The Handbook of Technology Foresight

PRIME SERIES ON RESEARCH AND INNOVATION POLICY

Series Editor: Philippe Larédo, *ENPC (France) and University of Manchester, UK*

The last decade has seen dramatic transformations in the configuration of national systems of innovation in Europe and in the way in which knowledge is produced. This important new series will provide a forum for the publication of high quality work analysing these changes and proposing new frameworks for the future.

In particular it will address the changing dynamics of knowledge production within the NBIC (Nano, Bio, Information, Cognitive) sciences and within different industries and services. It will also examine the changing relationship between science and society and the growing importance of both regional and European public authorities.

The series will include some of the best empirical and theoretical work in the field with contributions from leading and emerging scholars.

Titles in the series include:

Universities and Strategic Knowledge Creation
Specialization and Performance in Europe
Edited by Andrea Bonaccorsi and Cinzia Daraio

The Handbook of Technology Foresight
Concepts and Practice
Edited by Luke Georghiou, Jennifer Cassingena Harper, Michael Keenan, Ian Miles and Rafael Popper

The Handbook of Technology Foresight

Concepts and Practice

Edited by

Luke Georghiou
PREST, Manchester Institute of Innovation Research, MBS, University of Manchester, UK

Jennifer Cassingena Harper
MCST, Malta Council for Science and Technology, Malta

Michael Keenan
PREST, Manchester Institute of Innovation Research, MBS, University of Manchester, UK

Ian Miles
PREST, Manchester Institute of Innovation Research, MBS, University of Manchester, UK

Rafael Popper
PREST, Manchester Institute of Innovation Research, MBS, University of Manchester, UK

PRIME SERIES ON RESEARCH AND INNOVATION POLICY

Edward Elgar
Cheltenham, UK • Northampton, MA, USA

Published by
Edward Elgar Publishing Limited
The Lypiatts
15 Lansdown Road
Cheltenham
Glos GL50 2JA
UK

Edward Elgar Publishing, Inc.
William Pratt House
9 Dewey Court
Northampton
Massachusetts 01060
USA

Paperback edition 2009
Paperback edition reprinted 2015, 2016

A catalogue record for this book
is available from the British Library

Library of Congress Control Number: 2008922102

MIX
Paper from
responsible sources
FSC® C013604
www.fsc.org

ISBN   978 1 84542 586 9 (cased)
        978 1 84844 810 0 (paperback)

Printed and bound in Great Britain by the CPI Group (UK) Ltd

Contents

Figures

Tables

Acronyms

BMBF	German Federal Ministry for Education and Research
BMFT	German Federal Ministry for Science and Technology
CAB	Andrés Bello Agreement
CAF	Andean Development Corporation
CEE	Central and Eastern Europe
CGEE	(Brazil) Centre for Management and Strategic Studies
CNAM	(France) Conservatoire National des Arts et Métiers
COLCIENCIAS	Colombian Office of Science and Technology
CYTED	Science and Technology for Development Network
DASTI	Danish Agency for Science, Technology and Innovation
Defra	(UK) Department for Environment, Food & Rural Affairs
DTI	(UK) Department of Trade and Industry
EC	European Commission
ECLAC	United Nations Economic Commission for Latin America and the Caribbean
EFMN	European Foresight Monitoring Network
ERA	European Research Area
EU	European Union
FOI	Swedish Defence Agency
FOREN	Foresight for Regional Development Network
ICT	Information and Communication Technologies
IDA	Danish Society of Engineers
IPTS	(EC) Institute for Prospective Technological Studies
IVA	Royal Swedish Academy of Engineering Sciences
JRC	(EC) Joint Research Centre
MDIC	(Brazil) Ministry of Development, Industry and Commerce
MERCOSUR	Common Market of the South
MEXT	(Japan) Ministry of Education, Culture, Sports, Science and Technology
NAE	(Brazil) Nucleus of Strategic Issues of the Presidency
NICe	Nordic Innovation Centre

NISTEP	(Japan) National Institute of Science & Technology Policy
NUTEK	Swedish Board for Industrial & Technical Development
OAS	Organisation of American States
OCCT	Cuban Observatory of Science and Technology
OECD	Organisation for Economic Co-operation and Development
OPEC	Organization of the Petroleum Exporting Countries
OSI	(UK) Office of Science and Innovation
OST	(UK) Office of Science and Technology
OTA	(USA) Office of Technology Assessment
PACs	(EU) Pre-Accession Countries
PDVSA	Venezuelan State-owned Oil Corporation
PIU	(UK) Performance and Innovation Unit
PREST	Policy Research in Engineering Science & Technology
RCN	Research Council of Norway
RTD	Research and Technology Development
SCOPE	Scenarios for RTD Cooperation with Europe
SECYT	(Argentina) Secretary for Science and Technology
SELF-RULE	(Academic Network) Strategic Euro-Latin Foresight Research and University Learning Exchange
SENA	(Colombia) National Training Service
SITRA	Finnish National Fund for Research and Development
SPRU	(UK) Science Policy Research Unit
STEEPV	Social-Technology-Economics-Environment-Politics-Values
STI	Science, Technology and Innovation
TEP	Hungarian Technology Foresight Programme
TF	Technology Foresight
TFCEE/NIS	(UNIDO) Technology Foresight Programme for Central and Eastern Europe and Newly Independent States
TFLAC	(UNIDO) Technology Foresight Programme for Latin America and the Caribbean
TULIO	Finnish Graduate School in Future Business Competencies
UCV	Central University of Venezuela
UFRJ	Federal University of Rio de Janeiro
UNASUR	Union of South American Nations
UNIDO	United Nations Industrial Development Organisation
UNIVALLE	(Colombia) El Valle University
VINNOVA	Swedish Governmental Agency for Innovation Systems
VTT	Technical Research Centre of Finland

Contributors

W. Bradford Ashton is a senior programme manager at Battelle and has over 15 years of experience in advanced technology development. He is an industrial engineer with research and teaching expertise in capital investment planning, technology assessment and research and development (R&D) management. He has managed technology development efforts in computer technology and information systems, and has technical experience in energy conversion and industrial automation systems. He has developed original methods for technology monitoring and for using patent information in business decisions.

Rémi Barré holds a Civil Engineer diploma (Ecole des Mines) and a PhD in economics from EHESS (Paris). He is Director of the *Observatoire des Sciences et des Techniques (OST)* and Visiting Professor at CNAM in Paris. His current work relates to research policy decision-making and its relationship to evaluation, foresight and technology assessment. He has also been a member of the Executive Committee of ESTO (European S&T Observatory).

Jennifer Cassingena Harper is the Director of Policy within the Malta Council for Science and Technology (Office of the Prime Minister), with responsibility for National Research and Innovation Strategy and Foresight. Dr Harper is a member of the JRC Board of Governors, CREST and the FP7 Social Sciences and Humanities Advisory Group of DG Research. She was a member of the EU Regional Foresight (2001) Expert Group, and rapporteur of the Key Technologies (2005) and Agricultural Research (2006) Goups. Dr Harper is active in EU projects: ERANET ForSociety, INNFORMED, FUTURREG, EFMN and Forlearn. She was a member of the international evaluation panel for the German FUTUR Programme and the Hungarian Technology Foresight Programme. She is a graduate of Keele University and the London School of Economics. Her PhD research on the internationalisation of S&T Policy was carried out at the University of Malta and Sussex University (SPRU). She lectures in foresight within the University of Malta's Masters Programme in Innovation and Creativity.

Patrick Crehan has worked on various RTD collaboration projects in Information Technologies. In 1998 he left the European Commission to work as an independent consultant in international RTD, industrial and economic development policy. He worked as an advisor to OCYT – an agency in the Cabinet of the Prime Minister of Spain. In 1999 he founded Crehan, Kusano & Associates (CKA). CKA prospects for, evaluates and invests in new technology-based business opportunities. It designs and manages processes of strategic reflection that support innovation related decision-making for enterprise, government and administration.

Kerstin Cuhls studied Japanology, Chinese Studies and Economics/Business Administration at the University of Hamburg and one year in Japan (Kansai Gaikokugo Daigaku in Hirakata-shi near Osaka). In 1990, she spent four months at the National Institute for Metrology in Beijing, China. Since 1992, she has been at the Fraunhofer Institute for Systems and Innovation Research (ISI). In 1993, she spent four months at the National Institute of Science and Technology Policy (NISTEP) in Tokyo, Japan. In 1997 she received her PhD at the University of Hamburg. She has been the scientific project coordinator for different German–Japanese foresight projects, especially the German national foresight study Delphi '98, and monitoring Futur – The German Research Dialogue. She also coordinates the foresight/future studies projects in Fraunhofer ISI. Her research interests include: technology foresight; Delphi Methodology and combinations; Innovation Strategies; Management of Research and Development; Comparison of Japanese and German Technology Policy; and Japan and Asia in general.

Annele Eerola is Deputy Technology Manager and Senior Research Scientist at VTT Technical Research Centre of Finland. As leader of the Technology Foresight and Technology Assessment team she has been actively involved in developing cross-border foresight cooperation and in designing and carrying out technology foresight studies at Nordic and European levels, including Nordic H2 Energy Foresight (2003–2005) and Nordic ICT Foresight (2005–2007). She was a member of the Executive Committee of the European S&T Observatory (ESTO) in 2000–2003 and has been a member/substitute member of the Scientific Steering Committee of the European Techno-Economic Policy Support Network since 2005. Her working experience includes academic research and teaching at the Helsinki Swedish School of Economics and Business Administration, as well as industrial project management, systems planning/analysis at Jaakko Pöyry Companies. She holds a Doctor of Economics degree from the Helsinki Swedish School of Economics and Business Administration and a Licentiate of Technology and MSc (eng.) degrees from Helsinki University of Technology.

Luke Georghiou is Professor of Science & Technology Policy and Management and Associate Dean of Research within the Faculty of Humanities at the University of Manchester. Luke is also a Director of PREST, Manchester Institute of Innovation Research, and has been on its staff since 1977. His research interests include evaluation of R&D and innovation policy, foresight, national and international science policy, and management of science and technology. Recent projects include several studies of industry–science relations, policy for international scientific co-operation, evaluation of foresight, public procurement for innovation and changes in public sector research institutions. During 2006 he chaired the evaluation of the EUREKA Initiative and was member and rapporteur of the Aho Group report on Creating and Innovative Europe. He has recently chaired committees on rationales for the European Research Area on behalf of the European Commission, and the Evaluation of Futur – the German Foresight programme and TEP – the Hungarian Foresight Programme. He is an elected member of the Board of Governors of the University of Manchester and a member of the Board of Directors of Manchester Science Park Limited.

Attila Havas is a Senior Research Fellow at the Institute of Economics, Hungarian Academy of Sciences, and regional editor of *International Journal of Foresight and Innovation Policy*. His academic interests are in economics of innovation, theory and practice of innovation policy, and technology foresight as a policy tool. He holds a PhD in economics, from the Hungarian Academy of Sciences, 1997. He has participated in a number of EU-funded research projects on STI policies, innovation and transition, as well as on foresight and prospective analyses, and has been invited speaker at a number of EU and UNIDO conferences on technology foresight. In 1997–2000 he was Programme Director of the Hungarian TF Programme (TEP).

Birte Holst Jørgensen has been Managing Director of Nordic Energy Research since 2005. She holds an MSc in Business Economics from Copenhagen Business School and a PhD in Political Science from the University of Copenhagen. She is an expert and reviewer at DG Research, the Research Council of Norway, the Danish Strategic Research Council and the Swedish Gas Technology Centre. She is also involved in bridging Nordic Energy Research to the wider European Research Area (ERA), as the vice-chair of the Mirror Group of the European Technology Platform for Hydrogen and Fuel Cell and as core member of two ERAnets. Her research is focused on science, technology and innovation policies and administration, strategic management and technology foresight, in particular in the field of new energy technologies.

Ron Johnston is Executive Director of the Australian Centre for Innovation at the University of Sydney. He is also fellow of the Australian Academy of Technological Sciences and Engineering. His research interests include: the knowledge economy, e-commerce, the future of manufacturing, scenario planning for risk management, the future of financial services, strategies under globalisation, the new rules of the knowledge economy and effective knowledge management. He consults regularly to industrial firms across the manufacturing and service sectors, to research organisations and to government departments. He is also a member of the International Advisory Board of the APEC Technology Foresight Centre in Bangkok.

Michael Keenan is Lecturer at PREST, Manchester Institute of Innovation Research, University of Manchester. His research activities focus mostly upon policy analysis and advice, with particular emphasis upon foresight and evaluation studies. He has been an active player in shaping the foresight community in Europe, running projects on foresight capacity-building, writing a number of foresight guides for different target groups (for example, the EC's *Practical Guide to Regional Foresight* – co-authored with Ian Miles), and building a foresight knowledge platform. He was a member of the EC's High-Level Expert Group on Regional Foresight, as well as a member of international panels for several national foresight programmes. He is responsible for postgraduate teaching modules on foresight and science, technology and innovation policy, and is director of PREST's annual short course on foresight for organisers and practitioners. In addition, he is regularly retained by UNIDO and the EC to design and deliver foresight courses to decision-makers, and is a member of the International Board of the Finland Futures Academy.

Terutaka Kuwahara is Deputy Director General of the National Institute of Science and Technology Policy (NISTEP), Ministry of Education, Culture, Sports, Science and Technology (MEXT) since 2005. He also serves as Visiting Professor at the National Graduate Institute for Policy Studies (GRIPS). He started working for NISTEP in 1989 after administrative work including life sciences, nuclear energy and S&T information in the ministry. He has conducted the fifth, sixth, seventh and eighth Japanese Delphi surveys as well as an international Delphi study with Germany. He has organised international foresight conferences and took part in the activities of the APEC Technology Foresight Center. He also served as Planning Director for Research and Analysis, Council for S&T Policy, Cabinet Office (CSTP). His interest lies in prioritisation policy and in evaluation of scientific activities by bibliometric analysis. He received an MSc from the Graduate School of the University of Tokyo.

Javier Medina has been the Manager of the Technology Foresight Programme of Colombia (2003–2007). He holds a BSc in Psychology from the Universidad del Valle (1992), a Masters in Business Administration, Universidad del Valle (1997) and a Doctorate in Social Sciences at Pontificia Università Gregoriana (2001), with a focus on Human and Social Foresight. He has been a consultant for the World Bank, the Peruvian Government, the Economic Commission for Latin America (ECLAC), the Latin American Institute of Economic and Social Planning (ILPES), the National Planning Department, the Andean Development Corporation (CAF), UNIDO, and the Development Ministry of Colombia. He is a co-responsible of the S&T Programme for the Development of CAB countries.

Ian Miles is Professor of Technological Innovation and Social Change at the University of Manchester. He is based in the Manchester Institute of Innovation Research, Manchester Business School, and was appointed a CoDirector of PREST in 1990. His original training was as a social psychologist, and he spent almost twenty years at the Science Policy Research Unit, University of Sussex. Apart from being engaged in Foresight, his work has focused on innovation studies (especially with respect to services innovation); social aspects of Information Technology (especially with respect to working life and consumer activities); and research evaluation, science policy studies, and social indicators and quality of life issues. He has numerous publications – try Google.

Rafael Popper is a Researcher at PREST, Manchester Institute of Innovation Research, University of Manchester. He is also an Economist from the Central University of Venezuela (UCV). His research has focused mostly upon S&T policy analysis and advice, especially on foresight and evaluation studies. He has been an active player in shaping foresight practices in Europe and Latin America. He is often recognised for his contributions to foresight methodologies, for pioneering online Delphi and cross-impact analysis tools, back in 1999, and for raising the profile of Latin American foresight. Since 2003 he has regularly advised on the design and operation of the Colombian Foresight Programme, including its evaluation. In addition, he has led international projects and networks aimed at better connecting the Latin American foresight community to other parts of the world (e.g. SELF-RULE). He has also been a member of the European Commission's Expert Group on Rationales for the European Research Area. Apart from his research, he teaches postgraduate modules on foresight at the University of Manchester and he has been guest lecturer in universities in Finland, Russia, Colombia and Venezuela. He has also designed and delivered short courses for UNIDO, CAB, CAF and decision-makers in more than a dozen countries.

Alan Porter received a BSc in Chemical Engineering from Caltech (1967) and a PhD in Engineering Psychology from UCLA (1972). He served on the University of Washington faculty through 1974. In 1975 he joined the School of Industrial & Systems Engineering at Georgia Tech, where he now serves as Professor Emeritus. He also is Professor Emeritus of Public Policy and co-directs the Technology Policy and Assessment Center at Georgia Tech. He has taught short courses on Analysis of Emerging Technologies, Competitive Technological Intelligence, and Technology Forecasting & Assessment at Georgia Tech, various IBM locations, and in Brazil, Mexico, South Africa and Botswana.

Chatri Sripaipan is an electrical engineer with extensive experience in three broad areas–academic, public sector research and private sector consultancy and advisory. He has obtained a unique perspective on development and business issues through his wide-ranging experience and exposure to academia, industry and government agencies both in Thailand and in the international science and technology policy community. In 1976, he obtained a PhD in electrical engineering from the University of Hawaii, USA and became an associate professor in 1985. In 2000, he was promoted to be the Vice-President of NSTDA, supervising APEC CTF and the Science, Technology and Innovation Policy Research Department.

Preface

A paradox faced by those who wish to predict, understand or react to technological futures, is that they must comprehend drivers of change which go far beyond the domains of science and technology. In so doing, they have to draw upon a range of expertise involving not only the systematic study of science, social science, and even the humanities, but also make efforts to draw upon the experience of those more directly engaged with the economy and society in industry, government and beyond. More rarely they may also seek the opinions of the public at large. Finding the actors and the data is insufficient. They then need to create structures in which these actors can channel their knowledge and experience in a creative manner, unfettered by their short-term interests and prejudices. If this challenge is overcome, there still remains the issue of making sure that those who matter, including the participants, take notice of what has been produced and amend their actions accordingly.

It is the process of designing and implementing such an activity which is the subject of this book. It is an activity which we and others call foresight, though as we shall shortly see this term has a variety and complexity of meanings, contexts and modes of implementation. To attempt to capture the main elements of these in one volume has been a daunting task. For the editors there seemed to be two options: to collect the first-hand thoughts of leading practitioners and analysts or to attempt our own synthesis of the experience to date. The result seeks to combine these two approaches in a three–part structure. Part One aims to define foresight in terms of its rationales, roots and methods; Part Two hands the baton to the contributors, asking a range of leading experts from fourteen countries to assess the meaning of what has happened in their own countries or region. Regrettably not all nations could be covered and the significant contributions of Australia, New Zealand, Greece, Turkey, Ireland, Netherlands and Spain are but a few of those not here which deserve international attention. Part Three returns to cross-cutting issues, collecting under the general heading of policy and management the processes involved in learning about, planning and evaluating foresight before drawing to a conclusion.

Moving to the chapters, the starting point of our discussion is the problematic issue of definitions. A collective chapter by the editors notes the restriction here to the national level, while recognising the importance of regional and supra-national foresight. A consideration of several of the definitions offered leads to the conclusion that it is more important to identify what is distinctive about foresight than to dwell on semantics. This is as much about what it is not (broadly forecasting by experts) as what it is (broadly long term, engaged with decision-makers, drawing upon wide networks of expertise, and transdisciplinary). Through its evolution and a generational model the revealed rationales for foresight are explored and summarised as showing foresight as an approach extending the breadth and depth of the knowledge base for decision-making and creating new action networks. How critical foresight is as an element of the knowledge economy is a theme to which we shall return in the concluding discussion.

The future has a history which only later can be perceived as a series of key transitions. Chapter 2 is concerned with the origins of technology foresight (TF) and the specific characteristics which distinguish it from the more general approaches of future studies. In his overview of historical developments in future studies and the emergence and prominence of technology foresight in the beginning of the twenty-first century, Miles broadly identifies three phases of activity. In the first part of the twentieth century technological forecasting began to emerge as a systematic effort – beyond the work of individual visionaries in science fiction and speculative analysis. Technological change and the disruptive impact of the second world war which helped to raise recognition of the role of science in policy-making, and the rise of "*La Technique*" – technical rationality and tools ranging from operations research to systems analysis. This second phase of futures work was tied in the US to the emergence of futurology, with much innovation in forecasting methods, especially as applied to major military and aerospace programmes. European approaches tended to stress social alternatives more, and *La Prospective* was a (mainly French) crystallisation of such futures studies. One line of development of *La Prospective*, reflected in Godet's "Greek Triangle" of anticipation, appropriation and action, constitutes a striking precursor of TF, though without the emphasis on informing technology and R&D programmes. During this second phase of effort there was a lively debate on long-term global futures, catalysed by the Club of Rome's *Limits to Growth* report in 1972. But there was a downturn in futures activity in the 1980s, resulting from failures in economic forecasting and the rise of neoconservative governments. We are now in a third phase of activity, in which futures work in the form of TF is more widely institutionalised than ever before. The need for tools to assist with research priority-setting in the 1990s lead to a resurgence of TF activity, and a new emphasis on combining three elements: policy-related, prospective and participative activity. Miles concludes by identifying a number of tensions affecting TF, which may well take us beyond the third phase into new and uncharted TF territory.

An "unbiased" overview of foresight methods, practices and tools has to avoid the pitfalls of descriptive narratives, favoured techniques and the illusion of a prescriptive one-size-fits-all cure which is equally effective in all contexts. In Chapter 3 Popper provides a comprehensive and structured overview starting with the foresight process and a further elaboration of Miles' five-phase structuring of common practices in the sequence of pre-foresight, recruitment, generation, action and renewal. The potential contribution of the full range of methods to each phase is mapped providing a useful guide for selection and combination of appropriate approaches. Popper's presentation of three different frameworks for classifying foresight methods and practices provides important insights for distinguishing between varying rationales, approaches and uses. The first framework classifies 33 methods with detailed descriptions into qualitative, quantitative and semi-quantitative, whilst the second framework distinguishes between exploratory and normative types of approaches. Finally a third and novel framework classifies methods in a Foresight Diamond by type of knowledge source (creativity-based, expertise-based, interaction-based and evidence-based). The main addition here is the reference to more evidence-based methods for sourcing knowledge, a fact which has been confirmed in recent EFMN surveys, where literature review, scanning and other methods have been identified by practitioners as key foresight tools.

In Part Two of this volume we move from the general to the specific in terms of national experiences. Five countries are treated individually because of their seminal role in foresight practice. The remaining chapters look at regions or clusters of countries to draw out commonalities and differences.

In tracing the evolution of the United Kingdom Foresight Programme over three cycles, Keenan and Miles highlight in Chapter 4 the initial switch from a broadly priority-setting and networking exercise with a scientific focus to a distributed exercise of multiple foresight initiatives with a more business and social orientation in the second cycle, reflecting a more application-oriented and supply-chain-based approach. This less successful second cycle was to result in significant (negative) knock-on effects on the UK Foresight Programme as a whole, despite some lasting effects in terms of (sustained) ongoing initiatives such as the Foresight in the regions and the Young Foresight programmes. The reduced scale and scope of the third cycle has resulted in a programme which operates no more than two to three projects at a time and is predominantly science and technology focused. Keenan notes the marked difference in structure and organisation from its predecessors, as this rolling programme focuses on discrete projects concerned with a key issue or an area of cutting-edge science. The outlook for UK foresight is, however, largely positive with indications of foresight embedding across national and regional organisations and the emergence of a fourth cycle focused on horizon scanning, cross-government priority-setting and more evidence-based policy and strategy formation within the context of the new ten-year strategic framework for science and innovation.

The multi-level types of foresight which have emerged in France reflect the State-coordinated planning context and a response to an absence of inter-ministerial coordination and strong advisory structure. In Chapter 5 Barré reviews the critical mass of experience generated through three case studies: FutuRIS at the political (macro) level, AGORA 2020 at the programmatic (meso) level and INRA2020 combining the programmatic and research organisation (micro) levels. FutuRIS, aimed at correcting systemic failure and strengthening the system of innovation, achieved an important methodological breakthrough in translating scenarios into quantitative indicators. AGORA2020 focused on shared visions in transport and housing, managed the challenge of integrating a wide diversity of stakeholder visions through a complementary mix of methods. INRA2020, addressing the issue of relative weight to be assigned to four research axes: basic, food/nutrition, agricultural/agronomic, and environment research, was less successful due to the complexity of managing the cognitive aspects of research and the association of the exercise with the INRA Chairman. Barré's typology of the wide variety of foresights, relates to focus (e.g. technology areas or strategic issues), main objectives, ways of engaging social players, nature of the cognitive activity involved and organisation. Barré is confident that new developments such as the new Law on research are opening up other new opportunities for foresight embedding at multiple levels of the French system of innovation in transition.

Current foresight challenges for German R&D policy orient round effective approaches for generating and handling concepts and visions of emerging socio-economic futures, balancing problem-based demand for science and technology with science-based creativity at the interface of emerging technologies. Dating back to the 1990s, national foresight activity in Germany emerged in response to economic and budgetary realities and the need for information on the future as an input for decision-making and priority-setting. Cuhls traces the evolving rationales and objectives of four foresight exercises in Chapter 6, highlighting growing demands on foresight to address a broad range of policy needs, from a more transparent and strict methodology for ranking technologies in the T21 (Technology at the Beginning of the 21st Century), to the testing of Delphi methodology for more robust information about future science and technology time horizons in the first German Delphi study. The broadening of the stakeholder base, firstly in the second Delphi study which responded to industry demand and in the ongoing Futur Programme oriented more to societal needs, has had its effects on future prospects and orientations in foresight activity. Broader scope and participation, needs-orientation, increased creativity, reflexive learning and continuous quality management are the key emerging features balanced with a more realistic set of objectives.

In contrast with Europe's growing engagement with foresight at national and regional level, foresight in the USA has tended to be less embedded in policy-making both at the federal or state level, due to an inherent scepticism of any form of centralised planning. Porter and Ashton identify a range of foresight-type activity on the part of various government agencies (OTA, NIH, EPA) over the years coupled with impressive US contributions to foresight methods, but counterbalances this with an evident disenchantment and detachment of US R&D policy-making/policy-makers with foresight and its potential impacts. This "anti-foresight situation", as termed by Porter and Ashton, can be understood within the US policy context as a conscious policy approach which perceives limited utility for foresight approaches in a system where science and technology are already handled through pluralistic processes and checks and balances. The poor fit between US S&T policy and foresight is rooted, according to Porter and Ashton in Chapter 7, in a lack of conviction of what foresight can realistically deliver in terms of tangible future indicators, the possibility of free-riding on the results of other foresight activity underway elsewhere, and an innovation system which already works well with its predisposition to radical disruptions. Future interest in foresight activity in the US is likely to focus more on foresight methods and tool development, in particular the potential of advanced data-mining and processing tools, the appropriate mix of methods for eliciting different opinions, and possible forms of electronic voting processes in foresight.

Japan's more general foresight culture is reflected in a range of future-looking activity by public and private organisations. The chapter focuses on foresight activities undertaken by the State to understand scientific and technological developments within a holistic framework. Foresight activity has been largely rooted in the Delphi method, particularly appealing in the Japanese context for its subtlety in reaching a consensus, whose successful adaptation to the local context lead its extensive emulation by other countries, including Germany and France. Foresight activity has thus evolved since 1969 through successive rounds of Delphi exercises, presented in Chapter 8 by Kuwahara, Cuhls and Georghiou as the underpinning element of a four-layer model of Japanese foresight, complemented by macro level surveys, and corporate group and company level exercises. Used to support Japan's catch-up strategy, the surveys were limited in scope and time-span and did not target a single policy or group. The surveys grew in scope and in relation to the range of fields and sub-fields covered and increasingly took into consideration socio-economic needs. The Delphi approach and methodology improved over time and were kept alive through shared international experiences and interactive relationships developed with other countries, in particular Germany, allowing a two-way flow of concepts and techniques. Japanese foresight is currently focused on closer engagement with policy-making and expansion of the foresight toolkit to include scenario analysis, bibliometrics and socio-economic needs analysis.

Foresight in Nordic countries has focused primarily on improving the coherence and efficiency of the national system of innovation system, and enhancing the anticipatory intelligence of the producers and users of knowledge. Certainly, they have the advantage of being able to build on geographical proximity, shared cultures and values, and the strong potential for exploiting economies of scale and scope and exchanging experiences and mutual learning. The review of national foresight activities in Sweden, Denmark, Finland, Norway and Iceland highlights a diversity of rationales, approaches, capacities and competencies, with the first three countries dominating the foresight landscape and Norway catching up fast. The SECI model of shared knowledge creation is identified by Eerola and Jørgensen in Chapter 9 as one of the key conceptual tools in use in Nordic countries, which provides a useful framework for comparative analyses of various foresight processes, being a complementary perspective to the innovation system framework. By explicating the dynamics of shared knowledge creation the model also complements the "multiple perspectives" and "knowledge-people-system-organisation" frameworks of foresight management. The national overview is complemented by an exploration of the rationales, approaches and impacts of recent efforts to launch Nordic level foresight. The experiences with the Nordic H2 Energy Foresight and other exercises highlighted the potential for cross-border learning which can be unleashed, triggering both new exercises and opening up possibilities for cooperation with other countries outside Europe.

Foresight in small countries has an altogether different array of challenges and concerns to contend with. Not only is foresight often considered an unaffordable luxury but it tends to succumb to extreme pressures to perform and deliver results according to high expectations. Crehan and Harper in their review of the foresight experiences of three small countries, balance this initially negative prospect with a more optimistic account of what can be achieved when small countries are brought together in a cooperative foresight learning mode. eFORESEE, an EU-funded project, provided the opportunity for fast-track hands-on foresight learning through the implementation of pilots in each of the partner countries, Cyprus, Estonia and Malta. Despite the need to focus on national priorities and themes and to address specific challenges related to national context, the three countries were able to find a common ground for exchanging experiences and sharing insights. The range of results generated from the project include tangible and intangible benefits, as the more formal political objectives combined with the informal social challenges, extending the remit of the foresight activity. In Chapter 10 Crehan and Harper highlight the need for smart creative approaches to foresight on the part of small states, adapted to their particular context but also exploiting their potential growth capacities in the long term. Their outlook for the increasing utility of foresight approaches tailored to small territories is naturally positive, given the current environment favouring regional and city foresight activity.

In Asian industrialising countries, the foresight experience has evolved in a three-stage process of imitation, assimilation and adaptive innovation. In Chapter 11 Johnston and Sripaipan's typology of foresight exercises in Asia based on evolution, level, focus and objectives, highlights a high level of foresight embedding in planning and policy in industrialised countries as compared to early stage foresight in industrialising countries though with some evidence of foresight embedding. The China case study underlines the role of foresight in the development of a series of National S&T Plans dating back to 1963. In Thailand the hosting of the APEC Centre for Technology Foresight (in 1988) lead to the start of an extensive foresight embedding process totalling ten projects at national, sectoral and organisational levels, and a wide diffusion of foresight methods in Thai society. A number of success factors may be derived from the Thai foresight experience including level of foresight experience, the foresight skills of those participating in the exercise, the support of the lead organisations and individuals and alignment with political priorities. It is possible to develop foresight capabilities in a short space of time as revealed by the China case study but this depends on cross-border sharing of experiences and tacit knowledge.

In Latin America, foresight has acquired a critical development role as a major catalyst in the productive and social transformation of the region. Popper and Medina highlight in Chapter 12 the fact that while foresight entered the LA policy environment as a tool for better anticipating futures, its interpretation and use has evolved differently in each country. This depends primarily on country size and resources, international sponsorship, the political will and the perceived needs and opportunities. A key influence shaping the institutional development of foresight in the region is the role of international organisations, in particular UNIDO, in providing the initial political impetus, together with financial, technical and logistical support for the exercises themselves. However in terms of shaping the form and methods of foresight activity itself, Popper and Medina note that the European experience has been highly influential and is likely to increase with EU-funded initiatives such as SCOPE and SELF-RULE. Latin American foresight has evolved in a number of organisational forms with national programmes as the prevailing form, complemented by international programmes, specialised centres and institutions, networks, project and consortia and one-off and sectoral exercises. The impressive and diverse range of activity, particularly in the larger countries (Brazil, Colombia and Venezuela) highlights the fact that despite substantial international influences, there is a particular LA flavour to foresight activity which is in its own right making an important contribution to foresight methods and practice.

For the countries of Central and Eastern Europe the transition to market economies and the consequent reforms of science and industry formed the backdrop to their engagement with foresight. Havas and Keenan in Chapter 13 see an inherent tension between the haste to catch up and the need to consider the longer term, even when fundamental change is a strong need. Among the roles for foresight has been that of a bridge between policy for science made in isolation and the needs of the economy and society for innovation. The experiences of six countries are summarised. Among the common features noticed by the contributors is a tendency to favour approaches that generate "solid", expert-based codified outputs rather than more open-ended scenario methods. They speculate that this may be a legacy of central planning and the mindset of a single future.

Moving to Part Three of the book, this recognition of the role of policy transfer and learning in foresight is treated in more detail in the following chapter where a clear distinction is made between the process and outcomes of transfer. The key message from Georghiou and Harper in Chapter 14 is that there needs to be less emphasis on the successes or failures of policy transfer and more study of the factors favouring such transfers where they occur and their impacts on the new policy context. The learning activity generated through joint foresight exercises involving more than one country and region is projected as a step beyond adaptive foresight in the Asian case towards a more collaborative learning process. In Europe there has been a wealth of this type of activity underway in recent years, providing important insights in ways of enhancing policy transfer and mutual learning. The three case studies focus on different aspects of transfer, methodological in the case of national Delphi surveys in the early 1990s, transnational mutual learning in the eFORESEE project, and the EU Enlargement Futures project. The success factors for effective transfer relate to the level of shared features and context, and similar levels of foresight expertise and resources together with international knowledge sharing rooted in learning as a policy and learning as a policy instrument.

Nowhere is the policy transfer and learning aspect of foresight more critical than at the start of a foresight exercise when the extent of copying, adaptation and (we hope) creativity in relation to previous or ongoing exercises is determined. In Chapter 15 Keenan and Miles argue that foresight exercises have their origins and roots and depend heavily for their quality and success on the initial scoping, design and planning phase which has proven a critical pre-foresight activity at national, regional or company/organisation levels. The elements of the scoping process, organised round the framework conditions and scoping variables, now form a compulsory part of European guidelines on how to conduct successful foresight. The clear articulation of expectations on the part of the sponsor, the stakeholders and the managers of the exercise regarding the focus, remit, approaches, options and the setting of realistic goals in terms results and impacts, emerge as key considerations. The case for undertaking foresight activity highlights a number of potential advantages to policy-makers including its support in connecting science and technology to the wider socio-economic framework, in priority-setting and in providing value for money. The authors however end with on cautionary note: foresight should not be seen as a "quick fix", nor should it be undertaken where there is limited or no political, economic or cultural leverage or where there is no possibility to tailor foresight to its context. Moreover the results tend to enter into a reservoir and the benefits take a long time to materialise, often well after an initial evaluation or assessment of the process.

When all is said and done the policymaker normally wishes to know what the investment in a foresight exercise of time, money, effort, and often political capital has yielded. This is the territory of evaluation. The approach to evaluation of foresight is contingent upon the type of foresight undertaken and the associated rationales and objectives. In Chapter 16 Georghiou and Keenan see an understanding of the interaction between foresight and the environment in which its outcomes are to be implemented as central to the question of impact. Accuracy of prediction turns out to be a rarely applied criterion, mainly because of the long timescales involved but also because foresight has more complex aims. They review three of the most detailed national experiences of evaluation and conclude that practice here is still immature but that important lessons both from these experiences and from the more general practice of evaluation may be applied.

At this introductory stage we do not propose to summarise the conclusions that we draw at the end of this volume in Chapter 17. This is because most of the lessons and insights are distributed throughout and even our own thoughts and reactions in the final chapter will be most appreciated (or disputed!) by readers who have shared the journey through the "many faces of foresight", the title we give to the first chapter but in some ways also the theme of the entire Handbook.

PART ONE

The Nature of Foresight

1. The Many Faces of Foresight

Ian Miles, Jennifer Cassingena Harper, Luke Georghiou, Michael Keenan and Rafael Popper

INTRODUCTION

It would be convenient to begin with a definition that provides a common thread through the complex fabric of foresight. However, that complexity is in many ways the essence of an activity which is better understood through observation and the examination of the communities in which it is practiced. In this chapter we will discuss some definitions, inclusions and exclusions, and then move on to ask how the different facets of foresight are embedded in particular problem and decision-making contexts. These in turn can be related to the rationales for doing foresight and help us to answer the key question of: what is it for?

In this book, our central concern is that of national "foresight", that is foresight initiated by central government or its agencies. This is already a significant restriction, though one we must follow because of lack of space. In an era of multilateral governance, foresight is just as likely to be embedded at regional or local levels. Trans-national institutions have also engaged in foresight to cast light upon their own particular responsibilities. Trans-national because, as we shall see in Chapter 14, this has been a critical vector in the spread of foresight between nations, and regional because this often sits in an interactive relationship with national activity. Beyond the public sector, there is another major venue for foresight in the world of business. Not surprisingly, given its role in supporting the competitive advantage of firms, this is less visible. Nonetheless it is important both for its contribution to the development of practice and because the knowledge it generates may be fed into national foresight through the participation of firms. Furthermore, in due course public programmes draw upon knowledge of the future held in firms when they are bringing together collective visions, creating a "virtuous circle of foresight".

Table 1.1   Chronology of foresight activities (1971–1999)

Year	Country	Exercise/Programme	Method(s)
Since 1971	Japan	1st to 4th STA surveys	Delphi
1991	Japan	5th STA survey	Delphi
	USA	Critical Technologies	Others
1992	New Zealand	Public Good Science Fund	Others
	Germany	BMFT, T 21	Others
1993	South Korea	Foresight Exercise	Others
	Germany	Delphi '93	Delphi
1994	UK	1st TF Programme	Delphi + Others
	France	Technology Delphi	Delphi
1995	France	100 Key Technologies	Others
1996	Japan–Germany	Mini-Delphi	Delphi
	Austria	Delphi Austria	Delphi
	Japan	6th STA survey	Delphi
	Australia	Matching S&T to futures needs	Others
1997	Spain	ANEP	Delphi + Others
	Hungary	TF Programme (TEP)	Delphi + Others
	Netherlands	Technology Radar	Others
	Finland	SITRA Foresight	Others
1998	South Africa	Foresight Exercise	Delphi + Others
	Germany	Delphi '98	Delphi
	Ireland	Technology Foresight Ireland	Others
	New Zealand	Foresight Exercise	Others
1999	UK	2nd UK Foresight Programme	Others
	Sweden	1st Swedish Foresight	Others
	Spain	OPTI Technology Foresight	Delphi
	South Korea	Korean Technology Delphi	Delphi
	Thailand	ICT Foresight	Delphi + Others
	China	TF of Priority Industries	Delphi + Others

Others include: scenarios, panels, roadmapping, critical technologies, etc.

Note: Dates given are point of significant activity rather than formal start or end

Table 1.2   Chronology of foresight activities (2000–2006)

Year	Country	Exercise/Programme	Method(s)
2000	Japan	7th STA Survey	Delphi
	Brazil	Prospectar	Delphi
	Brazil	TFP Brazil (UNIDO/MDIC)	Delphi + Others
	France	2nd 100 Key Technologies	Others
	Portugal	ET2000	Others
2001	Venezuela	TFP Venezuela 1st cycle	Delphi + Others
	Chile	TFP Chile	Delphi
	Germany	FUTUR	Others
	Czech Republic	TF Exercise	Others
2002	Turkey	Vision 2023	Delphi + Others
	Colombia	TFP Colombia 1st cycle	Delphi + Others
	UK	3rd UK Foresight Programme	Others
	Cyprus, Estonia, Malta	eForesee	Others
	Denmark	National TF Denmark	Others
	USA	NIH Roadmap USA	Others
2003	China	TF Towards 2020	Delphi + Others
	Greece	Technology Foresight Greece	Others
	Norway	Research Council 2020 studies	Others
	Sweden	2nd Swedish TF	Others
2004	Japan	8th Japanese Programme	Delphi + Others
	South Korea	Korea 2030	Delphi + Others
	Ukraine	Ukranian TF Programme	Delphi + Others
	France	FuturRIS	Others
	France	AGORA	Others
	Venezuela	TFP Venezuela 2nd cycle	Others
	Russia	Key Technologies	Others
2005	Colombia	TFP Colombia 2nd cycle	Delphi + Others
	Brazil	Brazil 3 Moments	Delphi + Others
	Romania	Romanian S&T Foresight	Delphi + Others
	Finland	Finnsight	Others
	Luxembourg	FNR Foresight	Others
	USA	21st Century Challenges GAO	Others
2006	Finland	SITRA Foresight	Others
	Poland	Poland 2020 – TF Programme	Delphi + Others

Others include: scenarios, panels, roadmapping, critical technologies, etc.

Note: Dates given are point of significant activity rather than formal start or end

In Tables 1.1 and 1.2 we set out a chronology of foresight activities. Although this is far from being comprehensive it does give some sense of the growth of national foresight in the 1990s. Various hypotheses can be attached to the reasons for this growth, including simple explanations such as diffusion through an "epidemic" model or fashion, through to more complex analyses about the emergence of new challenges to the role of S&T in a networked economy for which foresight seemed to offer an answer. We explore motivations and rationales further later in this chapter. In the more recent activities (Table 1.2) we see two phenomena – a "bow-wave" of foresight spreading to new countries, particularly but not exclusively driven by the challenges of accession to the European Union, and new iterations of activity by most of the original players, though often departing from the formats they used initially. These national stories are played out in the chapters of this book, while the dynamic of their interrelation is discussed in Chapter 14 which deals with policy transfer in foresight. Detail, including some additional exercises, can be found in Part 2 of this book. In a sense this mapping of progress indicates the historical opportunities for learning and transfer. There is a clear family tree in terms of the use of large-scale Delphi surveys which also spills over into the hybrid exercises (those combining Delphi with other methods). Another explicitly-related family tree is that of critical technologies exercises. Among the activities which use other methods (e.g. scenarios, panels and roadmapping), the linkages are more complicated. Broadly speaking the earlier exercises have been the most influential, partly because of their pioneering nature and partly because some of their key participants have become expert in the process of policy transfer itself.

DEFINITIONS AND RATIONALES

How NOT to Define foresight

The growth of the foresight literature in the last decade – evidenced, for instance, by the emergence of two journals with "foresight" in their titles – has been matched by the growing confusion as to what the term actually refers to. In many ways this is an inevitable result of the appropriation of an everyday term to describe a specialised activity or construct. The endless debates about the meaning of "intelligence" bear witness to this. While researchers are justly castigated for inventing new terminology at the drop of a journal paper, their reinterpretation of existing terminology opens the way for controversy. In this case the term "foresight", as well as its everyday uses – for which, consult any English-language dictionary – has much application in a range of academic areas. A quick search of journal articles bearing the term will yield many "hits" in fields such as artificial intelligence, economics of the firm, market analysis, and cognitive psychology, etc. Articles dealing with foresight programmes appear quite low down on the list![1]

The dictionary definition of foresight tells us about a fundamental human capability, and the broad application of the term in a range of scholarly disciplines and practical applications tells us about the importance of this capability.[2] But defining foresight in terms of the dictionary definition, or a sort of lowest common denominator of its use across the spectrum of published articles, is not going to progress us very far in terms of exploring common features of the wave of systematic studies designed to inform policies bearing on longer-term prospects for technology development.

While there is a tendency for commentators to discuss the evolution and social articulation of the human capacity for foresight – and this is indeed a topic worthy of sustained attention – the confusion in the recent literature also springs from another pair of sources. First, there is a strong desire on the part of many scholars to situate their contributions in the corpus of published literature. If they are writing about technology foresight or foresight in the context of futures studies more generally, they are thus inclined to search the specialised futures literature for uses and definitions of the term. There is remarkably little use of the term in the futures literature until the 1990s: it is possible to search the major texts and handbooks without any use of the term – or if it is employed, it is merely used in passing as a convenient synonym, and is not dwelt upon. There are few reference points for scholars attempting to ground the term in a futures tradition, then. Typically they will come up with early uses like that of H.G. Wells, in the 1930s (his call for "Professors of Foresight" is discussed in Chapter 2) and Joseph Coates in 1985. Coates, talking of government policy activities in the USA, is a rare case where a futurist has provided a definition of the term, elaborating this along the following lines:

> Foresight is a process by which one comes to a fuller understanding of the forces shaping the long-term future which should be taken into account in policy formulation, planning and decision making ... Foresight includes qualitative and quantitative means for monitoring clues and indicators of evolving trends and developments and is best and most useful when directly linked to the analysis of policy implications. Foresight prepares us to meet the needs and opportunities of the future. Foresight in government cannot define policy, but it can help condition policies to be more appropriate, more flexible, and more robust in their implementation, as times and circumstances change. It is therefore closely tied to planning. It is not planning – merely a step in planning (p. 343).

This definition has often been cited, perhaps because of its rarity; those citing it frequently note Coates' stress on the link between futures-orientation and planning. However, the emphasis is clearly on scanning and forecasting approaches, and not specifically oriented to science and technology forecasting.

There are, actually, very few uses of the term "foresight" – even in a casual way – in the "futures" literature up until the 1990s. Since then there has been an explosion of use of the term – such that many activities that went by the name of forecasting, scanning, strategy analysis, or prospectives are now relabelled foresight. (This was even the case in Coates and Jarret's consultancy firm, which renamed its technology scanning activities "technology foresight".) Ironically, the Japanese Technology Foresight Programme originally published its reports under the heading "Technological Forecasting"; only when its model was being emulated in Foresight Programmes round the world did it relabel itself!

The second source of problems stems from this rebranding of activities. Irvine and Martin (1984) introduced foresight as the term to describe strategic forward-looking technology analysis for policy-making. They used the term in counterpoint to Hindsight studies that had explored the origins of current technologies in historic discoveries. It proved a useful point of orientation for the wave of national Programmes that was to take off in the 1990s. It was the large scale and clout of these activities – and the scope for funding and prestige attached to them – that gave the term a resonance it had lacked before in the futures field.

Many futurists and forecasters have taken to describing their work as foresight. This has led to immense definitional confusion, because effectively all varieties of futures and forecasting work have been rallied under the foresight flag. Probably the most influential of the futurists who have been flying this flag is Richard Slaughter who has written books such as *The Foresight Principle*,[3] differentiated between individual, social and strategic foresight, and established the Australian Foresight Institute and a substantial presence on the Web. His influence has been amplified by his having been President of the World Futures Studies Federation, and compiler of a major handbook on futures studies, as well as a prolific author in his own right. His definitions of foresight are also numerous, with the following often being cited:

> the ability to create and maintain a high-quality, coherent, and functional forward view and to use the insights arising in organisationally-useful ways, for example, to detect adverse conditions, guide policy, and shape strategy and to explore new markets, products and services (1999, p. 287).

Slaughter here and elsewhere relates foresight to decision-making, especially in organisations; and his work asserts that institutionalised foresight is vital for the contemporary epoch. While he does discuss technological issues and choices, the examples of "foresight institutions" he gives in *The Foresight Principle* are notably not governmental technology foresight programmes. The emphasis of his work here is on articulating his own brand of critical futurism, which aims to challenge narrow and reductionistic futures work and worldviews more generally. Finally, a recent European study of "Advancing Foresight Methodologies" (COST Action A22)[4] quotes definitions along the lines of those above, but then goes on to identify foresight with what is more commonly known as "futures studies":

> Foresight emerged in decision-making contexts following the Second World War in fields such as US military strategic planning with the RAND Corporation, and in French spatial planning with DATAR (the national institute for spatial planning). In the 1960s, General Electric and Royal Dutch/Shell introduced Foresight techniques in their corporate planning procedures. In the 1970s, Foresight included scenarios of socio-economic and environmental futures accompanied the introduction of the first global models that attempted to address these issues in an integrated fashion. In recent decades, Foresight has also been adopted in many areas of public policy, policy analysis, technology assessment, and scooping studies for various sectors and industries … Foresight aims to identify opportunities and areas of vulnerability in complex strategic issues. Its application ranges from strategy development to the raising of the general public's awareness of developments that are likely to influence society's future. Common to all use of Foresight, however, is the structuring of knowledge about complex issues into manageable elements so that these issues can be understood better and more informed decisions can be made.[5]

This identification of foresight with the entire futures field is hardly helpful for understanding the nature and dynamics of technology foresight programmes as we now know them. Nor does the COST account provide much light when it comes to dealing with these programmes.

Foresight in Europe

Although foresight techniques have been actively applied since the late 1960s European foresight is arguably more diverse and fragmented both in terms of the actors involved and the methodology applied than in the USA. In the USA activities organised by the World Futures Society draw thousands of participants annually and organisations such as Global Business Network and SRI International are well known within the wider foresight community. In

Europe, many foresight initiatives are undertaken and they range from scenario development exercises for small and middle enterprises (SMEs), to regional and national foresight studies, to environmental assessments for European public policy. The methodology that is applied is diverse and leans on theory from various disciplines such as psychology, history, policy science, economics and business administration. "The diversity in European Foresight methodology makes for a rich pallet of techniques but it also leads to the reinventing of wheels and putting old wine in new bottles".[6]

But the reason for the rise of foresight as a specific way of defining futures work is that the term has been associated with a specific sort of futures work. The rationale for this Handbook lies precisely in the need to examine and explicate this new flavour of foresight, and definitions that dilute this flavour are not helpful for this mission.

The Distinctiveness of Foresight

Rather than search for an understanding of the essence of foresight in archives or bandwagons, it is surely more relevant to consider the ways in which those involved in the major Foresight Programmes and their related activities have gone about describing the activity. Indeed, an obvious place to start is with Irvine and Martin's *Foresight in Science* (1984), where the term was introduced to describe, in their own words: "Technology Foresight is a process which seeks to look into the longer term future of science, technology and economy and society with the aim of identifying the areas of strategic research and the emerging generic technologies likely to yield the greatest economic and social benefit."

In their later *Research Foresight* (1989, p. 3), Martin and Irvine went on to argue that technology foresight was "the only plausible response" to:

> resolving conflicts over priority-setting caused by escalating experimental costs, limited resources, complexity in scientific decision-making and pressures to achieve 'value for money' and socio-economic relevance. …Foresight provides, at least in principle, a systematic mechanism for coping with complexity and interdependence as it affects long-term decisions on research, in particular facilitating policy-making where integration of activities across several fields is vital.

When the UK government finally decided to launch a Foresight Programme, Martin (1995) updated his review of similar activities round the world, and defined foresight as:

...the process involved in systematically attempting to look into the longer term future of science, technology, the economy, and society with the aim of identifying areas of strategic research and the emerging new technologies likely to yield the greatest economic and social benefits.

This latter definition is often cited, together with that of another UK researcher who was heavily involved with the design and implementation of the first cycle of the UK Programme. For Georghiou (1996) technology foresight is:

...a systematic means of assessing those scientific and technological developments which could have a strong impact on industrial competitiveness, wealth creation and quality of life.

These definitions all date from before, or early in the process of, the international take-off of Technology Foresight Programmes. Once a number of Programmes had been accomplished effectively, we began to see efforts to consolidate the experience that had been gained and to bring key players together into formalised communities of practice. A "Practical Guide" was produced for the EC-funded FOREN Project, which had set out to explore the scope for using foresight approaches in a regional development setting. This provided a detailed account of foresight, which is worth citing at length:

Foresight is a systematic, participatory, future intelligence gathering and medium-to-long-term vision-building process aimed at present-day decisions and mobilising joint actions...

In recent years, the term "foresight" has become widely used to describe a range of approaches to improving decision-making. As the term implies, these approaches involve thinking about emerging opportunities and challenges, trends and breaks in trends, and the like. But the aim is not just to produce more insightful "futures studies", more compelling scenarios, and more accurate econometric models. Foresight involves bringing together key agents of change and sources of knowledge, in order to develop strategic visions and anticipatory intelligence. Of equal importance, foresight is often explicitly intended to establish networks of knowledgeable agents, who can respond better to policy and other challenges. This is made possible not only by the improved anticipatory intelligence they have developed, but also through the awareness of the knowledge resources and strategic orientations of other members of the network. The key actors involved can include firms, governments, business sectors, voluntary organisations, social movements and technical experts. The contexts in which foresight can be employed are equally wide-ranging: much work to date has focused on national competitiveness and especially the prioritisation and development of strategic goals for areas of research in science and technology.

But foresight can and does also deal with issues like demographic change, transport issues, environmental problems and other social, political and cultural factors. Indeed, one of the main lessons of foresight exercises to date is that science and technology issues are inextricably linked with a wider range of social factors – and vice versa. Social forces shape the development and use of science and technology and the social implications associated with this.

Foresight involves five essential elements (FOREN, 2001, pp. 3–4):[7]

1. It involves structured anticipation and projections of long-term social, economic and technological developments and needs;
2. Interactive and participative methods of exploratory debate, analysis and study, involving a wide variety of stakeholders, are also characteristic of foresight (as opposed to many traditional futures studies that tend to be the preserve of experts);
3. These interactive approaches involve forging new social networks. Emphasis on the networking role varies across foresight programmes. It is often taken to be equally, if not more, important than the more formal products such as reports and lists of action points;
4. The formal products of foresight go beyond the presentation of scenarios (however stimulating these may be), and beyond the preparation of plans. What is crucial is the elaboration of a guiding strategic vision, to which there can be a shared sense of commitment (achieved, in part, through the networking processes);
5. This shared vision is not a utopia. There has to be explicit recognition and explication of the implications for present day decisions and actions.

The first part of the definition here, seeing foresight as a systematic, participatory, future intelligence gathering and medium-to-long-term vision-building process aimed at present-day decisions and mobilising joint actions, has been widely used (though often this original source is not mentioned).[8] It has been employed widely in European Union websites and documentation, for example.

A definition more focused on technology foresight is provided by Barré (2001); this overlaps considerably with the FOREN one where foresight is perceived as a decision support process with the following common characteristics:

- Long-term perspective;
- Particular focus on changes;
- Interactivity among participants;
- Transparency, openness and bottom-up spirit;
- Appropriation of the process to the actors and stakeholders;
- Diversity of actors and inputs, with acknowledged diversity of visions;
- Interest in the S&T, and the social, economic and environmental realms;
- Concern for alternatives, identification and exploration of hypothesis and events of significance for the actors; and
- Strategy formation.

A last definition is one prepared by practitioners from national foresight programmes, brought together in the EU-funded network ForSociety to explicate features of technology foresight:[9]

- Structured anticipation and projections of long-term developments and needs;
- Institutionalised;
- Formal techniques;
- Guiding strategic visions;
- Commitment to the results;
- Forging new social networks;
- Communication among actors;
- Broad approach/interdisciplinary;
- Interactive and participative methods; and
- Implications for present day decisions and actions.

Five characteristics can be attributed to such kinds of exercise: (1) a focus on the long-term perspective, (2) communication among the actors, (3) co-ordination of the actors' strategies by means of interaction, (4) consensus in terms of shared visions of the future and (5) commitment to the results of the exercise. In the context of "ForSociety", two additional aspects should be underlined: First that the future is genuinely perceived as uncertain and that Foresight provides a forum to deal with this openness and potential future options. Second, Foresight activities aim at building networks on regional, national and trans-national levels and thus to overcome compartmentiarland [sic][10] thus to strengthen the innovation systems[11]

These definitions clearly draw on experience with Foresight Programmes – experience which is continuing to evolve, as we shall discuss soon.

Together, they present a striking picture of what technology foresight is – and is not. It is not:

- Only forecasting (let alone prediction) – though forecasting is an important component of foresight activities;
- "Ivory tower" futures studies, in which an expert academic or consultant group produces its vision of the future or of alternative futures.

Among features that do characterise it, in contrast, are:

- Long-term orientation, aimed at informing ongoing decision-making in the present (especially research and innovation policy decisions), and grounded in the assumption that the future is in many ways open and can be shaped in positive ways by improved understanding of opportunities and threats, driving forces and underlying processes of change;
- Use of a range of formal tools and techniques for developing long-term analyses – including survey methods like Delphi, scenario workshops, and more extrapolative trend analysis, and often drawing on the results of modelling, SWOT studies, and many other methods (see Chapter 3);
- Involvement of a wide pool of expertise, and often of stakeholders more generally, to access relevant knowledge, to engage more participants in the policy process, and to establish networks for ongoing coordination of action and sharing of information;
- Crossing disciplinary boundaries and professional compartments, to be able to address emerging real-world problems that know nothing of these impediments. This often requires extensive "translation" and fusion of knowledge from different sources.

DIFFERENT FORMS OF FORESIGHT

Institutions and Frameworks

We have been talking mainly of foresight programmes, and the most visible activities in technology foresight to date have been national technology foresight programmes. By a programme, we mean more than a single one-off foresight exercise, a study of a specific issue or set of issues designed to inform a particular decision. Such exercises are often quite rapid affairs.[12] They tend to have less emphasis on the network-building elements of foresight, and the structure established to run the exercise has a limited life-time.

Programmes, in contrast, are wider-ranging, longer-term affairs. They typically take several years to accomplish, and it is common for them to undertake a succession of activities – indeed, many national programmes are now in their second or third waves of work. There will usually be some kind of Foresight Office or Unit constructed within the main policy sponsor. The programmes usually cover a wide range of sectors or topics (mainly topics with high technology relevance), and are generally highly participative. Some more "focused" activities are undertaken within an ongoing Foresight Programme, investigating specific topics in great detail; the current third cycle of UK Foresight, for example, is now structured around providing in-depth studies of a small number (four or five) strategic topics.

Some countries have responded to the challenge of foresight by setting up Foresight Units that carry out studies – often contracting them out – in response to immediate policy requirements. These typically involve less emphasis on participation and network-building, more on gathering intelligence to inform particular debates or policies.

Finally, some foresight activities have more of a focus on capability building, on establishing training in tools and approaches – perhaps elaborating this with some pilot studies, though there may not be such a focus of work. The aim is to stimulate foresight and foresight-like activities across a range of actors – for instance, across government departments, or across stakeholders in a regional economy.

GENERATIONS OF FORESIGHT

Building on earlier accounts from Georghiou (2001), the development of technology foresight can be characterised in terms of five generations of activity:

1. *First Generation*: Foresight is here emerging from what are mainly technology forecasting activities, with the analyses driven mainly by the internal dynamics of technology.
2. *Second Generation*: Foresight projects seek to grapple with technology and markets simultaneously. Technological development is examined in terms of its contribution to and influence from markets; and there is a strong emphasis on matching technological opportunities with market developments (and also with nonmarket needs such as environmental and social problems).

3. *Third Generation*: Foresight's market perspective is enhanced by inclusion of a broader social dimension, involving the concerns and inputs of a broad range of social actors.13 The need to take into account complicated issues concerning social trends and alternative institutional arrangements means that the methods used and the knowledge bases drawn on have to be expanded to deal with such issues.

4. *Fourth Generation*: Foresight programmes have a distributed role in the science and innovation system, rather than being "owned" by a single policy sponsor. Multiple organisations sponsor and/or conduct exercises that are specific to their own needs, but are coordinated with other activities (e.g. sharing resources and results, having shared working groups).

5. *Fifth Generation*: A mix of foresight programmes and exercises, also distributed across many sites but in combination with other elements of strategic decision-making. The principal concern of these activities is either (a) structures or actors within the STI system or (b) the scientific and technological dimensions of broader social or economic issues.

These "generations" are somewhat idealised. Most national foresight programmes at any one time have contained elements from more than one of the generations. There is no inevitability, either, and individual countries need not necessarily traverse the generations in a sequential progress. What the model suggests, however, is a broad trajectory of development in the foresight community of practice. Tables 1.1 and 1.2 tend to bear this out.

The developmental model bears some analogy to the evolution of thinking about innovation. The original linear model of technology development and transfer has now been recognised – not least because of the failures of policies based on it – as a rare special case rather than the norm. Innovation Policy has accordingly moved through a number of generations of its own, with better knowledge of innovation processes racing to catch up with ongoing transformation in the way in which technologies are developed and deployed. (see LLA *et al.*, 2003 – who see foresight as one of the responses to such transformation). Second and Third Generation Foresight are more akin to the "new production of knowledge" hypothesis of Gibbons *et al.* (1994), who characterise Mode 2 research as being carried out in the context of a problem addressed through the joint production of knowledge. The Fourth Generation has its analogue in the concept of an ecology of industry (Coombs and Georghiou, 2002) or the open innovation system (Chesborough, 2003). Miles (2004) has likened the switch from Third to Fourth to a move from a single mountain of activity to a range of foothills – which nonetheless has larger total mass. In part a Fourth Generation is enabled by its predecessors, as a foresight culture is created through contact with a large-scale but more monolithic exercise.

The different generations of foresight imply that different types of knowledge and of policy issues are being addressed, and thus that the actors involved in the process are liable to change, too. First Generation Foresight is clearly in the domain of the technological experts and may even be confined to the professional futurists. In the Second Generation academics and industrial researchers and managers – especially those who can work across the institutional gap – are typically the key actors. Third Generation Foresight means more involvement from social stakeholders such as voluntary organisations, consumer groups, pressure groups, etc. In addition, there is a shift in the role of government, with the parts of the policy system responsible for science being complemented by Ministries responsible for matters such as health, safety and the environment. The Fourth Generation is likely to further extend the range of those taking part – well beyond those who would naturally be drawn to a national policy exercise. In the Fifth Generation, arguably domain expertise becomes more significant; thus in some sense we are liable to return to the experts (who have never disappeared from the scene, in reality) but now they will be working alongside stakeholders and those with foresighting skills.

It also follows that the generations are liable to feature different structures for foresight, for example in the way panels or surveys are delineated. The First Generation is prone to follow the disciplinary taxonomies of science and engineering. Second Generation Foresight is much more likely to be structured in terms of industrial and service sectors to provide a bridge to the economy. For the Third Generation, with a socio-economic problem-solving organising principle, the structure is thematic, reflecting the problems that are being addressed, and with panels characterised by interdisciplinarity. Both the Fourth and Fifth generations build their own structures in terms of the object of analysis, and this has led to some diversity in the ways in which foresight is organised in countries that have reached these stages.

What is Foresight for?

The generations of foresight may also be distinguished by the economic rationales which underpin them and by the approach which should be taken to evaluate them. First Generation Foresight rests in the domain of economic planning. The Second Generation is mainly within a market perspective but the main rationale for its existence is one of market failure. In particular the market failure rationale rests on the recognition that many firms have excessively short horizons, and that intervention is necessary to stimulate them to take a longer view and consequently to afford a higher priority to research. In Third Generation Foresight, the rationale moves to being one of system failure. The underpinning analysis is that there are insufficient bridging institutions in the socio-economic system: foresight provides an arena in which the necessary network connections can be made. The succeeding generations retain this focus on institutions.

The market/systems failure cases have been the subject of much discussion, especially as systems failure has been increasingly adduced as a rationale for innovation policies – but one that has been resisted by many trained in more restrictive types of socio-economic analysis. In the traditional framework of innovation policy a policy intervention should be seen to be correcting a market failure such as asymmetric information, high uncertainty or inability to appropriate the benefits (Arrow, 1962). Foresight may be justified as enabling creation and pooling of knowledge that reduces uncertainty. This requires public support because (some of) these benefits are publicly available (through published outputs in particular). Individual firms may be unable to develop this knowledge themselves, and/or unwilling to invest in creating "public goods". But the knowledge of where to invest in technological development is vital, and left to the market, players would dissipate their efforts over too wide a spread of activity without reaching critical mass. This makes a case for a priority-setting exercise or an effort to establish the critical technologies for an economy. Furthermore, network externalities can be achieved by bringing activity into the sphere of a Programme, as working within a technology is more likely to be successful if others are working on complementary aspects (so that, to take just one example, standards are compatible).

The market failure approach is still widely held to be the only case for public intervention in a policy field. But innovation researchers, in particular those grounded in a structuralist-evolutionary perspective, will now often see that there are additional insights as to justifications for public intervention that flow from the systems failure approach. In many ways this provides a more compelling rationale for foresight as public policy.

In this perspective the failures may be seen to arise in the rigidities and mistakes of innovation agents (firms, public agencies, etc.) and in the system itself through a lack of linkages and fragmentation between innovation actors (Smith, 2000). There is also a stress upon the interrelatedness between innovation and a series of other factors, such as the availability of trained people, access to finance, and capabilities marketing and production (often collectively termed "complementary assets"). This could be seen to emphasise the networking benefits of foresight. Typical problems addressed by foresight as an instrument of innovation policy under this rationale are:

- Helping to overcome a lock-in failure by introducing a firm to a new or extended technology or market area. In this case the firm, without the benefit of public intervention through foresight, receives signals only from its own market and existing technological networks and misses the major threats and opportunities coming from beyond these horizons;
- Building new networks or coordinating systemic innovations such as those requiring establishment of standards, either between firms or between firms and the science base. Again historical factors may have created certain configurations and relationships which could be deficient in supporting the emergence of a new idea or area. Foresight can be used to build the new social structures which bridge these gaps.

Working from a conceptual rationale like this is important, but rather more concrete examples of the rationales that are used in practice, to argue the case for, and inform the design and use of, foresight, are also vital. We find that numerous rationales have been and are being used to justify the national Programmes and more limited exercises described in this volume. These may be clustered into five main groups (though the categories are not necessarily exclusive):

Rationale 1: Directing or prioritising investment in STI (Setting general research directions by identifying previously unknown opportunities)

- Informing funding and investment priorities, including direct prioritisation exercises;
- Eliciting the research and innovation agenda within a previously defined field;
- Reorienting the science and innovation system to match national needs, particularly in the case of transition economies;
- Helping to benchmark the national science and innovation system in terms of areas of strength and weakness, and to identify competitive threats and collaborative opportunities;
- Raising the profile of science and innovation in government as means of attracting investment.

Rationale 2: Building new networks and linkages around a common vision

- Building networks and strengthening communities around shared problems (especially where work on these problems has been compartmentalised and is lacking a common language);

- Building trust between participants unused to working together;
- Aiding collaboration across administrative and epistemic boundaries;
- Highlighting interdisciplinary opportunities.

Rationale 3: Extending the breadth of knowledge and visions in relation to the future

- Increasing understanding and changing mindsets, especially about future opportunities and challenges;
- Providing anticipatory intelligence to system actors as to the main directions, agents, and rapidity of change;
- Building visions of the future that can help actors recognise more or less desirable paths of development and the choices that help determine these.

Rationale 4: Bringing new actors into the strategic debate

- Increasing the number and involvement of system actors in decision-making, both to access a wider pool of knowledge and to achieve more democratic legitimacy in the policy process;
- Extending the range of types of actor participating in decision-making relating to science, technology and innovation issues.

Rationale 5: Improving policy-making and strategy formation in areas where science and innovation play a significant role

- Informing policy and public debates in these areas;
- Improve policy implementation by enabling informed "buy-in" to decision-making processes (for example, so that participants in foresight activities are able to use the understanding acquired here to argue the case for change, and to bring it to bear in more specialised areas than the Programme as a whole has been able to).

Seen collectively these practical rationales emphasise the role of foresight as an approach which extends the breadth of the knowledge base and the depth of analysis available to decision-makers, and at the same time through its participative element creates new action networks. For this reason, we find many commentators arguing that foresight is a vital component of the "knowledge-driven economy".

As we shall see in later chapters, foresight has had its share of problems as well as its successes. This is only to be expected from such a rapidly developing and diffusing set of policy innovations. And, given that Foresight

Programmes are policy innovations, the future of foresight is likely to be contingent on more than the emergence of the knowledge economy. In particular, politics and fashion may pay rather large roles. The persistence of institutionalised foresight in some countries or government departments may be threatened, for example, by being identified with a particular party or political philosophy. (Experience suggests that this may be particularly an issue in certain developing countries.) Individual policy-makers, politicians, and bureaucrats, too, are liable to seek elevation by strapping themselves onto "the next big thing"[14] – and as foresight is successfully institutionalised, it may have less chance of being portrayed as such. It would indeed be ironic if foresight proved to be good at identifying big new things on the horizon, but itself began to look rather passé! Only by codifying and professionalising its practice can foresight be made proof against such political vagaries.

NOTES

1   A Google scholar search on 13 March 2006 found the first reference to "foresight" as we here conceive it as its ninth "hit" – Martin and Johnston's 1999 paper "Technology Foresight for Wiring Up the National Innovation System" in the journal *Technological Forecasting and Social Change*. The themes of the preceding eight citations are suggested by the journals they come from, which include *The Bell Journal of Economics, Economics Letters, Environmental Pollution, Quality and Safety in Health Care, International Economic Review, International Library of Critical Writings in Economics*, the *Journal of Political Economy*, and *Studies in Subjective Probability*. The strong interest of economists in the topic relates to the need to take account of the fact that economic actors cannot possess the perfect information ascribed to them by conventional theory.

2   It may be of interest that "foresight" rarely appears in dictionaries of quotation. But there is at least one rather alarming appearance of a character called Foresight in literature – in a play in fact, William Congreve's *Love for Love*, where the character is described as "an illiterate old fellow, peevish and positive, superstitious, and pretending to understand astrology, palmistry, physiognomy, omens, dreams, etc." The text is available as an ebook on the Project Gutenberg website.

3   A summary of this book is available at http://foresightinternational.com.au/ catalogue/resources/The_Foresight%20_Principle_Summary.pdf.

4   Available online at http://www.costa22.org/ mou.php.

5   See: http://www.costa22.org/mou.php accessed 13/3 2006.

6   See: http://www.costa22.org/mou.php accessed 13/3 2006.

7   A second edition of this Practical Guide was heavily rewritten, but the only change in the definition of Foresight was more cosmetic: "Foresight is a systematic, participatory process, involving gathering intelligence and building visions for the medium-to-long-term future, and aimed at informing present-day decisions and mobilising joint actions." (Miles and Keenan, 2002, p.14).

8 For example, this definition is used without citation in the STRATA report (2002).
9 For further information see Task 2.3 of the ForSociety initiative.
10 Elsewhere ForSociety talks of "compartmentalised activities".
11 http://www.eranet-foresociety.net/ForSociety, accessed 03/08/07.
12 There has been little effort to differentiate between activities along the lines developed here – which draw in part on FOREN (2001). Thus the terminology is fairly unstable, and one person's programme is another's exercise, and vice versa.
13 A similar concept has emerged in research policy more broadly, notably in the European Union's Fifth Framework Programme and its successors.
14 We would hesitate to suggest that academic colleagues have any such tendency, of course, but consultants are fair game for such an accusation, too.

REFERENCES

Arrow, K. (1962), 'Economic Welfare and the Allocation of Resources for Invention', in Nelson, R. (ed.), *The Rate and Direction of Inventive Activity*, Princeton: Princeton University Press, pp. 609–625; republished in Rosenberg, N. (1971), *The Economics of Technological Change*, Harmondsworth: Penguin, pp. 164–181.

Barré, R. (2001), 'Synthesis of Technology Foresight', in Tübke, A., Ducatel, K., Gavigan, J.P. and Moncada-Paternò-Castello, P. (eds) (2001), *Strategic Policy Intelligence: Current Trends, the State of Play and Perspectives*, Seville, Spain: Institute for Prospective Technological Studies (IPTS), IPTS Technical Report Series.

Chesbrough, H. (2003), *Open Innovation: The New Imperative for Creating and Profiting from Technology*, Boston: Harvard Business School Press.

Coates, J.F. (1985), 'Foresight in Federal Government Policy Making', *Futures Research Quarterly*, **1**, pp. 29–53.

FOREN Network (IPTS, PREST, CMI and SI) (2001), *A Practical Guide to Regional Foresight*, Seville, Spain: Institute for Prospective Technological Studies (IPTS), available at: http://forera.jrc.es/documents/eur20128en.pdf.

Georghiou, L. (1996), 'The UK Technology Foresight Programme', *Futures*, **28**(4), pp. 359–377.

Georghiou, L. (2001), 'Third generation foresight – integrating the socio-economic dimension', *Proceedings of the International Conference on Technology Foresight. The Approach to and Potential for New Technology Foresight*, NISTEP Research Material, Vol. 77, March.

Gibbons, M., Limoges, C., Nowotny, H., Schwartzman, S., Scott, P. and Trow, M. (1994), *The New Production of Knowledge*, London: Sage.

Irvine, J. and Martin, B. (1984), *Foresight in Science*, London: Frances Pinter.

LLA, PREST, ANRT (2003), *Innovation Tomorrow*, Luxembourg, European Commission, at: http://www.cordis.lu/innovation-policy/studies/gen_study7.htm.

Martin, B. (1995), *Technology Foresight 6: A Review of Recent Overseas Programmes*, London: Office of Science and Technology.

Martin, B. and Irvine, J. (1989), *Research Foresight: Priority-Setting in Science*, London: Pinter.

Martin, B.R. and Johnston, R. (1999), 'Technology Foresight for Wiring Up the National Innovation System', *Technological Forecasting and Social Change*, **60** (1), January, pp. 37–54.

Miles, I. (2004), 'Centrifugal Foresight: Foresight in the UK', Paper presented at NISTEP Workshop on Science and Technology Foresight Tokyo, Japan.

Miles, I. and Keenan M. (eds) (2002), *Practical Guide to Regional Foresight*, (available in various country versions with local editors), Brussels: European Commission DG Research, available at: http://cordis.europa.eu/foresight/cgrf.htm.

Slaughter, R. (1995), *The Foresight Principle. Cultural Recovery in the 21st Century*, London: Adamantine Press.

Slaughter, R. (1996), *New Thinking for a New Millennium*, London: Routledge.

Slaughter, R. (1999), *Futures for the Third Millennium: Enabling the Forward View*, Sydney: Prospect Media.

Smith, K. (2000), 'Innovation as a Systemic Phenomenon: Rethinking the Role of Policy', *Enterprise & Innovation Management Studies*, **1**(1), pp. 73–102.

STRATA High Level Expert Group report (2002), *Thinking, debating and shaping the future: Foresight for Europe European Commission*, Directorate-General for Research, Unit RTD-K.2.

2.   From Futures to Foresight

Ian Miles

INTRODUCTION

Technology Foresight (TF) became prominent just before the turn of the millennium. By the beginning of the twenty-first century, TF programmes were underway or in preparation in a large number of countries, in all regions of the world. But studies of the long-term future, of one sort or another, have been in existence for a very long time. How much of a change does TF mark from such futures studies (FS)? Is this more a change of labels than of substance? What accounts for the rise of TF in the recent period, and how can the activities known as TF be related to other types of FS? This historical perspective is intended to help clarify what TF is, and is not. It should also suggest some ways in which efforts to take the long-term future into account in policy-making may evolve in the coming decades.

"Foresight" is a capability that humans have exercised for time immemorial. The use of the term to describe systematic study of the future – especially of a future transformed through use of new technologies – dates back at least to H.G. Wells' work in the early twentieth century. Futurists occasionally deployed the term to describe one or other element of their activities over the following decades, but the term rose to prominence in the 1980s and especially the 1990s, initially in the context of national TF programmes. This chapter will use "foresight" to describe the activities involved in these programmes in particular. But we have to acknowledge that the impact of these programmes on the futures field has led to a process of "rebranding". The term is now widely used to describe a host of activities that are often in key respects much more limited than TF; and in some cases they are indeed only tangentially related to TF.

THE REDISCOVERY OF THE FUTURE

Futures Studies (FS) are largely a phenomenon of the period after the Second World War, but as with most social phenomena there are deep roots. In particular, technology-oriented FS can be related to utopian fiction, with elaborate non-fictional studies only appearing towards the end of the nineteenth century (and, before that century, most utopian fictions were situated in distant places, rather than future times).[1]

We can differentiate FS from the narrow sorts of forecasting that date back at least to the seventeenth century, when Sir William Petty (one of the founding fathers of descriptive statistics) projected the population of England and London for several centuries to come.[2] Population forecasting (and the more restricted forecasting of mortality that actuaries engage in), economic forecasting, and similar enterprises mushroomed in the nineteenth and especially the early twentieth centuries. This reflects a growing sense that the future was quantitatively, and perhaps also qualitatively, different from the present. Technological change was increasingly recognised to be a factor that shaped both sorts of development, and novelists (most prominently H.G. Wells) considered the implications of such change. Indeed, after a lifetime of "genius forecasting", Wells (1932) returned to a theme he had articulated decades earlier, when he called for Professors of Foresight:

> ...though we have thousands and thousands of professors and hundreds of thousands of students of history working upon the records of the past, there is not a single person anywhere who makes a whole-time job of estimating the future consequences of new inventions and new devices. There is not a single Professor of Foresight in the world.

It is notable that Wells related systematic foresight to technological change. This reflects not only his own pioneering efforts, over several decades, as a science journalist and science fiction (SF) writer of note. The rise of SF itself in the pre-war years reflected the rise of new social groups with economic and intellectual power based in knowledge of new technologies and techniques, and bearing (by and large) optimistic views of technological progress and technical rationality. This was pronounced in the USA, where the pulp magazine SF phenomenon (crystallised by Hugo Gernsback) typically envisaged a future in which amazing technological progress (especially space travel and robotics) was fused with remarkable persistence of mid-century American mores and social relations.

While many similar challenges of economic and political crisis and industrial and technological change were being confronted across the industrial worlds, a marked divergence was beginning in the nascent (but not yet named) FS field. In both cases there were radical efforts to place scientists and engineers at the heart of policymaking – for example in the technocracy movement in the United States, and in Europe as part of "scientific socialism". More generally, though, the US field tended to focus more on continuity, while Europe was confronting the position of substantial political change.

In the Depression-era United States, social scientists were fusing statistical enquiry with forecasting. The sociologist William F. Ogburn and his colleagues pioneered social indicators research by producing a series of studies of social trends charting the course of the New Deal (e.g. "Report of the President's Research Committee on Social Trends", 1933). Ogburn also applied techniques of trend extrapolation to technological forecasting (Tf), with such reports as *Technological Trends and National Policy* (National Resources Committee, 1937 – notably the full title goes on to specify including the "Social Implications of New Inventions").[3]

He discussed the consequences of agricultural mechanisation in the USA (displaced workers would move to cities…) and examined the track record of technological forecasts. As well as developing a model of cumulative invention, he proposed a (rather basic) base-superstructure model of technological and social change. The notion of cultural lag which he proposed here can be traced back to a Marxian account of forces of production shaping relations of production. Thus Ogburn analysed and attempted to predict technological trends, and to make what we would now call technology assessments; and colleagues like Gilfillan played important roles in the application of extrapolative methods to Tf.[4]

In Europe, many intellectuals reacted against the failures of capitalism to pull itself out of the Great Depression by looking to left-wing political alternatives (and of course the far right rose to power in several countries). In the UK, the analysis of scientific and technological development in the interwar period was influenced by Marxism of a different sort. Many scientists were influenced by a Marxian analysis of the relations between scientific and technological thinking and more general social conditions (Werskey, 1978) and the case for science policy began to be articulated by thinkers like Haldane and Bernal.[5] Other great figures, such as Keynes, similarly attempted to look beyond the current economic difficulties to think how new sorts of planning tool might be used to establish a more stable future – famously in Keynes' case, to bring about a world of plenty in the long term. But visioning the future was more of an activity for gifted individuals than the sort of expert team work that Ogburn and others pioneered in the US. The Second World War was to have a disruptive impact. On the one hand it put the brake on long term FS (though this did not stop intensive work designing the political institutions – such as the welfare state – for the post-war world).

On the other hand, it gave a major boost to the development of systematic ways of appraising problems that fall outside of (or across) conventional scientific and technological disciplines.

LA TECHNIQUE AND LA PROSPECTIVE

In the Allied countries in general, the Second World War was a vindication of what Ellul was to term "la technique" (1954), as methods such as operations research and statistical planning proved vital to the war effort. Figures such as Haldane and Bernal were intimately involved in such work in the UK, for instance. But the problems of post-war Britain did not make this a fertile base for FS (though the odd "genius forecast" continued to appear from the likes of Arthur C. Clarke – and notably UK science fiction was very vigorous in this period).

In the US, the situation was different. The wartime experience of big science and mission-led programmes – notably the Manhattan Project – was of great symbolic importance. (This eventually fused, under the impetus of the challenges of Sputnik and Gagarin from the USSR, with the "frontier" narrative, to underpin NASA's space programme.) Innovation (nuclear weapons, jet aircraft, and many more) and marshalling of mass production were seen as vital to winning the war, and military expenditure became *de facto* industrial policy. Shortly after the war a series of "think tanks", such as RAND, emerged, providing research and strategy services to the US military (and intelligence services). Such organisations became high-status centres of multidisciplinary research, where academics could work with representatives of both sides of the military–industrial complex, considering wide-ranging political, social and technological issues. Many of the familiar tools of modern FS were nurtured in this context.

The dominant figure on this US scene – though not the primary innovator of the underpinning methods – was Herman Kahn, who became well-known – if not notorious – in the 1960s and 1970s. Across a series of studies – with ambitions growing from the analysis of prospects for nuclear war (and for "winning" it) to scenarios for the Year 2000, to studies of the next two centuries – he presented scenario analyses, extrapolations and Delphi results. His work synthesized the results of numerous studies, and established many methods of "extrapolative" forecasting, within a highly techno-optimistic framework. The space race – itself stimulated in no small way by military interests – has also played a notable role in providing techniques of "normative" analysis for goal-directed futures research – relevance trees, morphological analysis, and so on.

Much of this work was focused on prospects for the USA and other advanced industrial countries. An important step in the move of FS out of the military arena towards more general social analysis was the work of The Commission on the Year 2000. This was set up by the American Academy of Arts and Sciences, on the initiative of Lawrence K. Frank – one of Ogburn's collaborators on *Recent Social Trends*. The case was that the implications of rapid social and technological change for America's way of life needed to be thoroughly examined. The work of the Committee was reported in an issue of the Academy's journal *Daedalus* in summer 1967 (over 1000 pages in length!), with contributions from Herman Kahn and many luminaries of American social science. This concentration of intellectual fire power did a great deal to legitimise the study of the future as a serious affair, as well as beginning to popularise ideas such as "post-industrial society" as ways of describing the emerging social order.

Authors like Kahn proclaimed the emergence of "futurology", as a way of expertly synthesizing the results of different forecasting studies and methods to provide wide-ranging visions. This terminology was less popular in Europe, where the technocratic orientation of much of this work was widely distrusted, together with what seemed to be an unduly North American flavour. This distrust also reflected suspicions about the role of organisations such as RAND in the era of the Cold War – the survivability of nuclear war looked a lot less plausible to Europeans, and a lot less probable for Europe, where US interventions in developing countries were finding wide disfavour.[6] In Japan, however, the approaches being developed in the USA were recognised to have considerable potential for helping that country's policy-makers identify areas of technology development where industry could do more than simply imitate Western innovations. This was to be the basis of the Japanese foresight effort, beginning in the early 1970s, as strategic futures tools were applied to the task of creating visions of future technological opportunities that could be used to inform and mobilise research priorities in government, and industry, in particular. We can interpret the development of TF in Japan as a Japanisation of tools largely developed in the USA – just as was the case in other Japanese innovations such as "just-in-time" organisation.[7] The OECD also played an important role in consolidating and disseminating information on Tf (Jantsch, 1967) and later (and probably less successfully) technology assessment (Hetman, 1972).

What in retrospect became known as the world futures debate was also being kicked off in the 1960s, as many more pessimistic commentators argued against the techno-optimism of Kahn and other think-tank futurists. Whereas the optimistic vision was one in which affluence and advanced technology would trickle down from the rich world to the poor, a stream of critical writers stressed persistent inequality in the world. Whereas many economists and policy analysts from developing countries took this as a starting point from which to build arguments for alternative development paths (appropriate technology, self-reliance, Marxist revolution, and numerous variants of these approaches were on the cards), a great deal of impact in North America and Europe was secured by books arguing that the main problems confronting the world are demographic. Poor countries are supposedly largely poor because there are too many people for the available wealth; the world as a whole is expected to meet catastrophe as the population bomb explodes and resources are unable to support a massively expanded population.

The 1960s saw an upsurge of FS in Europe, with key individuals in many countries articulating the case for systematic analysis of future prospects – Gaston Berger and Bertrand de Jouvenel in France, Robert Jungk in Austria, Johan Galtung in Norway, and so on. Underpinning a good deal of this work, and perhaps helping to explain its relative strength in France and weakness in the UK, was a desire to find a "third way" between US *laissez-faire* capitalism and Soviet command socialism. The European work tended to place less emphasis on technological change, and often drew on US studies for this element; but there was more emphasis on FS as being not only holistic (rather than focused on a few highly specific trends) but also open to alternatives, able to consider qualitative and structural changes as well as more quantitative and continuous evolution.[8]

This particular element of the philosophy was strongly voiced by many French futurists (see Chapter 5). Alongside futuribles (De Jouvenel, 1967, and the *Futuribles* journal and research group), a French tradition of FS, labelled strategic prospectives or *la prospective*, has been particularly influential. Following especially the work of Gaston Berger (1967), numerous authors have elaborated this approach, but a strong impact in particular has been made by Michel Godet – whose position as Professor of Strategic Prospective at the Conservatoire National des Arts et Métiers (CNAM) in Paris is among the closest approximations to Wells' Professors of Foresight in the world today. He is also director of the Laboratory for Investigation in Prospective Strategy and Organisation (LIPSOR),[9] and has been particularly assiduous in developing and promoting a number of specific futures tools, such as cross-impact matrices, MACTOR, and others (see Chapter 3), that are less frequently used in the Anglo-Saxon world. Godet and colleagues have undertaken prospective studies for organisations of many kinds, on various strategic issues: science, technology and innovation policies are a minority interest in this body of work.

Often in his books, and often at the outset, Godet sets out a particular view of *la prospective* (e.g. Godet 1987, 1994, 2001). The associated diagram even features as the cover to his *Creating Futures* (2001). This

"Greek Triangle" discussed in each of these books sees "strategic prospective" as a management tool enabling anticipation to be linked to action, through appropriation or "incarnation" in earlier versions (see Figure 2.1).[10] Anticipation is defined in terms of future awareness and prospective thought; action as strategic resolve, and planning; and incarnation/appropriation as joint commitment, collective mobilisation, and sharing of values.

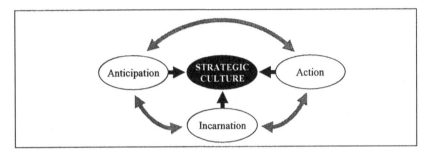

Source: Godet (1994) *From Anticipation to Action*

Figure 2.1   The Greek Triangle

Alongside interest in Tf, a body of work on technology assessment (TA) also began to be developed. This focused mainly on in-depth understanding of implications of, and issues surrounding, specific new and emerging technologies. The approach received a considerable boost with the founding of the Office of Technology Assessment in 1972 as a support agency to the US Congress. This mainly focused on preparing extensive – and often extremely high-quality – reports explicating technoeconomic issues and outlining policy options. The OTA survived Reagan's administration, only to be dismantled by Newt Gingrich's neoconservative Congress in 1995; it had been one of the most authoritative (and expensive – peak funding was around $20 million per year) groups examining future technologies to date.[11] TA approaches took root in a number of countries – notably Denmark, Germany and the Netherlands, and have continued to evolve as an input to decision-making (see, for example, Rip *et al.*, 1995; Vig and Paschen, 2000).

Futures activities were instituted in many countries, with government departments in charge with long-term analysis, with technology assessment agencies, and/or with strong professional groupings of futurists. For much of the 1960s and 1970s, the "futures scene" in the UK was very low key compared to that in most comparable countries. There was activity underway around long-range planning, and social and technological forecasting, in many large companies, but the coordination of, and public interest in, such work was generally muted. The Social Science Research Council (SSRC) did establish a Committee on the Next Twenty Five Years, and later funded a series of seminars on Social Forecasting (Freeman *et al.*, 1975). But these were low-key activities compared to the substantial research and consultative projects in other countries, with their Commissions on the Year 2000 and the like. Though the journal *Futures* was set up in the UK in 1968,[12] unlike France, even now the UK has no Professors of Foresight. A government body that did take a long-term, cross-departmental view, was the Central Policy Review Staff (set up in the early 1970s), but this was deemed unnecessary by Mrs Thatcher, who closed this down (in 1983), along with numerous other state activities that were suspected of trying to pre-empt the wisdom inherent in market forces.

LIMITS TO FUTURES

One set of researchers in the UK did attempt to engage with the futures and technology assessment movements, and it is notable that these were in many respects detached from the academic mainstream. The science and technology (S&T) policy studies community, building on the thinking of Haldane and especially Bernal, recognised that mainstream economics, failing to confront the many complicated issues surrounding research funding and technological change, was unable to provide more than the most anodyne insights for policy-making.[13] While there were a few university departments that focused on the history and philosophy of science, these were often preoccupied with rarefied questions whose relevance to contemporary problems was far from clear.[14] In some cases new S&T policy groups were established, while in other cases more traditional departments mutated in this direction. S&T policy and policy analysis became more detached from its largely socialist roots, and came to be seen as important tools for advanced economies in general. By the mid-1980s there were several research groups active in the area in existence in the UK, with counterparts springing up in many other European countries (Fagerberg, 2004).

Probably the most influential of these groups in the latter quarter of the twentieth century was the Science Policy Research Unit (SPRU) at the University of Sussex (where the Institute of Development Studies was urgently proposing an agenda for S&T for development). Beginning with studies of research expenditure and the dynamics of innovation, SPRU's inspirational founding leaders stressed that it was always important to be aware of the long-term objectives and implications of scientific discovery and technological change. Some staff had close links with the OECD, where TF and TA had been actively discussed. At the beginning of the 1970s, and impressed by the way in which some of the futures research community had articulated questions surrounding long-term technological trajectories, SPRU launched a research programme entitled Social and Technological Alternatives for the Future (STAFF). Almost as soon as this programme had been launched, the global futures debate was brought to the fore in a very striking manner. The Club of Rome, an elite organisation founded by an Italian industrialist, and bringing together many senior business and political leaders, funded a number of studies of what they termed the "global problematique", and achieved a major breakthrough with *The Limits to Growth* (Meadows *et al.*, 1972) – probably the most high-impact futures study ever (for a detailed account of the gestation and aftermath of *Limits*, see Moll, 1991). The massive publicity accorded *Limits* was in no small part due to assiduous marketing by the Club of Rome, who were able to pull many influential strings to secure media coverage and audiences with the great and the good. It also reflected the striking exposition of the underpinnings and results of a computer model applied to assess the future prospects for the world, taking into account a wide range of economic, demographic and environmental variables. At this time, computer modelling had considerable mystique. The impression was readily created that scientists had simply fed the best possible knowledge into a powerful computer, which produced a considered and objective set of conclusions. The impressive graphs that were reproduced in *Limits* were reprinted around the world. And they were especially impressive because they told a tale of impending doom – the world was heading for a Malthusian crisis of Biblical proportions, where population growth would outstrip agricultural and other resources, and catastrophe would result.

The STAFF team decided to take *Limits* as a starting point, since it was such a compelling and influential piece of work, and because its arguments were ones that needed to be taken seriously. While views were initially divided, fairly rapidly opinion within the group crystallised around a consensus that the Club of Rome study was heavily flawed. SPRU produced a detailed study that meticulously took apart *Limits*, and criticised its assumptions, presentation and conclusions (Cole *et al.*, 1973). A major element in the critique was that *Limits to Growth* failed to take any account of how S&T might be used to create a more sustainable future, in which developing countries could share in improved living standards.[15] The world futures debate raged on, as many influential economists and policy analysts took sides, new areas of controversy were staked out, new models were created, and futurists found themselves in demand as diagnosticians for a troubled, changing world.

The STAFF group continued its work though the 1970s. It prepared (fairly sceptical) methodological overviews of FS and reviews of social and technological forecasting in practice.[16] Arguably its most important contribution was a lengthy study which began with a detailed dissection of the debate about global futures that ranged from Kahn and techno-optimists through to prophets of ecological catastrophe. It moved on to consider the prospects for technology development, and to project a range of scenarios for the future of world development – these included futures characterised by high growth and global equity, as well as a range of more problematic futures (Freeman and Jahoda, 1978).

The engagement of FS and Innovation Studies was not just the preserve of SPRU, as other S&T policy groups also undertook futures analysis, technology assessment and similar lines of work. In many ways they were returning to the Ogburn and Haldane traditions, though they often did not recognise this; in some ways they were going beyond them, with more sophisticated analyses of process of change in S&T. The futures movement round the world entered something of a downturn during the 1980s, however. Some of the disfavour resulting from the widespread failure of economic forecasting to anticipate the rough times that the world economy was facing rubbed off onto FS more generally. More importantly, probably, the neo-conservative movement was gathering force, spearheaded by Thatcher in the UK and Reagan in the USA. Wary of government interference in matters that were best determined by "the magic of the market", and suspicious of the radical tenor of much of the world futures debate, neoconservatives had little time for FS. As we have seen, government units began to be dismantled.

While some futures and forecasting consultancies found that private business was still eager to use their insights, much of the policy interest in futures work evaporated. The STAFF team, for instance, moved on to new pastures – studying the growth of the service economy, modelling regional development or employment trends, and evaluation, and innovation studies dealing with new IT and the R&D programmes that supported this technology's development. These latter lines of work were important platforms for developing analyses of the systemic nature of innovation processes – ideas such as technoeconomic paradigms (new sets of problems and solutions, of institutional practices and cultural understandings, surrounding critical new technologies), revolutionary technologies (innovations that are so fundamental and so useful that they are employed to restructure huge swathes of social and economic practice), and innovation systems (the complexes of formal institutions, less formal networks, and flows of knowledge between researchers, financiers, innovators, users and policy-makers, that were posited as accounting for differences between countries, regions and economic sectors and clusters in the speed and shaping of technological change).

Though global FS were out of favour, there was a huge volume of speculation and some fairly systematic analysis of the implications of new IT – especially its significance for employment, but many other topics were addressed, in work that lies somewhere between FS and TA. S&T policy studies – or innovation studies as they became increasingly known – did mushroom during the 1980s. Innovation research centres proliferated round Europe in particular – because European countries were very commonly preoccupied by concerns that they were lagging behind not only the USA, but also new entrants like Japan and Korea, in terms of technology development and application.[17] Growing awareness of the strategic role of major new technologies (especially IT – but already people were identifying new biotechnology as a technological revolution in waiting), and acceptance that innovation was critical for national competitiveness, made innovation studies one of the most rapidly growing fields of social research. Policy debates about how to best encourage innovation often took this to be very much a matter of research policy. With the rise of neoconservativism, the idea of intervening in economic affairs through industrial policy became anathema. The role of government would henceforth be to correct market failures – by breaking up state monopolies and by funding precompetitive research – and to facilitate a more innovative environment. The question was one of how research funding (and organisation) could best be tailored to support innovation.

Research policy had in any case become a problematic issue. Especially as Western economies entered more turbulent times, it was apparent that the practice of regularly funding increases in the science budget, with the distribution of funds determined by an entrenched scientific elite, was unsustainable. If increases were not moderated, all of the workforce would be engaged in R&D; and if funds were not allocated to priority fields, European countries would continue to miss opportunities emerging in promising new fields. New ways of determining research priorities were required – and TF was among the solutions (other approaches included research evaluation, and critical technology analysis, all of these approaches drawing on the analysis of innovation as a systemic process).

An important step in the configuration of the present field of foresight studies came from SPRU, where two researchers undertook studies of the application of futures methods to S&T policy-making round the world – and applied the term "foresight" to these activities. Irvine and Martin introduced "foresight", according to Ben Martin, as a counterpoint to the retrospective analyses of the sources of technology development. Though these had been undertaken in several SPRU innovation projects, at the time the most famous example was the American "Project Hindsight". Their discussion of the Japanese experience in particular, brought "foresight" into play as the way to describe more forward-looking analysis. Other centres were meanwhile pursuing similar lines of work in the innovation studies area and in strategic decision-making for technology policy.

The issue of research prioritisation was the starting point for Irvine and Martin's reviews of foresight activities round the world: *Foresight in Science* in 1984, and then in *Research Foresight* (Martin and Irvine, 1989) Technology Foresight was seen as necessary for informed research policies, but also as intervening in a system marked by "complexity and interdependence". Multiple stakeholders would have to be mobilised to successfully mount major programmes, develop and apply disruptive technologies – and more generally, to achieve a better-functioning innovation system. After a number of false starts, the proposal to apply such approaches in the UK (and similarly in several other EU countries) was accepted. One result of the contribution of innovation research groups to thinking about and planning for UK Foresight was that an emphasis on systemic features of innovation, and on the institutional frameworks which are key components of these systems, was central to the design of the first Foresight Programme.

Irvine and Martin were particularly impressed by the Japanese system, where foresight processes were embedded within networks of academics and industrialists, policy-makers and entrepreneurs. Japanese decision-makers had taken FS techniques (most notably, Delphi surveys) and embedded them into their own cultural and institutional settings. Even in their earliest works, Irvine and Martin drew attention to the Japanese activities' process as well as the product elements, their combination of bottom-up inputs and systematic analysis of trends and efforts across a broad range of areas of science, and so on. Japanese foresight tools were designed, configured and used within a specific institutional environment, enabling the long-term perspectives that were created to be effectively linked to coordinated action. Whereas in Germany and France there were attempts to transplant the Japanese Delphi instruments with little modification other than translation (and dropping of questions dealing with what were seen as specifically Japanese concerns, such as earthquake prediction and prevention), in the UK a combination of newly designed technical tools and social networking was generated. The specific model of the first UK Foresight Programme was born out of these considerations. It explicitly built on the recognition that the tools that had been effective in Japan could not simply be transplanted without modification to the UK. Foresight would need to address the poor linkages in the UK innovation system, especially between the scientific research base and industry. Thus extensive networking across the parts of the system, via work in Panels and other meetings, was built into the Programme. Detailed accounts of the structure and orientation of the first period of UK Foresight are provided elsewhere (e.g. Georghiou, 1996; Keenan, 2003; Miles, 2005).

The UK Foresight Programme proved to be extremely influential, since it represented an exercise that involved priority-setting and networking, and tailored TF methods to fit the UK innovation system. In this it was seen as a better example than French and German attempts to reuse the Japanese Delphi survey, and the Dutch programme which spanned a wide area of fields of enquiry (not just technology) and avoided overall priority-setting. Many S&T policy and innovation research centres (typically based in universities or foundations) were important players in the first wave of Foresight Programmes across Europe (and in this they followed the Japanese model, though usually with more distance from government than the Japanese agency). As the term foresight gained a cachet, other established "futures" and forecasting groups – usually private consultancies – reacted by rebranding much of their activity as TF or more generally "foresight". As they began to displace the innovation studies community as perceived sources of foresight expertise, foresight has come to be more closely aligned to traditional FS – which may account for some of the mixed experiences in recent rounds of TF (e.g. the termination of the second cycle of UK Foresight, see Chapter 4).

In order to explicate just what was new about TF as it emerged in the mid-1990s, we have introduced the term "Fully-Fledged Foresight" to describe the combination of three elements:

- Prospective studies of long-term opportunities and alternatives;
- Participatory networking; and
- Policy orientation.

Figure 2.2, below, captures the essence of "Fully-Fledged Foresight" – and it will be clear that there are marked similarities to Godet's "Greek Triangle" (Figure 1.1), despite different terminology and nuances, and more of a focus on technological innovation. The similarities do suggest that TF has a great deal in common with *la prospective* (in Figure 2.2 "prospective" is used merely to cover Godet's "anticipation"). The balance of emphasis on the three elements varied considerably, but TF programmes in Europe from the mid-1990s on typically combined these policy-relevant, participative and prospective activities. The critical difference from more conventional FS is that these TF programmes are (a) linked to policy actions – often designed to feed into a process of priority-setting, in particular; and (b) draw on wider sources of knowledge than just an expert group, and often deliberately engage in networking not just to access broader sources of knowledge or to legitimate forecasts (and/or priorities) as stemming from a broad base, but also to enable shared knowledge to enter into the strategies of many organisations across the economy and society.

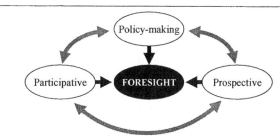

Foresight is a set of approaches to bringing longer-term considerations into decision-making, with the process of engaging informed stakeholders in analysis and dialogue being important alongside the formal products that can be codified and disseminated.

- Policy-making approaches adopt a longer-term perspective in the form of strategic planning, allowing flexibility and preparedness to deal with uncertainty, disruptive events and innovations. Another important role here is enabling greater integration and "joining-up" of discrete and compartmentalised lines of action. Foresight activities also bring material that can aid in prioritisation (setting priorities for R&D was a key goal of many national TF programmes). With increased need for coordination across policy areas, and mobilisation of effort across public and private actors, partnership can be fostered through the use of foresight in planning processes.

- Participative approaches involve interaction of wider ranges of stakeholders and experts in envisioning the future. This reflects several goals: (a) enlarging the knowledge base that is drawn upon, in recognition of the point that no single body encompasses all of the knowledge required to understand future opportunities and how to seize them – especially as the world grows more complex (through advances in S&T, through greater social differentiation, etc.). The technocratic rationale for participation lies in this distributed knowledge. But there is also a democratic rationale, (b) engagement, aimed at enhancing the democratic basis of future visions. This can give foresight processes and recommendations more legitimacy. Related to this is (c) enlistment, the mobilisation of those involved in the process as actors that can embed the messages of the programme into their own organisations/practices.

- Prospective approaches involve traditional forecasting efforts, using systematic methods to explore future dynamics, enabling development of coping strategies. A critical element of much foresight work is some matching of (present and forecast) opportunities and capabilities, framing a vision of desirable and feasible futures. These are the classic extrapolative and normative approaches, and between these are multiple scenario analyses, forging outlines of alternative development paths and possibilities.

Source: Miles (2005), drawing on Miles and Keenan (2002)

Figure 2.2   Three facets of Fully-Fledged Foresight

The original rationales for involving large numbers of experts through methods such as Delphi were (a) the desire to avoid domination of debate by a few particularly prominent voices; and (b) accessing a wider range of expertise than would otherwise be available. Arguably, anonymity played an important role in making Delphi approaches initially attractive in Japan. Such a problem is not restricted to Japan, of course, but the need to gain access to a wide knowledge base was a major driver in European Foresight. This point grew in importance as newly emerging technologies involved researchers active on many frontiers, required interdisciplinary work, and raised issues about new markets as well as new social and ethical concerns. Repeatedly, the 1990s TF programmes in Europe found themselves addressing unexpected social issues: this caused problems when the programmes had not been set up so as to systematically involve social science expertise, which could at least have mediated between critical social actors and the domain experts. More recent activities tend to recognise this requirement. The stress on the participative element has extended the scope of stakeholders beyond a narrow conception of users and suppliers in the innovation system, to a much wider social community. This marks a difference between TF in Europe (and other locations) and forecasting and critical technology analyses in the USA. The latter rarely involve the stress on broad participation developed in European work (especially when we see this as more than a matter of taking part in Delphi surveys, and see it as also including consultation exercises, public awareness workshops, consensus conferences, etc.). And participation extends beyond the technocratic imperative (gathering widely distributed intelligence) and the issue of implementing visions of desirable futures in societies with numerous power bases (distributed governance); there is also some prospect of widening democratic involvement in choice of technology and social development more generally.

As such, foresight now involves more than just anticipating the "impacts" of technological developments *à la* H.G. Wells, but a much more active process of shaping of technological development. Of course, even if foresight were to involve much higher levels of democratic participation than it currently does, it would be only one of a host of processes that shape social and technological futures. But it does add to the repertoire of methods – from SF and FS to TA and Tf (not to mention more conventional politics) – that are available for societies to reflect upon their trajectories.

This points to some of the major uncertainties surrounding the future of foresight. While the conditions that have given rise to the need for TF – rapid technological change across numerous frontiers, in more complicated and globalising societies, etc. – are likely to remain with us into the foreseeable future, there are several tensions visible in current TF practice. Among these we would highlight:

- Tension between three drivers: (a) the need to bring systematic appraisals to bear across a wide range of S&T fields that are candidates for RTD investment and innovation policy; (b) the need to explore the promises and problems of particular lines of development (breakthrough technologies like nanotechnology, or critical problems where one or other technology may be a large part of the solution, such as carbon sequestration and other ways of adverting disastrous climate change); (c) the need to understand and map out broader lines of social and economic development, which may influence adoption and implementation of specific technologies, and in turn be shaped by the pattern of use that results;

- Related to this, tension between the approaches and methods of TF developed by the innovation studies community, and the more general approaches of FS that have been rebranded as foresight by the wider FS community – notably lacking many of the emphases on system embedding and networking that have crystallised in TF programmes;

- Tension between, on the one hand, the need for expert knowledge of complicated issues, and the increasing applicability of sophisticated techniques of simulation, data mining, and statistical analysis to mapping out such issues, on the other hand, and the requirements for broad and often lay participation in foresight. There are problems of lay assessment and comprehension of cutting-edge analyses, and much scope for alienation and mystification – or for foresight to turn into a "public awareness of S&T" propaganda exercise;

- Tension between foresight activities conducted with different premises and missions in different institutional locations in a world of "multi-level governance". The problem is not that different institutions are undertaking their own activities, but that there may be overuse of the same experts and limited pools of knowledge, and problems in recognising the reasons for divergence and the steps required for synthesis of different types of exercise.

How these tensions will be worked out will be a matter of strategies and arguments, and the precise contours of the foresight of the future are hard to predict. What is likely is that we will see efforts to introduce an improved terminology to differentiate between distinctive approaches and combinations of approach. We can also anticipate increased use of approaches from TA and Tf within foresight, and more use of foresight-type approaches and philosophies in other fields.

NOTES

1   Thus Wordsworth's famous lines at the beginning of the nineteenth century, in reaction to the French Revolution: "Not in Utopia, – subterranean fields, / Or some secreted island, Heaven knows where!/ But in the very world, which is the world/ Of all of us – the place where in the end/ We find our happiness, or not at all!"

2   Around this time, Francis Bacon, another key figure in the scientific revolution, was promoting an expanded role for research in his utopian *The New Atlantis*.

3   It is well worthwhile taking the time to examine the original texts from this period, not least to see how much of current forecasting and indicators practice was anticipated in these studies; for a convenient overview, see Duncan (1964).

4   This tradition of work continued well into the 1940s, with studies of the future of the family, of aviation, even of atomic war. Gilfillan, like Ogburn, contributed to early social studies of invention and innovation processes.

5   J.B.S. Haldane had been one of the contributors to the publisher Kegan Paul's series of "genius forecasts", where prominent intellectuals gave views on the likely future of different areas of human activity. Haldane argued in *Daedalus* (1923) that scientific progress made for social progress – while Bertrand Russell focused on the use of science for oppression and destruction in his *Icarus* (1924). Haldane went on to argue about science policy and the social ends of science, with books like *Science and Ethics* (1928), *Science in Everyday Life* (1939) – and a great deal of practical activity. A science fiction utopia, *The Man with Two Memories*, went unpublished in his lifetime (Merlin Press in London eventually printed the text in 1976). John Bernal wrote important texts such as *The Social Function of Science* (1939) and *Marx and Science* (1952).

6   This was crystallised for social scientists by Project Camelot, designed to enhance the ability of the US to predict political developments in Latin America and to shape futures in its own interests (Horowitz, 1967). It was exposed when a European futurologist whose cooperation had been sought leaked details to Chilean politicians. On scholarly inputs to US involvement in Indochina, see Chomsky (1969).

7   Arguably, the implementation of Delphi method in Japan took a form that promoted and emphasised consensus, for example the way in which results were aggregated in providing feedback to respondents was one that tended to stress majority opinions. The Japanese foresight activities were originally described as "Technology Forecasts".

8   US FS did often produce multiple scenarios – but in Kahn's work, for example, these were variations on a single dominant trend, rather than dramatically different development paths. Qualitative change was implicit in visions of post-industrial society (from Daniel Bell and others, e.g. Bell, 1972) but these tended to assume convergence to a technocratic, "end of ideology" future, where the main alternative are means rather than ends.

9   He was earlier active with the European Commissions FAST Programme, and with the SEMA Prospective Consultancy.

10 Why the "Greek Triangle"? In *From Anticipation to Action* Godet quotes Thierry Gaudin as talking of Sophon (the spirit), Techne (the material world), and Poesis (emotion). In *Creating Futures*, the reference is to Logos (thought, rationality, discourse), Epithumia (desire in all its noble, and not so noble, aspects) and Ergo

(action or realisation). The latter recalls the psychologists' differentiation of cognition, affect and conation.

11 Its publications – many containing approaches that could well be emulated, even if the data are becoming out of date – available at http://www.wws.princeton.edu/ota/ns20/pubs_f.html.

12 There are several other well-established journals with more or less a UK base, including *Long-Range Planning*, the *International Journal of Forecasting*, and more recently *Foresight* (est. 1999). But there has been a dearth of university courses in futures studies, and until the turn of the century the futures movement was widely treated as less than respectable in scholarly quarters.

13 Thus innovation research came to be identified with evolutionary, neoSchumpeterian, and other heterodox traditions in economics.

14 Over time this relevance became increasingly clear. In particular, Kuhn's analysis of scientific paradigms and revolutions was to be instrumental in inspiring the notions of technoeconomic paradigms and technological revolutions, which were to become major conceptual rallying points in innovation studies (see Leydesdorff and van den Besselaar, 1994). More generally the fields of policy-related innovation studies and of more academic science studies remained fairly detached. The result was that when sociologists of science moved in to develop a sociology of technology, they could readily dismiss the innovation studies literature as irredeemably technologically deterministic, while laboriously rediscovering (and relabelling) many of the key results that it had contributed (see Woolgar, 1991).

15 Growing environmental crises might be (and often are) taken to mean that *Limits* was a vital early warning call; to criticise it was to help delay the inevitable realisation of, and responses to, these crises. But innovation researchers will point out that, unless we are to write off a large share of the world's population, "clean" (or at least "cleaner") technology has to be part of the solution. It was a major failing of *Limits* that it ignored this, stoking up despair.

16 Encel *et al.* (1974), Whiston (1979).

17 Innovation studies took off especially strongly and early in the UK – probably in response to the longstanding diagnosis of the country's economic difficulties reflecting an innovation problem. This formed a counterpoint to the (neo)conservative notion that these difficulties derived from the strength of the labour movement. It was a view that attracted more adherents as the power of trade unions was progressively dismantled under Thatcher's governments. Part of the problem was seen as being that public research priorities were ill-informed and inadequately linked to industry – while industrial managers were failing to exploit knowledge from public science or to funding their own R&D at levels similar to those of overseas competitors. S&T policy rose on the agenda in the UK, and in many other parts of the world, in the last quarter of the century, as concerns grew about limiting and redirecting scientific funding, and about taking into account revolutionary developments in new Information Technology (especially with the Japanese "Fifth Generation" programme at the beginning of the 1980s, which was seen as threatening to move Japan from its stereotypical position as brilliant imitator to pioneering innovator).

REFERENCES

Bell, D. (1972), *The Coming of Post-Industrial Society*, New York: Basic Books.

Berger, G. (1967), *Étapes de la Prospective*, PUF: Paris.

Chomsky, N. (1969), *American Power and the New Mandarins,* New York: Pantheon Books.

Cole, H.S.D., Freeman, C., Jahoda, M. and Pavitt, K.L.R. (1973), *Thinking About the Future*, London: Chatto & Windus; also published as *Models of Doom: A Critique of Limits to Growth*, New York: Universe Books.

De Jouvenel, B. (1967), *The Art of Conjecture*, New York: Basic Books (originally published in France, 1964).

Duncan, O.D. (ed.) (1964), *William F. Ogburn On Culture and Social Change*, Chicago: University of Chicago Press.

Ellul, J. (1954), *La Technique ou l'Enjeu du Siècle*, Paris: Armand Colin; reprinted (1990) Paris: Économica; English translation (1964) as *The Technological Society*, New York: Knopf.

Encel, S., Marstrand, P. and Page, W. (eds) (1974), *The Art of Anticipation*, London: Martin Robertson.

Fagerberg, J. (2004), 'Innovation: A Guide to the Literature', in Fagerberg, J., Mowery, D.C. and Nelson, R.R. (eds), *The Oxford Handbook of Innovation*, London: Oxford University Press.

Freeman, C., Jahoda, M. and Miles, I. (eds) (1975), *Problems and Progress in Social Forecasting*, London: Social Science Research Council.

Freeman, C. and M. Jahoda, (eds) (1978), *World Futures: The Great Debate*, London: Martin Robertson.

Georghiou, L. (1996), 'The UK Technology Foresight Programme', *Futures*, **28**(4), pp. 359–377.

Godet, M. (1987), *Scenarios and Strategic Management*, Butterworth.

Godet, M. (1994), *From Anticipation to Action*, Paris: UNESCO.

Godet, M. (2001), *Creating Futures: Scenario Building as a Strategic Management Tool*, Economica-Brookings.

Horowitz, I.L. (ed.) (1967), *The Rise and Fall of Project Camelot*, Cambridge, MA: The MIT Press.

Irvine, J. and Martin, B. (1984), *Foresight in Science*, London; Frances Pinter.

Jantsch E. (1967), *Technological Forecasting in Perspective*, Paris: Organization for Economic Co-operation and Development.

Leydesdorff, L. and Van den Besselaar, P. (eds) (1994), *Evolutionary Economics and Chaos Theory: New Directions in Technology Studies*, London: Pinter.

Martin, B.R. and Irvine, J. (1989), *Research Foresight: Priority-Setting in Science*, London: Pinter.

Meadows, D.H., Meadows, D.L. and Randers, J.W. (1972), *The Limits to Growth*, New York: Universe Books.

Miles, I. (2005) 'UK Foresight: Three Cycles on a Highway', *International Journal of Foresight and Innovation Policy*, **2**(1), pp. 1–34.

Miles, I. and Keenan M. (eds) (2002), *Practical Guide to Regional Foresight*, (available in various country versions with local editors), Brussels: European Commission DG Research, available at: http://cordis.europa.eu/foresight/cgrf.htm.

Moll, P. (1991) *From Scarcity to Sustainability: Futures Studies and the Environment*, Frankfurt am Main: Peter Lang.

National Resources Committee (1937), *Technological Trends and National Policy including the Social Implications of New Inventions*, A Report of the Sub-Committee on Technology, Washington DC: US Government Printing Office.

Report of the President's Research Committee on Social Trends (1933), *Recent Social Trends in the United States*, New York: McGraw-Hill.

Rip, A., Misa, T.J. and Schot, J.W. (eds) (1995), *Managing Technology in Society. The Approach of Constructive Technology Assessment*, London: Frances Pinter.

US National Resources Committee (1937), *Technological Trends and National Policy, Including the Social Implications of New Inventions*, Washington, DC: US Government Printing Office

Vig, N.J. and Paschen, H. (eds) (2000), *Parliaments and Technology: The Development of Technology Assessment in Europe,* Albany, State University of New York Press.

Wells, H.G. (1932), 'Wanted – Professors of Foresight', reprinted in *Futures Research Quarterly* (1987), **3**(1), (Spring), pp. 89–91; and in Slaughter, R. (ed.) (1989), *Studying the Future*, The Commission for the Future and The Australian Bicentennial Authority.

Werskey, G. (1978), *The Visible College: Scientists and Socialists in the 1930s*, Harmondsworth: Penguin.

Wesley, C.M. (1933), *Recent Social Trends in the United States: Report of the President's Research Committee on Social Trends*, Volume 1, New York: McGraw-Hill.

Whiston, T. (ed.) (1979), *Uses and Abuses of Forecasting*, London: Macmillan.

Woolgar, S. (1991), 'The Turn to Technology in Social Studies of Science', in *Science, Technology, & Human Values*, **16**(1), (Winter), pp. 20–50.

3. Foresight Methodology

Rafael Popper

INTRODUCTION

"Insanity: doing the same thing over and over again and expecting different results"
Attributed to Albert Einstein (1879–1955)

This chapter is an overview of foresight methods and common practices, outlining various ways in which a wide range of techniques can be classified as well as more practical tips on their selection and use. The main objective is to help foresight organisers and practitioners to improve the methodological designs of their projects and programmes, so that different and better results are achieved.

While there are plenty of discussions of foresight methods, many fall into a range between descriptive narratives of how methods are used in projects and thinly-disguised promotions of consultancy services and their "pre-packaged solutions". Some popular methods (e.g. Delphi, scenarios, SWOT and roadmapping) have attracted many articles and commentaries, though they typically fail to compare methods systematically (an exception is Scapolo and Miles, 2006) or to explicate the numerous decisions that need to be made in applying any tool and making use of its results. On this last point, Slaughter (2004) points out that "it is the depth within the practitioner that evokes depth and capacity in whatever method is being used". Part of this depth requires the acknowledgement of foresight as a process. For this reason, section one kicks off the discussion with a practical note on how methods relate to the foresight process and its phases.

The chapter also provides definitions for 33 methods in addition to various classifications and methodological frameworks with the overall purpose of raising the profile of forward-looking activities and setting the basis for a more structured research agenda for methodological developments in the foresight field.

FORESIGHT: THE PROCESS, PHASES AND APPROACHES

Foresight is "a process which involves intense iterative periods of open reflection, networking, consultation and discussion, leading to the joint refining of future visions and the common ownership of strategies … It is the discovery of a common space for open thinking on the future and the incubation of strategic approaches" (Cassingena Harper, 2003). A more systemic look into the *process* was made by Miles (2002) who outlined five complementary phases (Pre-Foresight; Recruitment; Generation; Action; and Renewal) which are here used to describe common practices (see Figure 3.1).

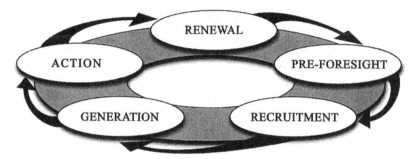

Figure 3.1 The five phases of the foresight process

Common Practices in the Pre-Foresight Phase

The *Pre-Foresight* or Design Phase (also known as *Scoping*) is the starting point of the process, where practitioners, together with the sponsor:

- Define the rationales and objectives;
- Assemble the project team; and
- Design the methodology.

For the definition of rationales and objectives the sponsor's participation is essential given that final outcomes should (ideally) inform future decisions. Some common objectives of foresight are: (a) fostering science, technology and innovation (STI) cooperation and networking; (b) orienting policy formulation and decisions; (c) recognising key barriers and drivers of STI; (d) encouraging strategic and futures thinking; (e) supporting STI strategy- and priority-setting; (f) identifying research/investment opportunities; (g) generating visions and images of the future; (h) helping to cope with "Grand Challenges" (e.g. climate change, natural disasters, terrorism, poverty, etc.) and (i) triggering actions and promoting public debate (Popper *et al.*, 2007).

Once objectives are clearly defined and (hopefully) shared, the project team should be assembled and a relevant methodological framework prepared, that is, creating a logic diagram of tasks and milestones. The work plan is normally sketched internally (sponsor + project team), but the thorough definition of milestones and its interconnections often requires an advisory group be created, with methodology experts and practitioners distributing tasks and defining work packages for teams and partners linked to the project. Choosing the right methodology is like choosing outfits: for formal meetings a suit with a tie may be appropriate; for informal meals then jeans and T-shirt could be fine; but, for scuba diving none of the above would help. In other words, methods must be chosen after the objectives (and the overall scope of the project, see Chapter 15) are defined and not the other way around.

Decisions about the methodological framework are also affected by resources: the budget, the availability of expertise, political support, technological and physical infrastructure, and time. For example, in a four-month project it would not be convenient to plan for large-scale Delphi surveys based on statements created by panels of experts, etc. Having valuable human resources (skilled and talented researchers) is essential. Such people do not necessarily need to be foresight specialists, but will often require intensive training courses (usually three to five days) in order to build internal capacities. Finally, technological support may facilitate the process: web-based communications and email are widely used and some tools allow for more work to be conducted online (but mastering these requires additional time and training).

After the methodology is designed, the team normally builds (or contracts out the building of) communication instruments, such as websites, leaflets, pamphlets and training material (especially important if the process involves methods such as benchmarking, modelling or Delphi). The Pre-Foresight phase does consume resources and also involves the use of certain methods. For instance, to identify the project objectives, there may be commissioning of literature reviews (on foresight itself or on the substantive issues), scanning, bibliometric or patent analysis, etc., from academics or research institutes.

Common Practices in the Recruitment Phase

Recruitment is normally ongoing with higher and lower intensity during the whole process. The core team (i.e. scientific/administrative coordinators and main sectoral/thematic/methodological experts) is usually built in the scoping phase; but additional members are often incorporated to the team (e.g. process facilitators, rapporteurs and expert panel members, among others).

However, enlisting key sources of information and stakeholders (who can supply not only knowledge but also act as "carriers" of foresight to wider constituencies) has to be undertaken at an early stage in the process. Identifying these key stakeholders and individuals, as well as enlisting their support and commitment, are the main activities of this phase. More or less formalised methods like stakeholder analysis, co-nomination surveys, brainstorming, and sounding the opinions of known contacts are widely used. Approaches such as bibliometric and patent analysis are occasionally used, especially for the identification of key researchers and groups (as well as giving insight into themes, technologies, etc.).

Common Practices in the Generation Phase

The *Generation Phase* is usually considered the heart of the process, where: (a) existing knowledge is amalgamated, analysed and synthesised; (b) tacit knowledge is codified; (c) new knowledge is generated; and (d) new visions and images of the future are created. New knowledge may result from the elucidation of emerging/prospective issues or from the amalgamation of existing knowledge but – whichever is the case – it should be relevant to the objectives defined in the *Pre-Foresight Phase*.

A sensible way of organising activities is to think of the *Generation Phase* as consisting of three main stages:

1. *Exploration* – understanding main issues, trends and drivers; and understating how 'key stakeholders' frame the context of the exercise;
2. *Analysis* – understanding how the context and main issues, trends and drivers influence one another; and synthesising knowledge generated in the exploration stage; and
3. *Anticipation* – considering previous analysis, this stage is aimed at anticipating possible futures and/or suggesting desirable ones.

Different types of knowledge sources (based on creativity, expertise, interaction and evidence) may be used to support these stages and design a comprehensive methodology (see Figure 3.3 and related discussion). These could also involve a number approaches reflecting various elements of the present situation and future contingencies. Among these are the following:

- Positioning or outward-looking approaches, that contrast the internal context with the external one. This approach is often used when foresight is stimulated by international (or industrial) competition, and thus aimed at comparing local and global trends, drivers and technologies. Its main challenge is the identification of commonalities. The need for comparable information makes methods such as, for example, Delphi, key technologies, SWOT analysis, benchmarking and expert panels widely used here;

- Networking or inward-looking approaches, that promote cooperation by linking stakeholders at different levels and areas. Interactive and participatory methods like stakeholder analysis, voting, citizen panels and workshops are relevant methods (see Slocum, 2003). Often these require development of shared perspectives on the present and future, as developed through the other approaches;
- Prospective or forward-looking approaches, which require that participants produce clear and plausible images of the future – often these will include desirable or aspirational options – while also if possible "thinking outside of the box" and avoiding "business as usual" perspectives. Common techniques draw on creativity and expertise, and include brainstorming, trend extrapolation, Delphi, key technologies and scenario workshops, among others.

Common Practices in the Action Phase

The *Generation Phase* should ideally produce "new knowledge" and compelling visions such that an *Action Phase* will follow it. But if intermediate outputs are not particularly new or simply do not stimulate the sponsor, then the process could finish just after contractual obligations are delivered. Using the wrong language may also result in lack of action (e.g. policy recommendations should be framed in a suitable lexicon, while business strategies may highlight market strengths and opportunities). A lack of focus can also hinder further use of results; for example, when an enormous amount of information is generated but there is insufficient synthesis. And it may be that the sponsor undergoes change such that they cannot act – there may be a regime change or political crisis, for instance.

But every effort should be made to ensure that foresight informs decisions, and, in so doing, the *Action Phase* may involve:

- Prioritisation and decision-making. Here methods like polling and multi-criteria analysis are used to aid decisions and broader strategies;
- Innovation and change. This may involve moving directly forward to promoting or transferring particular strategies, technologies or policy instruments, for example – or to changing attitudes and lifestyles. Here *expertise-based* creative methods (such as genius forecasts, technology roadmaps, backcasting, writing of narrative scenarios and

dramatising them through various media) are often useful to disseminate the visions of the future and underlying thinking achieved in earlier phases, while *interaction-based* creative methods (e.g. scenario workshops and role play) may help foster deeper awareness among smaller groups.

Common Practices in the Renewal Phase

The *Renewal Phase* involves constant monitoring and evaluation in order to assess whether the foresight process has helped to achieve its original objectives and how far results are being acted on. One main challenge here is the development of success indicators – a process on its own which may generate new questions and even turn into the *Pre-Foresight Phase* of a new process. But as pointed out by Georghiou (1998) it can take considerable time for project effects to become evident – in reference to a Norwegian Foresight study he indicates that some 12–15 years are needed for outcomes to become clear. A second challenge is how to maintain a systematic tracking of interconnected events and results. For follow-up activities, methods like roadmapping and relevance trees are rather useful. Ongoing elements of the *Action Phase*, including consciousness raising and cultural shaping activities such essays and scenario writing, and workshops and conferences are typical.

Here it is important to highlight that foresight should not only be about analysing or contemplating future developments but supporting actors to actively shape the future. Thus, foresight should only be undertaken when it is possible to make use of the outcomes. Both the Action and Renewal Phases are about "transformation". They look at possible implications and lessons that can be drawn for present-day decision-making, policy-making and strategy formulation (including knowledge and policy transfer). In essence, they centre attention on how the future could be shaped for the better. They take into account the practical steps necessary to implement the findings of the exercise, and consider how foresight as practice can become embedded in organisations and communities (Keenan and Popper, 2007, p. 16).

To conclude the discussion on common practices this section includes a table which provides an impressionistic view of the potential contribution that 33 methods might make to each phase of the foresight process. This "potential" is represented with bullets: one bullet [●] indicates little/no contribution, two bullets indicate [●●] some contribution, three bullets [●●●] indicate significant contribution and four bullets [●●●●] indicate major contribution to each Phase of the Foresight Process (see Tables 3.1 and 3.2). For example: Backcasting may have little contribution [●] in the Pre-Foresight, Recruitment and Renewal Phases, whereas significant contribution [●●●] in the Generation and Action Phases.

Table 3.1 Potential contribution of qualitative methods

Methods / Activities	Foresight Phases					Type of method
	Pre-Foresight	Recruitment	Generation	Action	Renewal	
1 Backcasting	•	•	•••	•••	•	Qualitative
2 Brainstorming	•••	••	••••	•••	•••	
3 Citizens Panels	••	•	•••	••••	•••	
4 Conferences/Workshops	••	••	•••	•••	•••	
5 Essays/Scenario Writing	•••	•	••••	••	•••	
6 Expert Panels	•••	••	••••	•••	•••	
7 Genius Forecasting	••	•	••••	••	•	
8 Interviews	••	••	•••	••	••••	
9 Literature Review (LR)	••••	••	•••	••	••	
10 Morphological Analysis	•	•	•••	•••	•	
11 Relevance Trees/Logic Charts	••	•	•••	•••	•••	
12 Role play/Acting	•	••	•••	•••	•	
13 Scanning	••••	••	•••	•••	••	
14 Scenarios/Scenario Workshops	•	•	••••	•••	••	
15 Science Fictioning (SF)	•	•	••••	•	•	
16 Simulation Gaming	•	•	•••	•••	•	
17 Surveys	•••	•••	••••	••••	•	
18 SWOT Analysis	••	•	••••	••••	••	
19 Weak Signals/Wild Cards	••	•	•••	••	•	

Legend of symbols: little/no contribution [•]; some contribution [••]; significant contribution [•••]; major contribution [••••]

Table 3.2 Potential contribution of quantitative and semi-quantitative methods

Methods / Activities	Pre-Foresight	Recruitment	Generation	Action	Renewal	Type of method
20 Benchmarking	●●●	●●	●●●	●●●	●●●	Quantitative
21 Bibliometrics	●●●	●●●	●●	●	●	
22 Indicators/Time Series Analysis (TSA)	●●●	●	●●●	●●	●●	
23 Modelling	●	●	●●●	●●●	●	
24 Patent Analysis	●●●	●●●	●●	●	●	
25 Trend Extrapolation/Impact Analysis	●●●	●	●●●	●●	●●●	
26 Cross-impact/Structural Analysis (SA)	●●	●	●●●	●●●	●●	Semi-Quantitative
27 Delphi	●	●●	●●●●	●●●	●●	
28 Key/Critical Technologies	●●	●	●●●	●●●	●●	
29 Multi-criteria Analysis	●●	●	●●●	●●●	●●	
30 Polling/Voting	●●	●●	●●●●	●●●●	●●●	
31 Quantitative Scenarios/SMIC	●	●	●●●●	●	●●	
32 Roadmapping	●●	●	●●	●●●●	●●	
33 Stakeholders Analysis/MACTOR	●●	●●●	●●	●●●	●●	

Legend of symbols: little/no contribution [●], some contribution [●●], significant contribution [●●●], major contribution [●●●●]

Note: the tables (above) provide an impressionistic view of the contribution that 33 methods might make to each phase of the foresight process. The "potential contribution" is represented with bullets. For example: *Backcasting* may have little/no contribution [●] in the *Pre-Foresight*, *Recruitment* and *Renewal* Phases, whereas significant contribution [●●●] in the *Generation* and *Action* Phases

FORESIGHT METHODS: CLASSIFICATIONS AND USES

This section presents three frameworks for classifying foresight methods and provides short definitions for 33 methods. The first framework is focused on types of techniques (qualitative, quantitative and semi-quantitative). The second focuses on types of approaches, distinguishing between exploratory and normative orientations. The third classifies methods by type of knowledge source (creativity, expertise, interaction and evidence).

Classification of Methods by Type of Technique

The distinction between qualitative and quantitative methods is commonplace in social research. In foresight, the former often refer to the use of more or less narrative and discursive texts, the latter to the analysis of trends and similar data. Here we introduce the category "semi-quantitative" to cover techniques that use more or less sophisticated probabilistic and statistical principles (e.g. SMIC and Delphi, among others) to manipulate judgements or tacit knowledge (i.e. weighting ideas, relationships, conjectures, opinions, etc.). In any case, the availability of more powerful computing capabilities means that there is increasing potential to represent qualitative material in numerical terms and process it using statistical tools. Quantitative analyses tend to follow a specific procedure quite meticulously, and for this reason they are often much easier to replicate than qualitative ones, where more tacit knowledge is required from the researcher. Some practitioners regard qualitative approaches as inherently superior (being able to deal with subjectivity and interpretations), others see them as a second-best set of tools that have to be used when hard data is not available. In many foresight projects, we are dealing with phenomena that have received little statistical attention, and have to apply qualitative approaches. And given that the available knowledge base often consists of a mixture of (partial) knowledge, assumption and ignorance, quantitative methods can be complemented with new qualitative approaches addressing aspects of uncertainty that are hard to quantify and were therefore largely underaddressed in the past (Van der Sluijs *et al.*, 2005). Van der Sluijs *et al.* highlight that although quantitative techniques are essential in any uncertainty analysis, they can only account for what can be quantified in a credible way, and thus provide only a partial insight in what usually characterises foresight activities, that is, a very complex mass of uncertainties. These authors also point out that quantitative methods address only the technical (inexactness) dimension of uncertainty in the knowledge base and that other key dimensions such as the methodological (unreliability), epistemological (ignorance) and societal (social robustness) may be better addressed with the help of qualitative techniques.[1]

The French mathematician René Thom noted that qualitative approaches can have real force: "Descartes, with his vortices, his hooked atoms, and the like, explained everything and calculated nothing; Newton, with the inverse square law of gravitation, calculated everything and explained nothing" (Thom, 1975). To make the point that qualitative approaches are not always second-best, Thom uses a simple example (reproduced below in Figure 3.2), concerning how we might explain the behaviour of a trend.

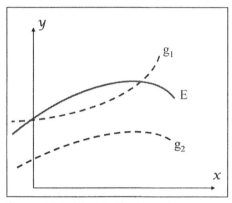

Source: Adapted from R. Thom (1975)

Figure 3.2   The qualitative vs. the quantitative

Suppose that a particular event E gives an empirical plot E (filled line). To explain the event we might have two theories, t1 and t2, leading to the plots g1 and g2, respectively (dotted lines). Neither of the plots fits the graph of the event E at all perfectly; the plot g1 fits better quantitatively in the sense that, over the interval considered, the difference between its predictions and the actual observations, $\int | E - g1 |$, is smaller than that of the other theory $\int | E - g2 |$. But the observer of the graph can see that the plot g2 has the same shape and appearance as E. For this reason, Thom points out the theorist would be likely to retain t2 rather than t1 even at the expense of a greater quantitative error. The rationale would be that t2, which gives rise to a graph of the same appearance as the experimental results, must be a better clue to the underlying mechanisms of E than the quantitatively more "exact" t1.

Of course, Thom's example does not demonstrate that qualitative is better than quantitative; it simply shows that the assumption that quantitative techniques are generally superior or more advanced than qualitative ones is at best dubious. Indeed, there are almost certainly qualitative judgements being made in the selection of quantitative methods, or the choice of which quantitative instrument to use. It is important to look beyond the received wisdom and prejudices of methodological fundamentalists, and consider when particular approaches can best be deployed. For example, many experiences, judgements and opinions are often better analysed using qualitative methods, and much effort has been devoted in recent years into outlining systematic ways of analysing qualitative data (see, for example, Miles and Huberman, 1994). If there is a general rule, it is that the best foresight methodologies are those that combine both quantitative and qualitative approaches, garnering the insights that come from each.

Table 3.3 presents some 33 methods often used in foresight in terms of the typology outlined above.

Table 3.3   Classification of foresight methods by type of technique

Qualitative	Quantitative	Semi-quantitative
Methods providing meaning to events and perceptions. Such interpretations tend to be based on subjectivity or creativity that is often difficult to corroborate (e.g. opinions, brainstorming sessions, interviews)	Methods measuring variables and applying statistical analyses, using or generating (hopefully) reliable and valid data (e.g. socio-economic indicators)	Methods that apply mathematical principles to quantify subjectivity, rational judgements and viewpoints of experts and commentators (i.e. weighting opinions or probabilities)
1. Backcasting 2. Brainstorming 3. Citizens Panels 4. Conferences/Workshops 5. Essays/Scenario Writing 6. Expert Panels 7. Genius Forecasting 8. Interviews 9. Literature Review (LR) 10. Morphological Analysis 11. Relevance Trees/ Logic Charts 12. Role play/Acting 13. Scanning 14. Scenario/Scenario workshops 15. Science Fictioning (SF) 16. Simulation Gaming 17. Surveys 18. SWOT analysis 19. Weak Signals/Wild Cards	20. Benchmarking 21. Bibliometrics 22. Indicators/Time Series Analysis (TSA) 23. Modelling 24. Patent Analysis 25. Trend Extrapolation/ Impact Analysis	26. Cross-impact/ Structural Analysis (SA) 27. Delphi 28. Key/Critical technologies 29. Multi-criteria analysis 30. Polling/Voting 31. Quantitative scenarios/SMIC 32. Roadmapping 33. Stakeholder Analysis/ MACTOR

It will be evident that the methods within each group cover a huge range – from modelling and simulation; through statistical analysis such as bibliometrics and patent analysis; to methods involving group discussions, networking and appropriation. For this reason, we need to examine some other ways of classifying methods that can provide a little more order to this wide variety of tools.

Qualitative methods

This section describes 19 qualitative methods. Most methods centre attention on the interpretation of perceptions with the aim of providing meaning to events. Typically, there is little ability to estimate the rate of occurrence or speed of change, other than in a highly impressionistic way. Such interpretations are often based on subjective judgements, and creative processes – the particular results may be difficult to replicate and/or corroborate (though this is not always the case). These techniques provide depth, rich discussions, and allow for the sharing of viewpoints and enhancement of understanding of different perspectives. Here they are listed in alphabetical order so as to facilitate their location in the text.

Backcasting is an approach that involves working back from an imagined future, to establish what path might take us there from the present. One version of backcasting involves simulation modelling – indeed, this method is much employed with planning models.[2] More commonly, backcasting is used in aspirational scenario workshops. Here, it involves the creation of a desired future and afterwards imagining (generally via brainstorming sessions) all necessary events, actions and milestones that must happen in order for that future to be achieved. It is considered by many as a less elaborated version of Roadmapping (explained below) which also requires the preparation of a timeline. The timeline usually includes key events and measurable goals that need to be reached. There may be some quantification (using probabilities) of the likelihood and feasibility of each event. A major objective of the technique is to identify possible policy and strategies required to approach a desired future (see Dreborg, 1996; Hojer and Mattsson, 2000).

Brainstorming is a creative and interactive method used in face-to-face and online group working sessions to generate new ideas around a specific area of interest. Aiming at removing inhibitions and breaking out of narrow and routine discussions, it allows people to think more freely and move into new areas of thought, and to propose new solutions to problems. The use of this technique was pioneered by Jungk and Müllert (1987) and was mainly applied in what they called "Futures Workshops". Brainstorming is usually conducted in groups that are undertaking open-ended discussions, but may also involve carefully prepared questionnaires and online approaches. The first step is to share and exchange views from a selected group of people. These views are gathered and made available for inspection as they arise, crucially without being criticised or discussed in depth (other than being used to trigger further ideas). Subsequently, all ideas are discussed and clustered into categories (in line with, for example, a framework like the STEEPV).[3]

Citizen Panels are groups of citizens (members of a polity and/or residents of a particular geographic area) dedicated to providing views on relevant issues, often for a regional or national government. The panel is more than a conventional opinion survey, since its members are encouraged to deepen their understanding of the issues involved. Activities carried out by such groups may include: completing questionnaires, discussing specific and cross-cutting issues affecting the community (e.g. environment, crime or local health services), and contributing to community planning (i.e. brainstorming on problems, social needs and possible solutions). They are normally established because of their effectiveness in widening participation. However, on occasions, panels are created to influence particular groups, in other words, they can be politically driven. Whichever the case, one major challenge for citizen panels is to decide how representative of the population the panel is to be, and how to achieve this – not only in terms of gender, age, ethnics or social status, but also in terms of ideologies and political orientation.[4] A second challenge is achieving commitment to investing energy in the process, which may require the organisers going beyond extracting information from the participants to more substantive consultation processes and the provision of coherent explanation of what has changed as a result of their contributions. Finally, we should recognise that Citizen Panels – just like Citizens' Juries – should be seen as a potential instrument to representative institutions, a way of bringing informed citizens' perspectives into the decision-making process (Smith and Wales, 2000).

Conferences/Workshops are events or meetings lasting from a few hours to a few days, in which there is typically a mix of talks, presentations, and discussions and debates on a particular subject. The events may be more or less highly structured and "scripted": participants may be assigned specific detailed tasks, or left very much to their own devices. Conferences are common settings for networking, knowledge exchange and consensus building. Participants comment on recent reports and take the opportunity to present the current status of their projects and research activities. Feedback is used to improve the scope of the foresight processes and/or validate its outcomes. The meetings may be used to spur people to action based on the results of earlier stages of foresight (see also Simon and Durant, 1995).

Essays/Scenario Writing involves the production of accounts of "plausible" future events based on a creative combination of data, facts and hypotheses. This activity requires insightful and intuitive thinking about possible futures, normally based on a systematic analysis of the present. It is often used in genius forecasting, but can also be implemented in systematic approaches, in desk or group work. Typically this will involve requesting the authors to examine each scenario in terms of a set of common features, e.g. what the scenario implies about business models, politics and environmental affairs. Essays may focus on one or a small set of images of the future, with a detailed description of some major trends promoting the evolution of the scenario, and/or of stakeholders' roles in helping to bring these about. They are usually fed by the outcomes of brainstorming sessions, SWOT exercises, Delphi, expert panels and many other activities; they may be prepared in or immediately after workshops, for example. Their main objectives are (a) to describe future situations resulting from the implementation of particular decisions, strategies or policies and (b) to make recommendations on those futures (see also Becker, 1983; Boucher, 1985; Schwartz, 1991).

Expert Panels are groups of people dedicated to analysing and combining their knowledge concerning a given area of interest. They can be local, regional, national or international. Panels are typically organised to bring together "legitimate" expertise, but can also attempt to include creative, imaginative and visionary perspectives. In many foresight programmes, when influence on decision-making is involved, panels are also expected to impact on the environment (disseminating results, building networks and/or reaching commitments). Methods like focused discussions and brainstorming are popular with panels; however thematic panels (e.g. biotechnology or nanotechnology) often use semi-quantitative methods like polling, Delphi and structural analysis, among others. Typical panel activities include: creating of networks and empowering existing ones; developing strategic intelligence; relating studies to much wider constituencies; diffusing results via blueprints, white papers, publications, declarations, and interviews; establishing priorities; and designing follow-up actions (see also Salo and Salmenkaita, 2002; Georghiou, 2003; Havas, 2003).

Genius Forecasting is an activity carried out by respected individuals that requires both expertise and creativity in relatively similar proportions. It involves the preparation of forecasts based on insights of a brilliant specialist, scientist or authority in a given area. There is inevitably a danger of one-sided views and special interest pleading – for example "scientific advertising", when people engaged in cutting-edge activities use media attention or political influence to claim that a certain technology is key or vital. However, it is undeniable that some individuals have particular insights and provocative ways of approaching topics, stimulating fresh thinking and ensuring that important perspectives and possibilities are being taken into account (see also Glenn and Gordon, 1999).

Interviews are often described as "structured conversations" and are a fundamental tool of social research. In foresight they are often used as formal consultation instruments, intended to gather knowledge that is distributed across the range of interviewees. This may be tacit knowledge that has not been put into words, or more documented knowledge that is more easily located by discussions with experts and stakeholders than by literature review. Interviews play an important role in the evaluation of foresight (e.g. assessing how well resources are being or have been used). They normally help in getting a sense of local experiences and understating of how studies are designed and carried out. Interviews may be more or less "open-ended", at one extreme taking a very exploratory form, at the other being much closer to a questionnaire survey that happens to be conducted in a face-to-face manner. More open-ended approaches, as long as the interview is structured with guidelines for major topic or thematic areas (in order to ensure some comparability across themes), areas or sectors, are more effective for pulling in knowledge, but organisation of the qualitative data that results from them can be very challenging (see also Ratcliffe, 2002).

Literature Review (LR) represents a key part of scanning processes (see below). Good reviews generally use a discursive writing style and are structured around themes and related theories. Occasionally the review may seek to explicate the views and future visions of different authors. LR involves the analysis of books, reports, journals or websites, and most often requires an expert in the topic concerned using their existing knowledge of the field to identify crucial contributions and synthesize their implications for the topic at hand (e.g. what they have to say about key drivers, alternative futures, policy instruments, etc.).

Morphological Analysis is closely related to relevance trees (see below), and to the soft-systems approach since it helps both complex problem-solving and management of change; it may be used in planning or scenario development. It maps promising solutions to a given problem and determines possible futures accordingly: the classic applications have involved systematically working through the entire range of conceivable technological solutions for a particular goal (such as attaining a manned mission to the moon). Working groups and/or desk working experts generally use the method to suggest new products or developments and build multi-dimensional scenarios.[5]

Participants consider possible events or hypotheses associated with each dimension. The combinations of developments with events create different paths or scenarios (see also Zwicky, 1969; Ritchey, 1998).

Relevance Trees and Logic Charts are methods in which the topic of research is approached in a hierarchical way. Each begins with a general description of the subject, and continues with a disaggregated exploration of its different components and elements, examining particularly the interdependencies between them. The activities may take place based on deskwork, or within a workshop where expert groups define a high-level goal. The goal is then connected to more specific second-level goals and to the possible means with which these could be achieved, and thus the third-level goals associated with these, and so on. The eventual result is a detailed diagram resembling a plant's root structure or an inverted tree. A relevance tree may be constructed so as to indicate the set of steps and stages required to bring about a desired end result. Logic diagrams are often employed in evaluation work, where they relate the various activities that are being evaluated to the criteria in terms of which the evaluation is to take place (see Chapter 16; Chaudhry and Ross, 1989; Grupp, 1993).[6]

Role Play/Acting requires reflection, imaginary interaction and creativity. The method tries to answer questions such as: If I were person X, how would I deal with problem Y? Or, if we were country X, what would be our position with regards to issue Y? Role play is an interesting and attractive method, though limited by the difficulty in suppressing one's own tendencies and emulating another party's beliefs, values, viewpoints, etc. In foresight work, gaming may be employed, in which several participants role play the responses of individuals or groups to unfolding situations.[7] Participants in role playing workshops are often provided with detailed profiles of the stakeholders involved in the acting. Such profiles may be circulated in advance, and themselves result from analyses such as literature reviews and targeted interviews (see also Young, 1998; Goodwin, 2002; Armstrong, 2002).

Scanning (often termed "environmental scanning") involves observation, examination, monitoring and systematic description of the technological, socio-cultural, political, ecological and/or economic contexts of the actor in question – a country, industry, firm, organisation, etc. Scanning techniques may be more or less formalised, systematic and comprehensive ways of searching for and collating information via literature reviews, SWOT, Web searching, bibliometric or patent analyses, etc. It is an activity often commissioned to academics or consultants, some of whom specialise in monitoring trends as reported in mass or scientific media (see also Defra, 2002; Lapin, 2004).

Scenarios/Scenario Workshops refers to a wide range of approaches involving the construction and use of scenarios – more or less systematic and internally consistent visions of plausible future states of affairs. Generally scenarios involve several features of the object of study, not just one or two parameters. They may be produced by means of deskwork, workshops or the use of tools such as computer modelling. Scenario workshops commonly involve working groups dedicated to the preparation of alternative futures. These groups generally focus on a particular subject or problem where resulting scenarios indicate (a) the views of experts in a particular field or (b) the views of a group of people carefully selected to represent a particular community, organisation or region. There are numerous ways of articulating and elaborating such scenarios – for example, using a 2*2 matrix cross-cutting key parameters; using "archetypal" scenarios such as "better than expected", "worse than expected", "different to expected"; selecting scenarios exemplifying key trends and drivers identified through STEEPV or similar approaches, and so on. But one can also find workshops aiming at the creation of an aspirational or "success scenario" (see Miles, 2005), for example elaborating a vision of a desirable and feasible aspirational future. Such a scenario requires the identification of specific objectives, targets and actions towards its achievement (see also Jantsch, 1967; Boucher, 1977, 1985; Miles, 1981; Schoemaker and Van der Heijden, 1992; Van der Heijden, 1996; Ringland, 1998; Andersen and Jæger, 1999; Roubelat, 2000; Krause, 2002; Berkhout and Hertin, 2002; Green *et al.*, 2005).

Science Fictioning (SF) is an activity that deals with stories assuming that possible events which have not yet materialised have taken place, usually at some point in the future, and elaborates on the consequences of this. Because it involves fictional narrative – and much commercial science fiction is driven more by the need to have adventure or surprise – the method is not very commonly linked to serious governmental or business policy-making. However, it is quite common for scenarios to be illustrated in reports by brief vignettes which use SF-like techniques to illustrate one or other point of the imagined future world. Such vignettes generally lack narrative drive, but may have considerable illustrative force. Commercial SF is often used, mostly informally, as a source of inspiration by people thinking about the future. The main limitation of generating new SF, as a technique, is the difficulty of finding people with inventive, novel and abstract mindsets. The main challenge in using published SF as a source of inspiration is locating the high-quality SF in the haystacks of pulp adventure and escapist fantasy (see also Livingstone, 1971, 1978; Miles, 1993; Steinmüller, 1997).

Simulation Gaming is one of the oldest forecasting and planning techniques, in that war gaming has long been used by military strategists. It is a form of role-playing in which an extensive "script" outlines the context of action and the actors involved. There have long been technological aids used here, such as model battlefields, and now computer simulations. Gaming can also be tied to computer modelling, with the software taking on the role of some actors, coordinating the effects of different actions with programmed rules corresponding to physical or virtual realities, and IT-based visualisation. The understanding of the process being "modelled" in the game – whether or not there are elements of computer simulation – is very important. Formal social scientific approaches such as game theory may be applied, or experts be asked to bring their tacit knowledge to bear. The technique may be used to build understanding and explore possibilities; its ultimate aim is often to propose action plans, cooperation instruments and provide material for developing roadmaps (see also Goodwin, 2002; Green, 2002).

Surveys, like interviews, are a fundamental tool of social research, and are widely used in foresight. A questionnaire is distributed or made available online, and responses drawn from what is usually hoped to be a large pool of respondents. High participation rates generally require attractive and clear design of the survey instrument. Most surveys – e.g. Delphi surveys – are mainly "close-ended", requiring respondents to reply by checking boxes; though surveys can also request much more qualitative responses, for example asking respondents to suggest key technological breakthroughs or socio-economic drivers. More quantifiable survey results can be used to examine the distribution of views across the respondent population, etc. (see also Popper and Miles, 2005).

SWOT Analysis is a method which first identifies factors internal to the organisation or geopolitical unit in question (resources, capabilities, etc.) and classifies them in terms of Strengths and Weaknesses. It similarly examines and classifies external factors (broader socio-economic and environmental changes, for example, or the behaviour of opponents, competitors, markets, neighbouring regions, etc.) and presents them in terms of Opportunities and Threats. This is then used to explore possible strategies – developing and building on strengths and overcoming or accommodating weaknesses, providing insight as to the resources and capabilities required to deal with changing environments and is a very widely used tool for strategy formulation and decision-making (see also Piercy and Giles, 1989; Klusacek, 2004).

Weak Signals/Wild Cards are types of analysis that are usually carried out by small groups of highly skilled people capable of combining expertise, examining data and creative thinking. The search for weak signals may be undertaken as part of the process of scanning. It involves the identification of "not necessarily important things" which do not seem to have a strong impact in the present but which could be the trigger for major events in the future (e.g. changes in public attitudes to one thing or another, an emerging pattern of concern about emerging health problems). Finding weak signals is one of the most challenging tasks in futures research and their analysis often leads to the identification of wild cards. The latter are surprising and unexpected events with low probability of occurrence but with very high impact (e.g. the 2001 attack on the World Trade Centre on September 11, sudden shifts in the dominance of nations or political ideologies, major disasters in environmental or technological systems). These are usually identified by such means as brainstorming, science fictioning and genius forecasting. It has been suggested that gaming and role playing may encourage participants to think of novel patterns of behaviour and responses (see also Ansoff, 1975; Rockfellow, 1994; Petersen, 1999; Cornish, 2003; Mendonca *et al.*, 2004; Steinmüller, 2004; Hiltunen, 2006; Ilmola and Kuusi, 2006).

Quantitative methods

Many quantitative methods are used in foresight, for providing an evidence base for futures thinking, or supplying forecasting tools themselves like trend extrapolation. When data can be quantified, powerful tools can be used to manipulate it: but the usefulness of this depends on the quality of the data in the first place. (For instance, how valid are the indicators for the subject at hand? How reliable is the statistical evidence in terms of the sample and measuring instruments involved?)

The impressive results that can be produced by these methods, and the scope for presenting these results in visually impressive graphs and charts – together, perhaps, with the mystique of statistical expertise and computerised data analysis – may lead to these tools having a particularly strong impact. Often there is insufficient effort to authenticate the reliability or validity of the data that are being used, or the extent to which conceptual misunderstandings or normative positions are being smuggled into the apparently technical assumptions required by the tools. The growing use of quantitative methods in foresight has been facilitated by the emergence of new information technology (IT) applications supporting rapid data acquisition and processing and providing effective visualisation techniques for further analysis.

Benchmarking is a method commonly used for marketing and business strategy planning and has recently become more popular in governmental and inter-governmental strategic decision-making processes. The main question here is what others are doing in comparison to what you are doing. It involves comparison of similar units of analysis in terms of common indicators (i.e. research capabilities of key sectors, market sizes of industries, potential for development and exploitation of technologies, capacity of human resources, etc.). Quite often, benchmarking studies are sub-contracted to specialised consultancy groups with access to relevant and up-to-date data about the countries, regions, industries or companies in question (see De la Porte *et al.*, 2001; European Commission, 2002; Lundvall and Tomlinson, 2002; Arrowsmith *et al.*, 2004).

Bibliometrics is a method based on quantitative and statistical analysis of publications. This may involve simply charting the number of publications emerging in an area, perhaps focusing on the outputs from different countries in different fields and how they are evolving over time. Impact analyses examine citations to assess, for example, the most influential pieces of work in specific areas. This involves using tools such as the Science Citation Index or text mining which requires: building algorithms for extracting multiword phrase frequencies and phrase proximities (physical closeness of the multiword technical phrases) from any type of large textual database; and using the interpretative capabilities of the expert human analyst (Kostoff *et al.*, 2001. See also Melkers, 1993; Narin and Olivastro, 1994).[8]

Indicators/Time Series Analysis (TSA) involve the identification of figures to measure changes over time. Indicators are generally built from statistical data with the purpose of describing, monitoring and measuring the evolution and the current state of relevant issues. As the term implies, they "indicate" features of the issue at hand, rather than provide a comprehensive description of it; famously when indicators are used to set planning targets, there are liable to be behavioural changes from those involved that subtly change what the indicator means and how it should be interpreted, and other desired goals may be neglected if they are not targets of this sort. Indicators can be economic (i.e. GDP, labour costs), social (literacy, infant mortality, etc.), environmental (i.e. gas emissions), scientific (i.e. RTD expenditure, HHRR, publications), technological (patents, inventions, innovations, etc.), among others. As for TSA (the analysis of a series of data points, measured normally at consecutive times, within (often) consistent intervals), it is possible to say that the method has become rather popular in doing economic forecasting, studying biological data and the like (see also Box and Jenkins, 1976; Harvey, 1989; Brockwell and Davis, 1996).

Modelling generally refers to the use of computer-based models that relate together the values achieved by particular variables. Very simple models may be based on statistical relations between two or three variables only – even extrapolation is an elementary form of modelling (in which time is one variable). More complex models may use hundreds, thousands, or even more variables; econometric models are routinely used in economic policy-making, for example, and are "calibrated" from economic statistics and statistical analyses of their interrelations. Models are frequently used in other sorts of planning activity, for example to estimate the effect of alternative land use or transport strategies. Many futures studies employ models that involve relations that are nonlinear and variables whose calibration is highly difficult: Systems Dynamics is a particular tool used for such modelling in many studies. There is considerable development in the field of modelling, as powerful personal computers (PCs) and tools such as spreadsheets can be used for this purpose, and as the frontiers of modelling are being extended with promising new approaches such as agent-based modelling (see also Forrester, 1971; Meadows *et al.*, 1972; Pagan, 2003).

Patent Analysis often resembles bibliometrics, but uses patents rather than publications as its starting point. It provides strategic intelligence on technologies, and can be used to indicate "revealed competitive advantage" based on leadership in technological development. It helps to understand who the leading technology providers are. It can be used to compare companies and countries, or different technology areas – for instance, fields where high levels of activity seem to be underway. Quantitative analysis utilises statistical methods to look at the number of patent registrations, assuming that increasing or decreasing registrations would (apparently) indicate, for example, low or high potential for technology developments in a specific area. More qualitative analyses may focus more on the contents of the patents. Normally this information is used to make strategic RTD and investment decisions as well as for the possible adaptations or even acquisition of patented technologies. One limitation is that, even in the most developed countries, patent information is usually two, three or more years out of date. Furthermore, some industries make little use of patents, leaving several technological developments and lines of innovation untracked (see also Narin and Olivastro, 1988; Ernst, 1997).

Trend Extrapolation/Impact Analysis are among the longest-established tools of forecasting. They provide a rough idea of how past and present developments may look like in the future – assuming, to some extent, that the future is a kind of continuation of the past. There may be large changes, but these are extensions of patterns that have been previously observed.

Essentially, it is assumed that certain underlying processes – which may or may not be explicated – will continue to operate, driving the trend forwards. In practice, of course, most, if not all, trends will confront limits and countertrends at some point in their evolution. Sophisticated trend analysis attempts to deal with such issues by fitting specific curves to particular phenomena (e.g. S-shaped logistic curves are often used to represent variables like population growth or the diffusion of a technology; envelope curves are used to depict trends in functionality that persist over several generations of different technologies, etc.). Recently, the concept of Megatrends has become popular to refer to macro level phenomena which include various (sometimes conflicting) sub-phenomena (e.g. globalisation, ageing, climate change). On the other hand, Impact Analysis aims to identify potential impacts that major trends or events would have on systems, regions, policies, people, etc. Impacts would then be described in terms of their likelihood (probable, possible or speculative), time of occurrence (short/medium/long-term), strength and expected consequences (positive, negative, neutral), for example (see also Porter *et al.*, 1980; Armstrong *et al.*, 2005; Armstrong, 2006).

Semi-Quantitative methods
This section describes eight semi-quantitative methods which apply mathematical principles to manipulate data derived from subjectivity, rational judgements, probabilities, values and viewpoints of experts, commentators or similar sources.

Cross-impact/Structural Analysis (SA) attempts to work systematically through the relations between a set of variables, rather than examining each one as if it is relatively independent of the others. SA requires that a set of key variables are determined in order to understand the system that is of concern. Usually, expert judgement is used to examine the influence of each variable within a given system, in terms of the reciprocal influences of each variable on each other – thus a matrix is produced whose cells represent the effect of a variable on each other.[9] Cross-impact has been also adapted to explore what groups of experts believe about the interaction between trends, stakeholders and objectives within a system (often called structural analysis). One limitation of the method – in addition to the tedious effort that the completion of a matrix can involve – is that it does not deal well with forms of causality that involve the interaction of several variables, or where nonlinear relationships may pertain. Structural analysis focuses on the identification and interpretation of correlations between variables (e.g. trends and drivers). Tools such as cross-impact matrices are applied – for example in the MICMAC method (promoted by Michel Godet) – to identify those drivers that are key in terms of their influence and dependency on other elements of a particular system. Nowadays structural analysis is often carried out with specialised computer software on a desktop or online basis (see also Godet, 2000; Popper, 2002).

Delphi is a well-established technique that involves repeated polling of the same individuals, feeding back (sometimes) anonymised responses from earlier rounds of polling, with the idea that this will allow for better judgements to be made without undue influence from forceful or high-status advocates. The technique was developed so as to circumvent "follow the leader" tendencies of face-to-face exchanges, and other problems such as the reluctance to discard previously stated opinions. Ideally Delphi will feed back explanations for people's initial decisions, so that participants can assess the strength of the case for "deviant" viewpoints; often this is only done to a limited extent, if at all, because of the effort required to produce, process and feed back this information. Delphi surveys are usually conducted in two, and less commonly three, rounds.[10] Delphis are most often employed to elicit views as to whether and when particular developments may occur, but the technique can be used for any sort of opinion or information – such as the desirability of specific outcomes, impacts of policies or technologies, etc. Likewise, Delphi is frequently used with a focus on the dominant views that emerge, but the technique may be oriented more to delineating different points of view. Delphi surveys are often carried out online, and findings are used to prepare recommendations, action plans, roadmaps, etc. But as Kuusi (1999) indicates, the active role of Delphi managers and synthesizers are critical for success (see also Loveridge *et al.*, 1995; Linstone and Turoff, 2002; Popper, 2003; Popper and Miles, 2005a).

Key/Critical Technologies methods involve the elaboration of a list of key technologies for a specific industrial sector, country or region. A technology is said to be "key" if it contributes to wealth creation or if it helps to increase quality of life of citizens; is critical to corporate competitiveness; or is an underpinning technology that influences many other technologies. However the method is implemented (expert panels or surveys, for instance) it implies some prioritisation process (such as voting, multi-criteria and/or cross-impact analysis). The exercise is most often oriented to emerging technologies, but may involve more familiar ones; it typically involves experts in the new fields but may also involve other stakeholders (such as politicians and entrepreneurs). Many critical technologies exercises are sponsored by the private sector (see also CSIRO, 1991; Bimber and Popper, 1994; Durand, 2003; Keenan, 2003; Wagner and Popper, 2003; Sokolov, 2006).

Multi-Criteria Analysis is a prioritisation and decision-support technique specially developed for complex situations and problems, where there are multiple criteria in which to weigh up the effect of a particular intervention.[11] The method works by asking participants to assess the importance of various evaluative criteria, and the impact of a series of options, policies or strategies in each of the criteria. Final scores are calculated based on these sets of judgements, and sensitivity analyses may be conducted (for example to evaluate effects of the weighting). The procedure requires many judgements to be made, and participants may feel that the underlying logic of their analysis is being lost,[12] but a convenient comparative summary of the views of different people (or groups) concerning different options and different impacts can be obtained in this way (see also Meyer-Krahmer and Reiss, 1992; Henriksen and Traynor, 1999; Salo *et al.*, 2003).

Polling/Voting refers to the use of voting or survey methods to gain an assessment of the strength of views about a particular topic among a set of participants. These may be members of a workshop, for example, who make a show of hands, place post-it stickers on one or other category on wall posters, enter views into a computer system, etc., to indicate how probable, uncertain, or important they consider events to be, which actions are priorities and how feasible alternatives are, and so on. It is often used as a way of gauging priorities for further analysis or for policies in late stages of a workshop. Polling may also be used to improve communication between different stakeholders and citizens, in which case the process could be done via Internet, telephone or (more recently) SMS voting (see also Paletz *et al.*, 1980; Witte and Howard, 2002; Cuhls, 2004).

Quantitative Scenarios/SMIC take various forms. One version involves quantification of the contingencies that bring about the scenario. Sometimes probabilistic analysis is established via expert opinion in order to build a system which evaluates the likelihood of occurrence of certain events. Such systems can be more or less simple (a short list of independent events) or complex (using a large matrix to interconnect events – e.g. cross-impact matrices are the foundation for the SMIC method proposed by Duperrin and Godet (1975) to identify how feasible scenarios are). A quite different sort of quantitative scenario is that deduced from survey analysis. For instance, it is possible to identify alternative points of view among a population that has been asked to complete a questionnaire dealing with future developments: these different viewpoints can be elaborated via factor analysis or cluster analysis, and contrasted in terms of the values that are assigned to the underlying variables by these methods (see also Godet, 2000).

Roadmapping is a method which outlines the future of a field of technology, generating a timeline for development of various interrelated technologies and (sometimes) including factors like regulatory and market structures. It is a technique widely used by high-tech industries, where it serves both as a tool for communication, exchange, and development of shared visions, and as a way of communicating expectations about the future to other parties (e.g. sponsors). Most often the technique is applied through a combination of group- and desk-work. It requires inputs from people with deep knowledge about the focus area. The method has occasionally been applied to topics other than technology development, and the term "roadmap" is used loosely to describe all sorts of forward planning accounts of expected or hoped-for stages of development. Although the elaboration of roadmaps rely less on numbers and statistics (i.e. synthesis and evaluation of ideas or documents), the actual outcome may be expressed in term of quantitative targets, such as, for example, the reduction of gas emissions by X per cent (see also Willyard and McClees, 1987; EIRMA, 1997; Kostoff and Schaller, 2001; Saritas and Oner, 2004; Phaal *et al.*, 2004).

Stakeholder Analysis/MACTOR are strategic planning techniques which take into account the interests and strengths of different stakeholders, in order to identify key objectives in a system and recognise potential alliances, conflicts and strategies. These methods are quite common in business and political affairs. In futures work, there are techniques such as MACTOR that take this further, systematically considering whether stakeholders are in favour of or against particular objectives, and representing the situation in terms of matrices that can be formally analysed. Such information is then used to build scenarios, plan strategic actions and determine stakeholders' strategies. Ideally the method requires reliable information on stakeholders' interests and the strength of their attitudes; there may also be difficulties in taking into account possible changes in viewpoints, compromises and new strategies associated with alliances, and the emergence of new stakeholders and/or issues. But, as De Jouvenel (1967, p. 45) noted, knowing the network of reciprocal commitments may trap the future and moderate its mobility, thus reducing uncertainty (see Burgoyne, 1994; Godet, 2001; Elias *et al.*, 2002; Cassingena Harper and Georghiou, 2005).

Classification of Methods by Type of Approach

A long-established classification of forecasting and futures tools distinguishes methods in terms of their "exploratory" or "normative" orientation. In many ways this terminology is misleading (see Miles and Keenan, 2002) because some value choices are implicit even in the most empiricist exploratory studies, and many foresight activities intertwine elements of each. In the received formulations, a method is said to have exploratory orientation if, based on what is known today, it examines what are the various possible futures. The sorts of methods commonly used to address these questions are: brainstorming, SWOT analysis, scenario workshops, Delphi, genius forecasting, trend extrapolation and systems analysis. The following examples show *exploratory*-type questions:

- What sort of socio-economic opportunities and threats for your country would emerge from the technological developments in a particular area (e.g. nanotechnology) in the next 10–20 years?
- What would be the impact of a technological trend (e.g. the uptake and increasing functionality of embedded systems) on your country's economy or on specific industrial sectors in the next 10–20 years?

On the other hand, a method is said to have normative orientation if, based on how the future is expected or desired, it examines how a particular scenario could be reached or avoided. Here there are similar questions but with *normative* orientation:

- Considering the evolution of a given technology area (e.g. new types of microprocessor architecture) in the next 10–20 years: what would need to be done today in order to achieve a high level of success for a country's research and industrial establishments in this field?
- Considering the expected future impacts of a technological development (e.g. systems of personal identification and location): what would need to be done today in order to reach a situation where the risks associated with the technology were well under control?

Questions with normative orientation may also be addressed with methods like brainstorming, scenarios, Delphi, genius forecasting, science fiction, and surveys, among others (see Coates, 1999). Many methods are not inherently exploratory or normative, but can be used more or less in one or other way. Some methods are more intrinsically normative, such as roadmapping, relevance trees, logic charts, morphological analysis and backcasting, for example. Overall, normative activities aim at the formulation of policies and strategies and they are seen as useful instruments for decision-making. However, they need to be fed by information and knowledge generated by exploratory techniques.

The exploratory/normative divide is only one of the ways in which people differentiate among approaches to foresight. For instance, a framework posed by Voros (2005) classifies methods into evolutionary and revolutionary ones.

Here the main arguments are that "evolutionary methods seek to develop or evolve forward in time relatively continuously from a distinct starting point or configuration (usually in the present) while revolutionary methods seek to project or jump forward largely discontinuously into some distinctly different (future) state of being, without necessarily a clear connection to the prior state" (ibid. p. 44). Again, we would expect most foresight programmes to involve a combination of these approaches, though in particular political and cultural contexts one or other element may need to be stressed.

Classification of Methods by Type of Knowledge Source

In the classification by type of technique we provided short descriptions for 33 methods. While this information certainly helps practitioners realise the wide range of options available for methodological designs, the sole understanding of their use is not enough information to select them and, furthermore, to design the methodological framework of a project. So, here we focus on the intrinsic properties for which methods are usually selected and combined. That is, the type of knowledge or information source. In other words, this section describes the predominant purposes of the methods.

In Cameron *et al.* (1996) a triangular structure was introduced to arrange ten methods around three main characteristics of foresight: creativity, expertise and interaction. One of the authors later argued that "neither creativity nor expertise can flourish without a constant flow of information from monitoring processes, which should be seen as an implied background activity" (Loveridge, 1996, p. 9). However, the triangle gave the impression that more formal exploratory methods (such as literature reviews or scanning, techniques that rely on historical data) are not part of the foresight process. These issues, together with emerging developments in foresight practices, have led the author of this chapter to introduce the Foresight Diamond, which takes account of evidence-based methods (e.g. trend extrapolation, literature review, benchmarking and patent analysis) and provides a more comprehensive portrait for methods used in the French and Latin American practices, often called *La Prospective*[13] (i.e. structural analysis, MICMAC, MACTOR and SMIC).

The Foresight Diamond is a practical framework for mapping the 33 methods considered in this chapter in terms of the core *type of knowledge source* each method is mainly based upon (see Figure 3.3). There are three font styles in the Diamond which indicate the type of technique: *qualitative* (using normal style), *semi-quantitative* (using strong style), and *quantitative* (using italic style). Arguably, a comprehensive foresight process should try to use at least one method from each pole. Exactly how methods are located will be to some extent contingent on particular forms of use.

For example, Delphi surveys are probably becoming less of an exclusively expert-oriented activity, and being used more as part of consultation processes for gathering views from wider pools of knowledge and practice; similarly, they are used to explore normative possibilities as well as to forecast "when" particular things might happen.

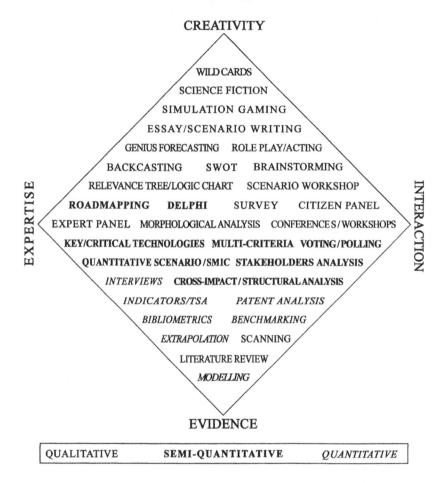

Figure 3.3 The Foresight Diamond

With regards to the type of knowledge source (based on creativity, expertise, interaction or evidence) it is important to emphasise that these domains are not fully independent from one another. However, it is helpful to consider characteristics that can be assigned to each of them, as indicated below:

- *Creativity-based* methods normally require a mixture of original and imaginative thinking, often provided by technology "gurus", via genius forecasting, backcasting or essays. These methods rely heavily on (a) the inventiveness and ingenuity of very skilled individuals, such as science fiction writers or (b) the inspiration which emerges from groups of people involved in brainstorming or wild cards sessions. As Albert Einstein once stated: "The only real valuable thing is intuition ... Imagination is more important than knowledge. Knowledge is limited. Imagination encircles the world" (Einstein as noted by Viereck, 1929).

- *Expertise-based* methods rely on the skill and knowledge of individuals in a particular area or subject. These methods are frequently used to support top-down decisions, provide advice and make recommendations. Common examples are expert panels and Delphi, but methods like roadmapping, relevance trees, logic charts, morphological analysis, key technologies and SMIC are essentially based on expertise. A warning note about expertise is sounded by Arthur C. Clarke (1962, p. 14): "If an elderly but distinguished scientist says that something is possible, he is almost certainly right, but if he says that it is impossible, he is very probably wrong".

- *Interaction-based* methods feature in foresight for at least two reasons – one is that expertise often gains considerably from being brought together and challenged to articulate with other expertise (and indeed with the views of non-expert stakeholders); the other is that foresight activities are taking place in societies where democratic ideals are widespread, and legitimacy involves "bottom-up", participatory and inclusive activities, not just reliance on evidence and experts (which are liable to be used selectively!). Scenario workshops, voting and polling are among the most widely used methods here; of course these often require some sort of expertise to apply the method and inform the interactions. Other methods like citizen panels and stakeholder analysis are becoming popular because of their potential contribution to further networking activities. But it is not always easy to encourage participation and the anonymous saying accurately states that "the world is ruled by those who show up".

- *Evidence-based* methods attempt to explain and/or forecast a particular phenomenon with the support of reliable documentation and means of analysis. These activities are particularly helpful for understanding the actual state of development of the research issue. For this reason, quantitative methods (e.g. benchmarking, bibliometrics, data mining and indicators work) have become popular given that they are supported by statistical data or other types of indicator. They are fundamental tools for technology and impact assessment and scanning activities (see Porter *et al.*, 1980). These methods can also be employed to stimulate creativity (sometimes by challenging received wisdom). And while supporting workshops, evidence-based information is quite useful to encourage interaction and getting feedback from participants. A word of warning here, for both practitioners and users, may be the well-known quote attributed to Benjamin Disraeli by Mark Twain (1924): "There are three kinds of lies: lies, damned lies, and statistics"[14] – which basically points out that sometimes statistics are used to mislead the public.

Information technology (IT) tools are being applied to most of these approaches, especially interaction- and evidence-based activities. Many applications are available now to support modelling, data mining, scanning, participatory processes, and visualisation – there are even tools designed to facilitate creativity. Use of IT does not always mean more effective application of foresight techniques, however. Salo and Gustafson (2004) identified five factors which need to be met in order to make good use of IT here: a clear mandate from the sponsoring organisation; high-quality process and technical facilitation; presence of senior representatives; presentation of unequivocal information inputs; and sufficient time for informal debate.

The important role of evidence-based methods in foresight was revealed in the first examinations of foresight practices in several hundred cases (mostly from Europe), by Popper *et al.* (2005, 2007) and Keenan *et al.* (2006). Figure 3.4 (below) outlines the use of 27 methods in 785 exercises. The findings roughly show four categories: most widely used; commonly used; less often used; and emerging and/or rarely used methods:

- Literature review is used in 52 per cent of the cases. In second place are expert panels followed closely by scenarios. These three (mainly qualitative) methods are the *most widely used.*
- The group of *commonly used* methods is led by futures workshops, trend extrapolation, brainstorming, megatrend analysis, surveys, key technologies, interviews, Delphi and SWOT analysis.
- A third group of *less often used* methods includes roadmapping, essays, scanning, modelling and backcasting.
- Finally, the fourth group of *emerging and/or rarely used* methods such as stakeholder analysis, citizen panels, structural analysis, multi-criteria analysis, simulation gaming, bibliometrics, relevance trees and morphological analysis.

Note that the fifth most often selected option is "other methods", thus indicating that methodological frameworks are very diverse and versatile.

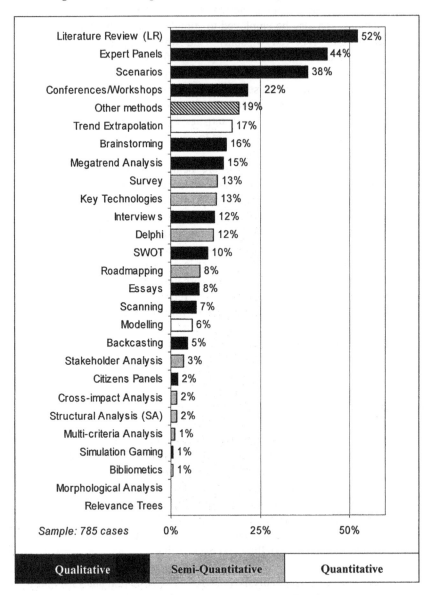

Source: Adapted from Popper *et al.* (2005, 2007) and Keenan *et al.* (2006)

Figure 3.4   Common foresight methods

METHODOLOGICAL FRAMEWORKS

The methodological framework used in a foresight project should be tailored to meet the specific objectives of the project and the resources and capabilities that are available. While the previous section provides advice with respect to the use and selection of methods, this section draws attention to *inter alia* the articulation and combination of methods.[15] Many of the methods described above can be used at different stages in a foresight process and practitioners should take into account (a) the contribution of each in the context of the study as a whole, and (b) the ways in which individual methods can be combined and synthesised to positive effect.

There is no "ideal" methodological framework providing the "best" combination of methods. In fact, there is no "ideal" number of methods to be used in a project. Popper *et al.* (2005) took a sample of 130 cases from 15 countries (Austria, Belgium, Czech Republic, Denmark, Estonia, Finland, France, Germany, Italy, Netherlands, Spain, Sweden, Turkey, UK and the USA) and found an average of five to six methods per exercise. Countries such as Turkey[16] and the UK demonstrated a high propensity to mix several methods, whilst others (e.g. Denmark and the USA) tended to exhibit greater conservatism in terms of methodological scope. Whilst this is interesting, the reliability of the results should not be taken at face value given the relatively small numbers of exercises considered. However, if we assume for a moment that, on average, foresight projects will combine six methods, then, with all 33 methods above as eligible options, a question we might wish to address is "what number of possible permutations (i.e. a selection of methods in which the order of the methods matters) exists?" In other words, how many ways can six methods from a set of 33 be combined in order to generate a methodological framework? The answer is simple, using the permutation formula,[17] there are nearly 800 million ways of combining six methods from a set of 33 to build a methodological framework. Having said this, it would be remarkable to find practitioners selecting methods from the vast range available in a random fashion: it is of course always the case that expertise and accumulated know-how in the use of certain methods will provide a rational justification for the selection of a particular combination (see also Popper *et al.*, 2007, pp. 25).

To illustrate this, the following charts (Figures 3.5–3.8) plot four idealised frameworks showing how different techniques may be combined within overall methodological frameworks X and Y (both using six methods only). Each methodology will be described in two ways:

1. Forward (combining methods in one sequence); and
2. Backward (combining methods in reverse order of sequencing).

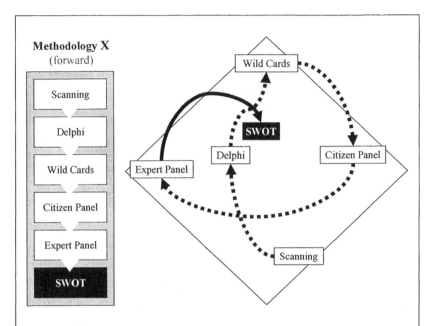

The following bullets indicate the sorts of application that may be scheduled for each of the techniques, illustrating how these may be used if *Methodology X* is carried out in a *forward* sequence:

- Scanning: detailed analysis of main issues around a particular sector/theme (sub-contracted);
- Delphi: large-scale exploratory study assessing the likelihood of occurrence and possible impacts of issues highlighted by the scanning activity;
- Wild Cards: workshop-type activity aimed at identifying events which may challenge the occurrence of "highly probable" situations in the future;
- Citizen Panels: conference-type activity aimed at identifying major public concerns on critical issues;
- Expert Panels: reduced group of key stakeholders looking at future implications of findings;
- SWOT: internal activity for synthesising outcomes in terms of current strengths/weaknesses and future opportunities/threats.

Figure 3.5   The role of methods in Methodology X (forward)

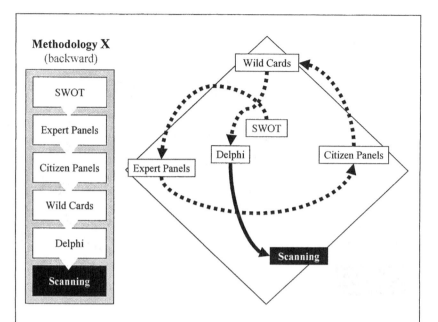

The following bullets indicate the sorts of application that may be scheduled for each of the techniques, illustrating how these may be used if *Methodology X* is carried out in a *backward* sequence:

- SWOT: large-scale workshop aimed at identifying strengths, weaknesses, opportunities and threats related to a specific sector or industry;
- Expert Panels: groups of experts looking at future implications of SWOT findings and clustering main issues into broader dimensions, such as social, technological, economic, etc.;
- Citizen Panels: regional task forces contextualising key issues and evaluating public acceptance;
- Wild Cards: internal activity aimed at identifying disruptive trends and events;
- Delphi: large-scale normative study aimed at formulating policy recommendations;
- Scanning: internal activity aimed at identifying similar policies being implemented in comparable contexts.

Figure 3.6   The role of methods in Methodology X (backward)

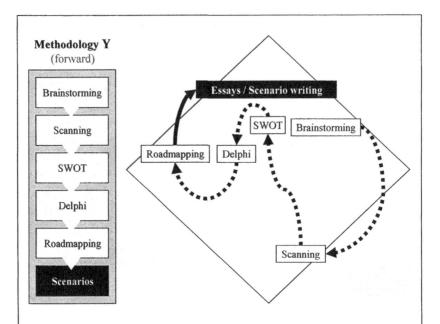

The following bullets indicate the sorts of application that may be scheduled for each of the techniques, illustrating how these may be used if *Methodology Y* is carried out in a *forward* sequence:

- Brainstorming: large-scale activity aimed at identifying key issues around particular dimensions (e.g. STEEPV);
- Scanning: a desk-research activity aimed at describing and expanding relevant ideas from the brainstorming exercise in terms of current developments and policies;
- SWOT: a workshop with approx. 20 experts from each of the following sectors (public, private, academic and civil society);
- Delphi: an exercise targeting a selected group of experts assessing the stage of development of particular technologies;
- Roadmapping: a panel-based activity looking at market needs and potential linkages between products and technologies, as well as future regulatory developments;
- Scenarios: same panel elaborating a vision of a desirable and feasible aspirational future.

Figure 3.7   The role of methods in Methodology Y (forward)

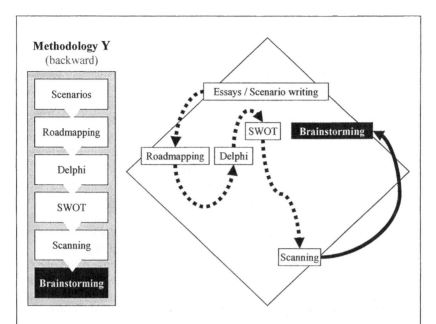

The following bullets indicate the sorts of application that may be scheduled for each of the techniques, illustrating how these may be used if *Methodology Y* is carried out in a *backward* sequence:

- Scenarios: business as usual, negative and positive scenario (desk-research);
- Roadmapping: workshops with targeted experts preparing time-line and discussing market needs and technology linkages for each scenario;
- Delphi: a large-scale activity aimed at identifying STEEPV impacts of suggested "action plans" resulting from the roadmapping activity;
- SWOT: internal activity looking at the strengths, weaknesses, opportunities and threats of suggested "action plans";
- Scanning: a parallel process (possibly outsourced) mapping the market penetration of products and services related to initial scenarios;
- Brainstorming: workshops with key stakeholders identifying new cooperation instruments and exploiting existing ones.

Figure 3.8   The role of methods in Methodology Y (backward)

CONCLUSION

This chapter has reviewed and classified a large number of commonly used foresight tools. If nothing else, it has indicated that there is no one method that provides all of the answers – indeed, there appears to be no single technique that is used in all, or even in a very large share of, foresight exercises. Nor is there one ideal combination of methods that is appropriate for all foresight exercises. The selection and detailed implementation of tools is highly dependent on the objectives of the programme, and only consultants with a particular interest in promoting their own expertise will claim that a given way of conducting foresight is the "right" way.

More generally a combination of different types of tools is most likely to yield a set of rich resources that can appeal to a range of stakeholders and satisfy a variety of intertwined objectives. The ways in which different techniques can be combined are diverse, and the crucial ingredients for success are bound to be the ability to manage the combination of techniques and to synthesise the results emerging from them so as to effectively inform and engage sponsors and other users. The proposed Foresight Diamond (Figure 3.3) illustrates the importance of the combination of methods that rely on different sources of knowledge (i.e. creativity, expertise, interaction or evidence).

Finally, the reader should be reminded that (a) the methodology must be chosen after objectives are defined and not the other way around, and (b) the selection of methods may be affected by resources, such as the budget, availability of expertise, political support, technological and physical infrastructure, and time. Having valuable human resources is essential and although such people do not necessarily need to be foresight specialists, they will often require intensive training courses in order to build internal capacities and know-how.

NOTES

1   For more discussion see, for example, Giorgi, 1970; Irvine *et al.*, 1979; Cassel and Simon, 1994; Funtowicz and Ravetz, 1990; and Bauer and Gaskell (2000).
2   As happens so often, the term has been used in several different ways. In addition to that discussed in the text, "backcasting" can be used to describe the use of a forecasting tool to "predict" the past. This can be done with computer simulation calculating whether a feasible path can be drawn between the present and the future (given the data and assumptions fed into it). For example, Cole *et al.* (1973) ran the systems dynamics model of the world that inspired *Limits to Growth*, in reverse. It turned out that the inherent structure of the model

necessarily depicted history as a succession of Malthusian crises (of decreasing magnitude, as successively more resources were exhausted).

3   STEEPV is an acronym for Social, Technological, Economic, Environmental, Political, and Values. This is simply a framework designed to prompt people to look for trends, drivers, impacts, etc., beyond those they normally restrict themselves to. For instance, participants in a workshop may be asked what important influences they can identify under each of the headings. There are numerous similar frameworks, of which PEST is a well-known example.

4   This may not necessarily require that the representativeness has to mirror the larger population precisely!

5   These are normally based on particular dimensions, i.e. social, economic, etc.

6   Logic diagrams are often structured around levels. For example: level 1 might describe the overall vision or rationale, level 2 the mission, level 3 the general objectives, level 4 the more specific objectives associated with these, level 5 for activities which are aimed at meeting these objectives, level 6 for the specific outputs anticipated from the activities, level 7 for their impacts and level 8 for broader outcomes.

7   Role Play is more commonly used in training people to deal with future situations and/or to empathise with the experience of others. This sort of approach has also been used in military forecasting, for developing alternative scenarios and for speculating on stakeholders' strategies.

8   Less systematic analysis may be conducted rapidly using web search engines such as Google Scholar.

9   For instance, the extent to which the occurrence of one event makes the occurrence of another event more likely may be examined.

10 It is argued that third or additional rounds are effective only when the questions and arguments of the Delphi are redesigned on the basis of the previous results, thus transforming the survey into a dynamic expert forum.

11 For example, capture and storage of greenhouse gases might be good in terms of climate change and ability to move the energy economy away from fossil fuels slowly; but it might cause local environmental problems and distribute economic costs in a skewed fashion.

12 For example, not everyone is convinced that a simple multiplication of the desirability of an impact by the intensity or probability of that impact will capture the essence of each situation. Some valuations may not be simply commensurable, for example when a particular impact is absolutely disastrous it may be misleading to treat it in the same way as a more mundane problem.

13 Please note that *La Prospective* should not be thought as a single method between the interaction and the creativity poles (as suggested in Loveridge's triangle in 1996). The richness of French practices include a wide range of methods, from creative (relevance trees) to interactive (MACTOR) to expertise-based ones (SMIC) to evidence-based (structural analysis & MICMAC).

14 Note that *La Prospective* should not be thought as a single method between the interaction and the creativity poles (as suggested in Loveridge and Van der Meulen's triangle in 1996). The richness of French practices include a wide range of methods, from creative (relevance trees) to interactive (MACTOR) to expertise-based ones (SMIC) to evidence-based (structural analysis & MICMAC).

15 Foresight guides and mapping reports such as those by Miles and Keenan (2003), Miles *et al.*, (2003), Keenan and Popper (2007) and Popper *et al.* (2007) provide further advice on the selection of methods.

16 Turkey's analysis is mainly based on the "Vision 2023" project led by the Scientific and Technological Research Council of Turkey (TÜBİTAK). Further information can be found at http://www.tubitak.gov.tr.

17 If *tm* = total number of methods and *sm* = number of selected methods, then the Permutation formula would be P = (tm!)/(tm - sm)! In the example above *tm* = 33 and *sm* =6, thus making P = (33!) / (27!), resulting in P = 33*32*31*30*29*28 = 797,448,960.

REFERENCES

Andersen, I. and Jæger, B. (1999), 'Danish Participatory Models. Scenario Workshops and Consensus Conferences: Towards More Democratic Decision-making', *Science and Public Policy*, **26**(5), pp. 331–340.

Ansoff, I. (1975), 'Managing Strategic Surprise by Response to Weak Signals', *California Management Review*, **18**(2), pp. 21–33.

Arrowsmith, J., Sisson, K. and Marginson, P. (2004), 'What can "Benchmarking" Offer the Open Method of Coordination?', *Journal of European Public Policy*, **11**(2), pp. 311–328.

Armstrong, J.S. (2006), 'Findings from Evidence-based Forecasting: Methods for Reducing Forecast Error', *International Journal of Forecasting*, **22**(3), pp. 583–598.

Armstrong, J.S., Collopy, F. and Yokum, J.T. (2005), 'Decomposition by Causal Forces: A Procedure for Forecasting Complex Time Series', *International Journal of Forecasting*, **21**(1), pp. 25–36.

Armstrong, J.S. (2002), 'Assessing Game Theory, Role Playing, and Unaided Judgment', *International Journal of Forecasting*, **18**(3), pp. 345–352.

Bauer, M. and Gaskell, G. (eds) (2000), *Qualitative Researching with Text, Image and Sound*, London: Sage.

Becker, H. (1983), 'Scenarios: A Tool of Growing Importance to Policy Analysts in Government and Industry', *Technology Forecasting and Social Change*, **23**, pp. 95–120.

Berkhout, F. and Hertin, J. (2002), 'Foresight Futures Scenarios: Developing and Applying a Participative Strategic Planning Tool', GMI newsletter.

Bimber, B. and Popper, S. (1994), 'What is a Critical Technology?', RAND paper DRU-605-CTI, Santa Monica.

Boucher, W.I. (ed.) (1977), *The Study of the Future: An Agenda for Research*, Washington, DC: National Science Foundation.

Boucher, W.I. (1985), 'Scenarios and Scenario Writing', in Mendell, J. (ed.), *Non-extrapolative Methods in Business Forecasting*, Westport, CT: Quorum Books.

Box, G. and Jenkins, J. (1976), *Time Series Analysis: Forecasting and Control*, San Francisco, CA: Holden-Day.

Brockwell, P. and Davis, R. (1996), *Introduction to Time Series and Forecasting*, Springer.

Burgoyne, J. (1994) 'Stakeholder Analysis', in Cassel, C. and Symon, G. (eds), *Qualitative Methods in Organisational Research: A Practical Guide*, London: Sage Publications.

Cameron, H., Loveridge, D., Cabrera, J., Castanier, L., Presmanes, B., Vasquez, L. and Van der Meulen, B. (1996), *Technology Foresight: Perspectives for European and International Co-operation*, Final Report to DG XII of the European Commission, Manchester: PREST, The University of Manchester.

Cassel, C. and Symon, G. (1994), 'Qualitative Research in Work Contexts', in Cassel, C. and Symon, G. (eds), *Qualitative Methods in Organisational Research: A Practical Guide*, London: Sage Publications.

Cassingena Harper, J. and Georghiou, L. (2005), 'The Targeted and Unforeseen Impacts of Foresight on Innovation Policy: The eFORESEE Malta Case Study', *International Journal of Foresight and Innovation Policy*, 2(1), pp. 84–103.

Cassingena Harper, J. (ed.) (2003), *Vision Document*, eFORESEE Malta ICT and Knowledge Futures Pilot. .

Chaudhry, S. and Ross, W. (1989), 'Relevance Trees and Mediation', *Negotiation Journal*, 5, pp. 63–73.

CSIRO (1991), *CRIRO Priority Determination 1990 – Methodology and Results Overview*, Canberra, Australia: Commonwealth Scientific and Industrial Research Organization.

Clarke, A.C. (1962), *Profiles of the Future: An Inquiry into the Limits of the Possible*, New York: Harper & Row, p. 14.

Coates, J.F. (1999), 'Normative Forecasting', in Glenn, J. (ed.), *Futures Research Methodology*, Washington, DC: AC-UNU.

Cole, H.S.D., Freeman, C., Jahoda, M. and Pavitt, K.L.R. (1973), *Thinking about the Future*, London: Chatto & Windus; also published as *Models of Doom: A Critique of Limits to Growth*, NY: Universe Books.

Cornish, E. (2003), 'The Wild Cards in our Future', *The Futurist*, 37, pp. 18–22.

Cuhls, K. (2004), 'Futur – Foresight for Priority-setting in Germany', *International Journal of Foresight and Innovation Policy*, 1(3–4), pp. 183–194.

Defra (2002), *Defra's Horizon Scanning Strategy for Science*, London: Defra.

De Jouvenel, B. (1967), *The Art of Conjecture*, transl. from French by N. Lary, London: Weidenfeld and Nicolson.

De la Porte, C., Pochet, P. and Room, G.J. (2001), 'Social Benchmarking, Policy-making and New Governance in the EU', *Journal of European Social Policy*, 11, pp. 291–307.

Dreborg, K. (1996), 'Essence of Backcasting', *Futures*, 28(9), pp. 813–828.

Duperrin, J.C. and Godet, M. (1975), 'SMIC 74 – A Method for Constructing and Ranking Scenarios', *Futures*, 7(4), pp. 302–312.

Durand, T. (2003), 'Twelve Lessons Drawn from "Key Technologies 2005", the French technology foresight exercise', *Journal of Forecasting*, 22(2-3), pp. 161–177.

European Commission (2002), *eEurope 2005: Benchmarking Indicators*, Brussels.

EIRMA (1997), *Technology Roadmapping – Delivering Business Vision*, Working group report, European Industrial Research Management Association, Paris, 52.

Elias, A.A., Cavana, R.Y. and Jackson, L.S. (2002), 'Stakeholder Analysis for R&D Project Management', *R&D Management*, 32(4), pp. 301–310.

Ernst, H. (1997), 'The Use of Patent Data for Technological Forecasting: The Diffusion of CNC-technology in the Machine Tool Industry', *Small Business Economics*, **9**, pp. 361–381.

Georghiou, L. (2003), *Evaluating Foresight and Lessons for its Future Impact*, Second International Conference on Technology Foresight; Tokyo, 27–28 Feb.

Georghiou, L. (1998), 'Issues in the Evaluation of Innovation and Technology Policy', *Evaluation*, **4**(1), pp. 37–51.

Giorgi, A. (1970), *Psychology as a Human Science: A Phenomenologically Based Approach*, New York: Harper & Row.

Glenn, J. and Gordon, T. (eds) (1999), *Futures Research Methodology*, Washington: American Council for the United Nations, The Millennium Project.

Godet, M. (2000), 'The Art of Scenarios and Strategic Planning: Tools and Pitfalls', *Technological Forecasting and Social Change*, **65**(1), pp. 3–22.

Godet, M. (2001), *Creating Futures: Scenario Planning as a Strategic Management Tool*, London: Economica, p.75.

Goodwin, P. (2002), 'Forecasting Games: Can Game Theory Win', *International Journal of Forecasting*, **18**, pp. 369–374.

Green, K. (2002), 'Forecasting Decisions in Conflict Situations: A Comparison of Game Theory, Role Playing and Unaided Judgement', *International Journal of Forecasting*, **18**, pp. 321–344.

Green, L., Popper, R., and Miles, I. (2005), 'IST Success Scenario and Policy Priorities', in Compano, R., Pascu, C., and Weber, M. (eds), *Challenges and Opportunities for IST Research in Europe*, Bucharest: Publishing House of the Romanian Academy.

Grupp, H. (ed.) (1993), *Technologie am Beginn des 21. Jahrhunderts*, Heidelberg: Physica (2nd ed 1995).

Forrester, J. (1971), *World Dynamics*, Cambridge, MA: Wright-Allen Press.

Funtowicz, S. and Ravetz, J. (1990), *Uncertainty and Quality in Science for Policy*, Dordrecht: Kluwer.

Harvey, A. (1989), *Forecasting, Structural Time Series Models and the Kalman Filter*, Cambridge, MA: Cambridge University Press.

Havas, A. (2003), 'Evolving Foresight in a Small Transition Economy', *Journal of Forecasting*, **22**(2-3), pp. 179–201.

Henriksen, A. and Traynor, A. (1999), 'A Practical R&D Project-Selection Scoring Tool', *IEEE Transactions on Engineering Management*, **46**(2), pp. 158–170.

Hiltunen, E., (2006), 'Was It a Wild Card or just our Blindness to Gradual Change?', *Journal of Future Studies*, **11**(2), pp. 61–74.

Hojer, M. and Mattsson, L. (2000), 'Determinism and Backcasting in Future Studies', *Futures*, **32**(7), pp. 613–634.

Ilmola, L. and Kuusi, O. (2006), 'Filters of Weak Signals Hinder Foresight: Monitoring Weak Signals Efficiently in Corporate Decision-making', *Futures*, **38**(8), pp. 908–924.

Irvine, J., Miles, I. and Evans, J. (eds) (1979), *Demystifying Social Statistics*, London: Pluto.

Jantsch, E. (1967), *Technological Forecasting in Perspective*, Paris: OECD.

Jungk, R. and Müllert, N. (1987), *Future Workshops: How to Create Desirable Futures*, London: Institute for Social Inventions.

Keenan, M. (2003), 'Identifying Emerging Generic Technologies at the National Level: the UK Experience', *Journal of Forecasting*, **22**(2–3), pp. 129–160.

Keenan, M., Butter, M., Sainz, G. and Popper, R. (2006), *Mapping Foresight in Europe and Other Regions of the World: The EFMN Annual Mapping Report 2006*, report to the European Commission, Delft: TNO.

Keenan, M. and Popper, R. (eds) (2007), *Practical Guide for Integrating Foresight in Research Infrastructures Policy Formulation*, Brussels: European Commission.

Klusacek, K. (2004), 'Technology Foresight in the Czech Republic', *International Journal of Foresight and Innovation Policy*, 2(1), pp. 84–103.

Kostoff, R., Toothman, D.R., Eberhart, H.J. and Humenik, J.A. (2001), 'Text Mining Using Database Tomography and Bibliometrics: A Review', *Technology Forecasting and Social Change*, 68(3), pp. 223–253.

Kostoff, R. and Schaller, R. (2001), 'Science and Technology Roadmaps', *IEEE Transactions on Engineering Management*, 48(2), pp. 132–143.

Krause, P.H. (2002), 'The Proteus Project – Scenario-based Planning in a Unique Organization', *Technological Forecasting & Social Change*, 69 (5), pp. 479–484.

Kuusi, O. (1999), *Expertise in the Future Use of Generic Technologies*, Helsinki: Government Institute for Economic Research (VATT).

Lapin, J. (2004), 'Using External Environmental Scanning and Forecasting to Improve Strategic Planning', *Journal of Applied Research in the Community College*, 11(2), pp. 105–113.

Linstone, H. and Turoff, M. (eds) (2002), *The Delphi Method: Techniques and Applications*, available at: http://www.is.njit.edu/pubs/delphibook/.

Livingston, D. (1978), 'The Utility of Science Fiction', in Fowels, J. (ed.), *Handbook of Futures Research*, Westpost, CT: Greenwood Press, pp. 163–178.

Livingstone, D. (1971), 'Science Fiction Models of Future World Order Systems', *International Organization*, 25, Spring, pp. 254–270.

Loveridge, D. (1996) 'Technology foresight and models of the future', PREST Ideas in Progress, Number 4, September, Manchester.

Loveridge, D., Georghiou, L. and Nedeva, N. (1995), *Technology Foresight Programme: Delphi Survey*, Manchester: PREST, The University of Manchester.

Lundvall, B. and Tomlinson, M. (2002), 'International Benchmarking as a Policy Learning Tool', in Rodrigues, M.J. (ed.), *The New Knowledge Economy in Europe*, Cheltenham: Edward Elgar.

Meadows, D.H., Meadows, D.L. and Randers, J.W. (1972), *The Limits to Growth*, New York: Universe Books.

Melkers, J. (1993), 'Bibliometrics as a Tool for Analysis of R&D Impacts', in Bozeman, B. and Melkers, J. (eds), *Evaluating R&D Impacts: Methods and Practice*, Boston: Kluwer, pp. 43–61.

Mendonca, S., Cunha, M.P.E., Kaivo-Oja, J. and Ruff, F. (2004), 'Wild Cards, Weak Signals and Organisational Improvisation', *Futures*, 36(2), pp. 201–218.

Meyer-Krahmer, F. and Reiss, T. (1992), 'Ex Ante Evaluation and Technology Assessment – Two Emerging Elements of Technology Policy', *Research Evaluation*, 2, pp. 47–54.

Miles, I. (1981), 'Scenario Analysis: Identifying Ideologies and Issues', in UNESCO, *Methods for Development Planning: Scenarios, Models and Micro-studies*, Paris: UNESCO Press, pp. 31–54.

Miles, I. (1993), 'Stranger than Fiction. How Important is Science Fiction for Futures Studies?', *Futures*, 25(3), pp. 315–321.

Miles, I. (2002), *Appraisal of Alternative Methods and Procedures for Producing Regional Foresight*, Report prepared by CRIC for the European Commission's DG Research funded STRATA – ETAN Expert Group Action, Manchester, UK: CRIC.

Miles, I. (2005), 'Scenario Planning', in *UNIDO Technology Foresight Manual*, Volume 1 – Organization and Methods, Vienna: UNIDO, pp. 168–193.

Miles, I. and Keenan, M. (2002), *Practical Guide to Regional Foresight in the United Kingdom*, Luxembourg: European Commission, EUR 20478, ISBN 92 894 4682 X.

Miles, I., Keenan, M. and Kaivo-Oja, J. (2003), *Handbook of Knowledge Society Foresight*, Report prepared by PREST and FFRC for the European Foundation for the Improvement of Living and Working Conditions, Manchester, UK: PREST.

Miles, M.B., and Huberman, A.M. (1994), *Qualitative Data Analysis: An Expanded Sourcebook*, Beverley Hills: Sage.

Narin, F. and Olivastro, D. (1994), 'Bibliometrics/Theory, Practice and Problems', *Evaluation Review*, **18**(1), pp. 65–77.

Narin, F. and Olivastro, D. (1988), 'Technology Indicators Based on Patents and Patent Citations', in Van Raan, A. (ed.), *Handbook of Quantitative Studies of Science and Technology*, North-Holland: Elsevier.

Pagan, A. (2003), *Report and Modelling and Forecasting at the Bank of England*, Bank of England Quarterly Bulletin, Spring.

Paletz, D., Short, J.Y., Baker, H., Cookman Campbell, B., Cooper, R.J. and Oeslander, R.M. (1980), 'Polls in the Media: Content, Credibility, and Consequences', *Public Opinion Quarterly*, **44**, pp. 495–513.

Petersen, J.L. (1999), *Out of the Blue-How to Anticipate Big Future Surprises*, Lanham: Madison Books.

Phaal, R., Farrukh, C. and Probert, D. (2004), 'Customizing Roadmapping', *Research-Technology Management*, **47**(2), pp. 26–37.

Piercy, N. and Giles, W. (1989), 'Making SWOT Analysis Work', *Marketing Intelligence & Planning*, 7(5), pp. 5–7.

Popper, R. (2002), *Cross-impact method for detecting key drivers in Peru*, Report of the foresight workshop organised by the Consortium Prospective Peru (CPP) 17–18 October, Lima, Peru: CPP.

Popper, R. (2003), *The Knowledge Society Delphi*, EUFORIA Project report submitted by PREST to the European Foundation for the Improvement of Living and Working Conditions (EFL), Dublin, Ireland: EFL.

Popper, R. and Miles, I. (2005), 'IST and Europe's Objectives – a Survey of Expert Opinion', in Pascu, C. and Filip, F. (eds), *Visions of the Future for IST, Challenges and Bottlenecks Towards Lisbon 2010 in an Enlarged Europe*, Bucharest: Publishing House of the Romanian Academy, pp. 87–101.

Popper, R. and Miles, I. (2005a), *The FISTERA Delphi: Future Challenges, Applications and Priorities for Socially Beneficial Information Society Technologies*, FISTERA report prepared by PREST, Manchester, UK.

Popper, R., Keenan, M. and Butter, M. (2005), *Mapping Foresight in Europe and other Regions of the World: The EFMN Annual Mapping Report 2005*, report prepared by PREST–TNO to the European Commissions' DG Research, Manchester, UK: The University of Manchester.

Popper, R., Keenan, M., Miles, I., Butter, M. and Sainz, G. (2007), *Global Foresight Outlook 2007: Mapping Foresight in Europe and the rest of the World*, The EFMN Annual Mapping Report 2007, report to the European Commission, Manchester: The University of Manchester/TNO.

Porter, A., Rossini, F.A. and Carpenter, S.R. (1980), *A Guidebook for Technology Assessment and Impact Analysis*, North-Holland, New York.

Ratcliffe, J. (2002), 'Scenario Planning: Strategic Interviews and Conversations', *Foresight*, 4(1), pp. 19–30.

Ringland, G. (1998), *Scenario Planning: Managing for the Future*, Chichester: John Wiley.

Ritchey, T. (1998), 'Fritz Zwicky, Morphologie and Policy Analysis', Paper presented at the 16th Euro Conference on Operational Analysis, Brussels.

Rockfellow, J. (1994), 'Wild Cards: Preparing for "The Big One"', *The Futurist*, 28(1), pp. 14–18.

Roubelat, F. (2000), 'Scenario Planning As A Networking Process', *Technological Forecasting and Social Change*, 65(1), pp. 99–112.

Salo, A. and Gustafsson, T. (2004), 'A Group Support System for Foresight Processes', *International Journal of Foresight and Innovation Policy*, 1(3/4), pp. 249–269

Salo, A., Gustafsson, T., and Ramanathan, R. (2003), 'Multicriteria Methods for Technology Foresight', *Journal of Forecasting*, 22(2–3), pp. 235–255.

Salo, A. and Salmenkaita, J. (2002), 'Embedded Foresight in RTD Programs', *International Journal of Technology, Policy and Management*, 2(2), pp. 167–193.

Saritas, O. and Oner, M. (2004), 'Systemic Analysis of UK Foresight Results: Joint Application of Integrated Management Model and Roadmapping', *Technological Forecasting and Social Change*, 71(1–2), pp. 27–65.

Scapolo, F. and Miles, I. (2006), 'Eliciting Experts Knowledge: A Comparison of Two Methods', *Technological Forecasting and Social Change*, 73(6), pp. 679–704.

Schoemaker, P.J.H. and Van der Heijden, C. (1992), *Integrating Scenarios into Strategic Planning at Royal Dutch/Shell Case Study*, Planning Review, pp. 41–46.

Schwartz, P. (1991), *The Art of the Long View: Planning for the Future in an Uncertain World*, New York: Doubleday.

Simon, J. and Durant, J. (1995), *Public Participation in Science: The Role of Consensus Conferences in Europe*, London: Science Museum.

Slaughter, R.A. (2004), 'Road Testing a New Model at the Australian Foresight Institute', *Futures*, 36(8), pp. 837–852.

Slocum, N. (2003), *Participatory Methods Toolkit. A Practitioner's Manual*, Bruges, UNU-CRIS available at: http://www.cris.unu.edu/.

Smith G. and Wales C. (2000), 'Citizens' Juries and Deliberative Democracy', *Political Studies*, 48(1), pp. 51–65.

Sokolov, A. (2006), 'Russian Critical Technologies 2015', European Foresight Monitoring Network Brief, 79, available at: http://www.efmn.eu

Steinmüller, K. (1997), 'Science Fiction and Science in the Twentieth Century', in Pestre, D. and Krige, J. (eds), *Science in the Twentieth Century*, Harwood, pp. 339–360.

Steinmüller, K. (2004), 'The Future as a Wild Card. A short introduction to a new concept', in Brockett, S. and Dahlström, M. (eds), *Spatial Development Trends: Nordic Countries in a European Context*, Stockholm.

Thom, R. (1975), *Structural Stability and Morphogenesis*, New York: Benjamin Addison Wesley, pp. 4–7.

Twain, M. (1924), *Mark Twain's Autobiography*, New York and London: Harper Brothers, vol. I, p. 246.

Van der Heijden, A. (1996), *Scenarios: The Art of Strategic Conversation*, Chichester, UK: John Wiley.

Van der Sluijs, J.P., Craye, M., Funtowicz, S., Kloprogge, P., Ravetz, J. and Risbey, J. (2005), 'Combining Quantitative and Qualitative Measures of Uncertainty in Model-Based Environmental Assessment: The NUSAP System', *Risk Analysis*, **25** (2), pp. 481–492.

Viereck, G. (1929), 'What life means to Einstein: An interview by George Sylvester Viereck', The Saturday Evening Post, on 26 October.

Voros, J. (2005), 'A Generalised "Layered Methodology" Framework', *Foresight*, **7**(2), pp. 28–40.

Wagner, C. and Popper, S. (2003), 'Identifying Critical Technologies in the United States: A Review of the Federal Effort', *Journal of Forecasting*, **22** (2/3), pp. 113–128.

Willyard, C.H. and McClees, C.W. (1987), 'Motorola's Technology Roadmap Process', *Research Management*, **30** (5), pp. 13–19.

Witte, J. and Howard, P. (2002), 'The Future of Polling: Relational Inference and the Development of Internet Survey Instruments', in Manza, J., Cook, F.L. and Page, B.I. (eds), *Navigating Public Opinion: Polls, Policy and the Future of American Democracy*, New York: Oxford University Press, pp. 272–289.

Young, H.P. (1998), *Individual Strategy and Social Structure*, Princeton: Princeton University Press.

Zwicky, F. (1969), *Discovery, Invention, Research – Through the Morphological Approach*, New York: Macmillan Publisher.

PART TWO

Foresight Experience Around the World

4.   Foresight in the United Kingdom

Michael Keenan and Ian Miles

INTRODUCTION

The UK Foresight Programme has been in existence since 1993 and is managed by the Office of Science and Innovation (OSI).[1] The Programme is now in its third cycle, having produced hundreds of reports and involved tens of thousands of people over the last decade or so (Georghiou, 1996; Miles, 2005). Its explicit aim is to bring together scientists, technologists, businesses and consumers to discuss the future. In its latest incarnation, it seeks to identify potential opportunities for the economy or society from new science and technologies; it also considers how future science and technologies could address key future challenges for society. It does this through the establishment of dedicated discussion forums, the use of questionnaire surveys, consultation documents and events, and the deployment of a range of futures techniques, such as scenarios. The overall aim is to increase the UK's exploitation of science.

In this chapter, we will provide an account of the UK Foresight Programme, describing each of its three cycles and explaining its evolution over time. Although there is plenty of other foresight activity going on in the UK,[2] the focus will be solely upon the national Programme, since this is the longest established exercise and its evolution illustrates well the shifting rationales for public support of foresight exercises.

HISTORICAL ANTECEDENTS

Although the UK Technology Foresight Programme (TFP) was not initiated until 1993, the idea of conducting a national technology foresight programme in the UK had been thought about since the mid-1980s, when two researchers at the Science Policy Research Unit (SPRU), John Irvine and Ben Martin, were asked by the main national science and technology (S&T) advisory body to the Government to review technology foresight experiences in other countries and to consider the benefits for the UK of using foresight. The resulting report was published as a book (Irvine and Martin, 1984), but its message was largely rejected by the Government, which mistook foresight as a new form of public sector selection of technologies for commercial exploitation. This was an era when UK Government activities were shaped by a neo-liberal orthodoxy, which espoused the merits of market mechanisms, including for the selection of technologies. The Government was therefore suspicious of any activity, including foresight, which seemed to suggest that non-market actors and signals should be involved in "picking winners" in technology.

The idea of a national UK foresight exercise therefore lay dormant for almost a decade, to be resurrected in 1992 with a change in the political and institutional climate. Although the Conservative Government remained in power, the neo-liberal programme had been toned down. At the same time, a new science ministry was created in mid-1992 – the Office of Science and Technology (OST) – with the aim of better coordinating the UK's extensive public spend on S&T. This new ministry was looking for some "big ideas" to include in a major policy document that would establish its mandate and set out its future plans for UK S&T (Flanagan and Keenan, 1998). Although foresight lay dormant, it had not completely gone away. Irvine and Martin were still strong advocates and had sympathisers in various parts of Government.

Other academics, such as Luke Georghiou and Denis Loveridge (PREST), also played important roles in keeping foresight on the agenda. In the meantime, various small-scale foresight-type work was being carried out by some of the Research Councils and by the Department of Trade and Industry's (DTI) Long-Term Studies Unit. The latter pushed particularly hard for a national technology foresight exercise to be launched and had funded some pilot work in early 1992 that was jointly carried out by academics from PREST and SPRU.

The newly-created OST later became involved in this work and was convinced of the benefits of foresight. In particular, the fact that both Japan and Germany – at the time, important reference points for policy learning by UK policy-makers – were already using technology foresight to inform national S&T policy was an important driver in the UK adopting the approach.

THE FIRST CYCLE

In 1993, the OST launched its White Paper – *Realising Our Potential* – which constituted a blueprint for British S&T for the coming decade (OST, 1993). It announced major structural changes to the Research Councils and a review of the organisation of science in state ministries. The new Research Councils were, for instance, given mission statements and instructed to organise panels of research users (including industry) to inform their funding strategies. At the heart of the White Paper was the announcement of the OST's intention to organise a national Technology Foresight Programme (TFP).

The rationale for the TFP was multi-faceted, referring most prominently to the need for prioritisation in the UK's S&T efforts. According to the White Paper:

> No one nation can afford to sustain a significant independent presence in all of the burgeoning fields of scientific research. The Government must therefore work closely with the scientific and industrial communities to determine the appropriate mechanisms for setting priorities both in terms of the areas of research to support, and the level of funds to be committed to them (OST, 1993, p. 2).

A systematic, well-informed assessment of the match between potential research outputs and the likelihood of their appropriation by firms and organisations would be required, it was argued. As for the priorities identified, it would be for "individual firms and organisations to decide for themselves what use, if any, to make of the results of this Programme" (ibid.). The White Paper also highlighted the TFP's role in developing networks and partnerships.

> The TFP was going to benefit the UK, "not only as a means of gaining early notice of emerging key technologies but also as a process which will forge a new working partnership. This partnership, and the networks which develop from it (...) should lead to greater common understanding of the trends and uncertainties involved in future technological developments" (ibid. p. 17).

In fact, the White Paper anticipated that the TFP would be most valuable if it succeeded in generating informal interaction. The Government believed that university Vice Chancellors already had adequate contacts with Company Chief Executives and saw no need for further formal committees or boards. The vision was of opportunities being generated on a much larger scale through "interaction between scientists and businessmen involved in the day-to-day business of selling in competitive markets" (ibid, p.18).

Shortly after the publication of the White Paper, a Steering Group, chaired by the Government's Chief Scientific Advisor, was established to oversee the running of the TFP. Fifteen sector panels (later 16) were subsequently set up, with experts and stakeholders drawn from business, Government and academia appointed as panel members (see Table 4.1). Among other things, these panels were charged with identifying key trends and drivers, benchmarking their sectors, developing scenarios, consulting widely with their communities through a Delphi and workshops, and constructing priorities and recommendations for action. The use of Delphi in the earlier Japanese and German national foresight exercises had prompted the OST to include the approach in the UK Programme. Its use was further rationalised by the need to engage a far broader base of expertise than could be accommodated on the sector panels. Thus, the Delphi survey was sent to around 10,000 people, with almost 3,000 responses received. The Delphi generated some extremely rich data (Loveridge *et al.*, 1995), but due to timing problems, the results were not fully used by the panels. This led to criticisms of the approach, sufficiently damaging the reputation of Delphi to ensure that the method has not been used extensively in UK Foresight since.

Table 4.1   Sector panels from the first Technology Foresight Programme[3]

Agriculture, Horticulture & Forestry	IT, Electronics & Communications
Chemicals	Leisure & Learning
Construction	Manufacturing, Production & Business Processes
Defence & Aerospace	Marine
Energy	Materials
Financial Services	Natural Resources & Environment
Food & Drink	Retail & Distribution
Health & Life Sciences	Transport

In total, it is thought that some 10,000 people were consulted during 1994, both through the Delphi and through regional and national workshops. The work of the Programme in 1994 was extremely intensive, with many sector panels meeting on a monthly or more frequent basis, with huge pressure to get the Delphi questions, the interim and final panel reports, and presentations for various workshops, prepared to meet tight deadlines. But these deadlines were met. In 1995, the panels published their findings in reports that aimed to identify the likely social, economic and market trends in each sector over the next ten to twenty years, and the developments in science, engineering, technology and infrastructure required to best address future needs. Some 360 recommendations for action were suggested across the Programme, with the Foresight Steering Group identifying a further 28 generic science and technology and 18 infrastructure priority areas (OST, 1995).

The sector panels were subsequently retained as the "hubs" of dissemination and implementation activity, continuing to meet regularly up until 1999 in order to coordinate and/or catalyse follow-up actions on their priorities. Other organisations, including Government Departments, Research Councils, trade associations and professional bodies also took up the mantle of dissemination and implementation. Between 1994 and 1999, over 600 Foresight events were held and 130,000 copies of the Foresight panel recommendations distributed. It is impossible to accurately estimate the impacts all of this effort had, but there is evidence that several hundred million pounds worth of research spending was at least "aligned" with (if not a direct result of) the priorities and recommendations to emerge from the Foresight Programme (see POST (1997) for a comprehensive discussion of the impacts of the Programme on Research Council spending). Moreover, many new fora emerged from the Programme, for example, in the form of new "virtual" centres, which brought together stakeholders on key issues, such as traffic management, health informatics and mobile telecommunications (Keenan, 2000). Finally, inspired by the national Programme, scores of organisations went on to conduct their own foresight studies, helped along by the publication by the OST of methodological guidelines and successful case studies, as well as by the availability of funding to support some limited new work in the regions.

The Programme also attracted a great deal of attention in Europe and more widely, being seen as a particularly successful experience. Its results were also widely disseminated, being readily available on the web. The perception was that this Programme effectively combined priority-setting (product) and networking (process) elements, wide and high levels of participation with technical sophistication, and policy influence with vision.

This apparent success should be viewed against an ever-changing background of developments and events. First, the OST had been relocated from its central position in the Cabinet Office[4] into the DTI in 1995, around the time that the Programme was presenting its major outputs. In the view of some commentators, this was not only disruptive, but also created implementation difficulties for the Programme, since it was no longer viewed by other ministries and agencies as a "neutral" player but rather as an advocate of an agenda set by a competing department (DTI). Also around this time, Sir William Stewart left his post as Chief Scientific Advisor and head of the OST. He had been particularly important in establishing the Foresight Programme, chairing the Steering Group, and making it a central part of the OST's work. His departure was just part of a wider draining away of the people best informed about the Programme from its management and implementation structures. For instance, the technical secretaries and facilitators to the original Panels mostly departed after the panels had accomplished their main tasks. Moreover, panel members were not expected to serve more than three years, so that by 1998, perhaps as few as 20 per cent of the original participants were still involved in the Programme. This was to have ramifications for the design and management of the second cycle, as discussed below.

Also during the first cycle, the incoming Labour government in 1997 came with a vision of the "knowledge-based economy" – that was rapidly reflected in studies and strategies developed by the DTI, among other parts of the state. An audit was commissioned into the Foresight Programme as an early step, and this resulted in a largely favourable appraisal of its activities – and since the audit was conducted on an interdepartmental basis, it actually raised the profile of the activity across government. However, the Programme continued to receive criticism from several quarters on its relationship with and impacts on industry (for example, see evidence presented by the Confederation of British Industry to the House of Lords Select Committee on Science and Technology (1997) in its review of the Foresight Programme). This saw the Programme partially re-launched in 1996 with a newly reinforced business focus (OST, 1996). At the same time, funding was made available to regional government offices to disseminate the results of the Programme to businesses in their region – especially small and medium-sized enterprises (SMEs). For the most part, these efforts seem to have met with little success, due to funds being spread too thinly and a lack of communication between OST and the regions. On the other hand, this limited regional activity did catalyse the establishment of long-standing foresight exercises in a few regions, such as North-East England and Northern Ireland.

THE SECOND CYCLE

Extensive consultation about what was to replace the first cycle of Foresight was undertaken in 1997 and 1998, resulting first in a formal consultative document and then in a "blueprint" – adding detail to the earlier proposals (OST, 1998). The innovation researchers who had been very influential in designing the methodologies of the first cycle were, for the most part, not active in the second cycle, though Ben Martin remained on the Steering Group (see Miles, 2005). The enhanced business focus of the latter stages of the first cycle was carried over to the second, along with greater emphasis upon quality of life issues.

The second cycle was launched by the OST in April 1999, explicitly aiming to:

- Build on the perceived success of the first cycle, while taking account of the changed conditions at the end of the 1990s;
- Provide more visionary and better integrated outputs (the high time pressures of the first round were viewed as having restricted the exercise);
- Involve a wider variety of participants -- including more small and medium-sized enterprises (who had been hard to enlist in the first cycle, even though there were "outreach activities"), and voluntary and public sector representatives;
- Put more stress on quality of life issues (some members of the Labour Party were suspicious of the supposed emphasis on wealth creation, and the new Government had social inclusion high on its agenda) – this meant examining social issues from the outset, not just as secondary factors or obstacles to technological innovation.

In two major respects the second cycle showed continuity with the first, and in two respects it marked a severe break. First, the combination of a Steering Group (composed of policy-makers, industrial managers, and academic and industrial researchers) and panels was retained, with panels still at the heart of the Programme. However, they were constituted around new themes. Ten panels dealt with sectoral and technology themes – fewer, and typically smaller, groups than in the first cycle. In addition, three thematic panels were established to tackle broad issues cutting across science and technology and sectoral boundaries, and that had been highlighted as general challenges by participants in the first cycle. These were "ageing population", "manufacturing in 2020" and "crime prevention".[5] All panels were also asked to consider the implications of their findings for another two generic issues – education, skills and training, and sustainable development, and appointed members who would pay particular special attention to "underpinning technologies" (identified as information technology and biotechnology). The panels were also requested to interact more intensively than had been possible in the first cycle, among themselves and more broadly.

Table 4.2   Panels in the second cycle

Sector Panels	Thematic Panels
Built Environment & Transport	Ageing Population
Chemicals	Crime Prevention
Defence, Aerospace & Systems	Manufacturing 2020
Energy & Natural Environment	
Financial Services	
Food Chain & Crops for Industry	
Healthcare	
Information, Communications & Media	
Marine	
Materials	
Retail & Consumer Services	

Second, networking remained a priority. Three new ways of furthering this were introduced. Task forces, extending beyond Panel members, were formed to address specific issues (sometimes highly specific ones, sometimes cross-panel concerns). Some 65 of these were set up, though most of these were not very effective – however, a few really did take off. Some of these took on the task of implementing and further developing panel recommendations from the first cycle. Over 500 people were involved in panels and task forces. In addition to the task forces, Associate Programmes to take Foresight practice to a wider community were encouraged: these were run by organisations outside the OST on topics of particular interest to their members. Finally, the Knowledge Pool was intended to be a common resource, a web-based collection of information and reports to support work in panels and task forces, and to enable dialogue and discussion with a wider community. Using software developed for a newspaper organisation, and requiring a substantial investment in design and maintenance, this pool initially contained results and other documents from Foresight and S&T policy activities in the UK and elsewhere.

But, third, while the panels were expected to produce consultation documents and work towards final reports for November 2000, these reports were regarded primarily as further ingredients in a wide dialogue, involving as many stakeholders as possible. Though the panel reports were meant to relate to each other, enabling coherent messages to be extracted and critical actions to be identified, there was not a clear target in terms of a policy statement or strategic document that was to be informed. The priority-setting elements that had been so important in the first cycle of Foresight were much less evident here.

Fourth, there was no Delphi, nor any similar systematic methodology that could help provide an integrated view across panels. While the Delphi study of the first cycle had found many users, many panel members had been somewhat alienated by the haste in which work was required, and the consequent difficulty in absorbing the results of something that had consumed a great deal of their efforts.

The second cycle of Foresight involved a large number of meetings, and covered an impressively wide-ranging spectrum of topics. Its website featured a valuable array of material on prospective issues and scenarios, presented attractively and accessibly. But the Knowledge Pool, while widely used (and not just in the UK), did not manage to:

- Serve as the hoped-for platform for wide discussion and debate. It proved difficult for a government-sited website to make itself open to inputs from a very wide constituency: questions of intellectual property and "off message" inputs could not easily be resolved;
- Establish a framework for integrating the diverse range of issues, perspectives, and approaches arising from the many groups reporting onto it;
- Provide a clear focus on challenges for S&T policies and priorities. It would be possible to explore details of topics such as demography without seeing clear links to S&T topics.

These two last points led to "ownership" difficulties for the sponsoring OST, and a new Chief Scientific Adviser (Sir David King), on reviewing the panel reports, asked pertinent questions about aims and objectives. A review of the Programme was subsequently instituted (OST, 2001). This concluded that:

- Foresight's objectives were unclear or too broadly defined. Objectives needed to be tightened and made more realistic;
- Panel reports provided limited value-added, mostly on account of lack of novelty. Reports therefore needed to contain more in depth analysis behind panel recommendations;
- There was a need to narrow down the focus of UK Foresight, with more emphasis upon technology;
- The structure of the Programme needed to be more flexible in order to accommodate emerging issues. A rolling programme, with possibly 3–4 projects running at any one time, was suggested; and
- Foresight's audience was too broad and it attempted to cover too many topics at any one time. Therefore, the number of issues being examined needed to be reduced.

The second five-year cycle had started in April 1999 and was due to come to an end in 2004, thereby following a similar pattern to the first cycle, with a 1–2 year consultation phase followed by a 3–4 year implementation period. But the uneven quality of the panel reports, coupled with the results of the review, led to the decision to cut short the second cycle. Thus, virtually all of the panels and task forces established under the second cycle were wound up shortly after publication of their reports.

Reflecting on the experience of the second cycle, its early closure was due to perceptions that it had failed to deliver against many of the expectations invested in it by the Government and other participants. It is interesting to ask how this could have come about, and in this regard, we propose a few hypotheses. Firstly, it should be acknowledged that the second cycle was informed by the experiences of the first. However, this learning was largely restricted to the experiences of the implementation phase rather than the consultation phase. This was simply because the personnel involved in the consultation phase of the first cycle had long moved on and were not involved with the organisation of the new Programme. This meant that the Programme had lost much of its institutional memory around how to organise the consultative phase of a foresight exercise. Instead, Programme designers took the implementation phases, with which they were familiar, as their reference point.

With this in mind, it is worth examining the thinking behind the implementation phases of the first cycle and its salient features. This was marked by a desire to "diffuse" and communicate the results (and process) of the Programme as widely as possible in the hope that this would lead to implementation successes. At the same time, it was also marked by a "directed" strategy, particularly towards key organisations and individuals, who were viewed as "critical enablers" for implementation follow-up, as well as multipliers (communicators) for the Programme's results and processes. Both this extensiveness and focus on key actors strongly influenced the organisation of the second cycle. For example, the desire for extensiveness can be seen in the establishment of over 60 task forces and the support given to more than 30 associate programmes. Whilst admirable on paper, such extended activity was extremely difficult to support with finite resources. The focus on involvement of key actors manifested itself most prominently in the appointment of (mostly company) chief executives as Panel chairs. Whatever their qualities, their commitment and engagement were inevitably less than that of their predecessors: the result was that panels mutated from active working units to rather formalised bureaucratic units meeting infrequently (as in the implementation phase of the first cycle).

A second explanation for the problems faced by the Programme centres upon a belief that it had to be more inclusive than the first cycle, in terms of both its subject coverage and its participation. Thus, cross-cutting themes were introduced and the process was opened up considerably to a wider variety of actors. The Programme was now viewed more as a social process than one of techno-rational analysis and calculation. This situation arose partly as a result of suspicion in expert-based tools, such as the Delphi. But it was also informed by a growing science and society agenda across Government in the wake of well-publicised controversies around issues like BSE. It led to a Programme that was perhaps overly concerned with ensuring representation of interests and perspectives whilst less interested in analysis and calculation. None of this is to say that the attempts for better inclusiveness and diffusion were unsuccessful in every respect. In fact, the extended nature of the Programme meant that many organisations became familiar with foresight practice, something that has subsequently led to an explosion in foresight activity across the UK. The problem for OST lay in the difficulty of being able to account for such activities given their wide distribution. This is the basis of our third hypothesis: that the diffuse and extended nature of the Programme's social process, whilst having untold benefits, was difficult to account for, leaving OST with few concrete results to point towards as indicators of success. Without these, the Programme was always going to be open to criticism, particularly from those who might not have been directly involved in the process.

A fourth and final hypothesis concerns developments external to the Programme. In 1998, the Performance and Innovation Unit (PIU) was established in the Cabinet Office (the former home of OST) with the aim of improving the Government's capacity to address long term and/or cross-cutting strategic issues. As part of this work, guidelines were created for futures thinking in the public sector, a Strategic Futures Team was created, and several future-oriented studies were conducted. In 2002, the PIU was merged with the Prime Minister's Forward Strategy Unit (FSU) to create the Strategy Unit (SU). As with the PIU, this was located in the Cabinet Office and continued to report directly to the Prime Minister. These developments unfolded during the era of the second cycle of the OST's Foresight Programme, an exercise that also sought to work across Government with a futures perspective. Whilst the authors are unaware of any evidence to suggest institutional competition between the PIU/SU and the Foresight Programme, clearly there was overlap. Moreover, the wide scope of the Foresight Programme was often resented by other ministries and agencies, which sometimes challenged OST's authority to deliberate on topics under their jurisdiction. The PIU/SU no doubt faced similar problems on occasion, but the fact that they were located in the Cabinet Office and reported directly to the Prime Minister meant that their involvement was often welcomed as a means of raising political awareness around a topic. Given its institutional location, the OST's Foresight Programme could never offer anything similar. Under these circumstances, it is not difficult to see why ministers and the new Chief Scientist decided that it was time to narrow the scope of the Programme.

With the second cycle wound up early, many of the civil servants in Foresight Directorate left and were not replaced. From a high number of almost 50 full-time officials working on the Programme in 2000, by 2002, this number had fallen to below 20. The early termination angered many of the Programme's participants, who had given their time freely to contribute to panel and task force meetings. Some vowed never to work with OST again. For certain commentators, Foresight was dismissed as an exercise that had served its time. This was therefore a time of great uncertainty for the future of the Programme. The task facing the OST was to set about reinventing the Programme in a bid to make it more useful and relevant for a new environment.

THE THIRD CYCLE

Launched in April 2002, the third cycle of UK Foresight is largely the outcome of the review of the second. Thus, in taking account of the review's recommendations, both the scope and scale of the Programme have been much reduced. In practice, this translates into a Programme predominantly focused upon science and technology and where no more than 3–4 "projects" are ongoing at any one time. This means the third cycle is markedly different from its two predecessors, both in terms of its structure and organisation. Moreover, whilst earlier generations of UK Foresight were principally targeted on the wealth creation agenda, the third cycle has seen a shift towards public policy concerns. Therefore, instead of panels chaired by industry with a government department representative, we now have a model where the main stakeholder is in Government at a political level, supported by very senior members of the civil service. This emphasises the fact that the key constituency is Whitehall, i.e. the British Government.

As already mentioned, rather than organising UK Foresight around panels, the Programme takes "projects" as its starting point. Projects are concerned either with a key issue where science holds the promise of solutions (e.g. flood and coastal defence) or with an area of cutting-edge science where potential applications and technologies have yet to be considered and/or articulated more broadly (e.g. cognitive systems). The third cycle operates through a flexible rolling programme of 3–4 such projects at any one time, with each estimated to last between 12–18 months. In this way, the Programme can respond quickly to emerging issues.

The process of selecting Foresight projects involves wide consultation with the science base, government departments, devolved administrations, business sector and others. Foresight maintains a fluid long list of issues that come up from one or more sources. When particular issues begin to develop clear momentum, the OSI may choose to scope them more formally. At any one time, the fluid long list contains about 40 issues. These range from topics such as the impacts of climate change for agriculture, to the possibilities of and implications of being able to control rain or hurricanes, and the future of the deep oceans. Every three years or so, public consultation exercises are conducted to identify new issues for inclusion in the long list. The first projects (described at the time as pilot projects) were launched in April 2002. They covered Flood and Coastal Defence (FCD) (OST, 2004) and Cognitive Systems (CS). In the absence of a formal selection process at that time it may be noted that they were topics reflecting the interests of their respective directors, Sir David King and Dr John Taylor (Director–General of Research Councils at that time). This meant that the projects had high level interest and backing from the start. Further topics were developed through a more systematic process. Initially an intensive workshop ("Hothouse") was held in which a group of senior scientists identified 12 potential project topics. These were put out to consultation in September 2002. The consultation paper invited comments and ideas for further topics and two suggested additional topics were subsequently added to the list. Six criteria for project selection were announced, indicating that projects would need to tackle issues that:

1. Require looking ahead at least ten years, in areas where the outcomes are uncertain. This typically occurs where the future direction of change is rapid, current trends are uncertain or different trends may converge;
2. Have science and technology as the main drivers of change or are capable of impacting substantially on future scenarios;
3. Have outcomes that can be influenced, to an extent that is significant for one or more of the economy, society and the environment;
4. Are not covered by work carried on elsewhere. However, they must build from areas of active research;
5. Require an inter-disciplinary approach to the science, and bring together groups from academia, business and government. They must not be capable of resolution by a single group; and
6. Command the support of the groups most likely to be able to influence the future and be owned by a lead government department.

Over one hundred responses were received to the consultation and other contacts were made in relation to the proposed selection. One of the externally suggested projects (on Tackling Obesities) has recently been launched as a Foresight Project. The scope of some projects was also refined in part as a result of the consultation. The selection itself was made using a process of scoring against a conflated set of the above criteria and with structured ranking information. A key factor was the ability to obtain senior stakeholder commitment.

The first projects to emerge from this process were Cyber Trust and Crime Prevention (CTCP) and Exploiting the Electromagnetic Spectrum (EEMS). CTCP was led by Professor King, with the support of the Home Office, and EEMS by the Director General of the Innovation Group in the Department of Trade and Industry (DTI). The next three projects were all short-listed for the future at this stage – Brain Science, Addiction and Drugs (BSAD), Detection and Identification of Infectious Diseases (DIID), and Intelligent Infrastructure Systems (IIS). Two new projects were launched in late 2006: Sustainable Energy Management and the Built Environment (SEMBE) and Mental Capital and Wellbeing (MCW).

Table 4.3   Foresight Projects and their "sponsors" (2002–2007)

Foresight Projects	"Sponsors"
Flood and Coastal Defence	Chief Scientist (OST) and Defra
Cognitive Systems	Director General of Research Councils
Cyber Trust and Crime Prevention	Home Office
Exploiting the Electromagnetic Spectrum	Innovation Group, DTI
Brain Science, Addiction and Drugs	Department of Health
Detection and Identification of Infectious Diseases	Defra
Intelligent Infrastructure Systems	Department of Transport
Tacking Obesities: Future Choices	Department of Health
Sustainable Energy Management and the Built Environment	Department for Communities and Local Government, Defra and DTI
Mental Capital and Wellbeing	Department of Innovation, Universities and Skills

Once a project has been selected, an initial seminar is usually organised to include expert speakers from related fields. This informs participants about each other's work, and enables the scope and objectives of the project to be refined. Each project has a dedicated project team in the Foresight Directorate at OSI, who are assisted by scientific experts. These teams have access to the latest research information and are skilled in futures techniques in order to best capture and explore future environments. Each project draws on inputs and insights from a network of external experts and scientists and leaders in their fields (see Figure 4.1 below).

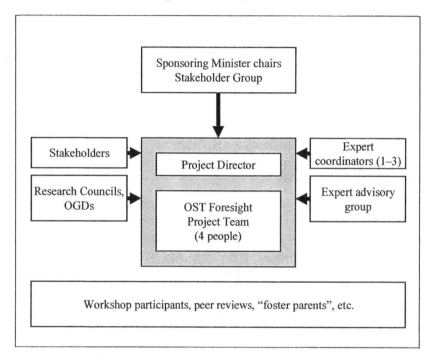

Figure 4.1   Extended project team structure

Each project should deliver analysis and current information about developments in science and technology and how these will pan out in a global context. The aim is to create visions of the future that will reflect the impact these developments will make. In line with earlier Programme cycles, each project is also supposed to deliver recommendations for action – by research funding agencies, business, Government and others – and to enable the creation of networks of those able to take forward actions. The generic stages for projects are shown in the table below.

Table 4.4   Master plan stages and key issues/actions

Stages	Key Issues/Actions
Scope	Subjects
	Stakeholders
	Consultation
	Relevance
	Value-added
	Planning
Review	Review the state-of-the-art of science
	Roadmap technologies
Analysis and Synthesis	Scenarios, trends and other futures
	Modelling
	Testing hypotheses
Engage	Broadening networks
	Engaging the influential with findings
	Draft action plans
	Test strategies
Launch and Action	Link actions to owners
	Launch plan and supporting outputs
	Track

From recent evaluation work carried out by the authors, it can be concluded that the third cycle of the Programme has operated a niche strategy and has in general succeeded in being regarded as a neutral interdepartmental and interdisciplinary space in which forward thinking on science-based issues can take place (PREST, 2005). More generally this model can be described as a licence to think outside the box. At the same time, given the public policy focus of most of the projects, industry has not been in any prominent role in respect of positioning of the Programme and has been downstream of the projects. However, this could change with the incorporation of the DTI's Innovation Unit into the old OST structure to create the new OSI. At the time of writing, a new process for topic generation has been launched (in addition to another "Hothouse" process in 2006), involving input from a new horizon scanning activity. This activity is focused within a newly established Horizon Scanning Centre, set up in 2004 in the OSI's Foresight directorate. The Centre's aims are:

- To inform departmental and cross-departmental decision-making;
- To support horizon scanning carried out by others inside government;
- To spot the implications of emerging science and technology and enable others to act on them.

Again, it is clear that the emphasis remains upon Government. Accordingly, the work streams of the Centre are focused upon: (a) regular cross-Government strategy horizon scans, to underpin existing horizon scanning and inform cross-Government priorities; (b) project work with stakeholders: demand-led opportunities for joint working on specific issues with stakeholders (departments or groups of departments); and (c) the provision of tools and support to spread good practice in departmental horizon scanning, including coaching, providing advice, brokering agreements and creating synergies that make the best use of resources and facilitate capacity-building. It is interesting to note that the above-mentioned cross-Government scans, as well as the provision of foresight tools and coaching, are somewhat reminiscent of the activities of the first and second cycles. At least with regards to the scope of the activities, it is possible that developments are turning full-circle, with a wider set of "new" objectives being "reincorporated" into the Programme. It is too early to say how the OSI will manage this expanded role, but past experiences suggest that care will be needed to keep the Programme sufficiently focused and to avoid unnecessarily raising expectations.

EXPLAINING THE EVOLUTION OF UK FORESIGHT

Looking across the three cycles, the national programme has moved from a broadly-scoped exercise in priority-setting and networking aimed at allowing S&T policy to be better designed within a better-functioning innovation system (first cycle), to a broadly-scoped distributed exercise of multiple foresight initiatives aimed at upgrading awareness of a wide range of futures issues and building a more general foresight culture (second cycle), through to a programme of more focused science-based discrete projects aimed at reaching across government departments to improve the ability of policy-making to deal with S&T issues (third cycle). This evolutionary path of the UK Foresight Programme can be clearly attributed to learning effects between the cycles. But the types of lessons, the ways in which they were learned, and even their validity, varied substantially. In moving from the first cycle to the second the atmosphere was conditioned by the perception of a rare success. This was at a time when the Government's research and innovation policies were generally in a state of disrepute as a result of long-term underfunding and ideological objections to support for industrial R&D. Since this success had not been scrutinised by a fully constituted evaluation, some perceptions about what the underlying positive factors had been, were in all likelihood mistaken. The discontinuity in staffing caused by the exodus of key personnel and panel chairs at the end of the cycle reduced the ability of the programme management to set these inputs in context. The evaluation of the second cycle was short, but to the point; it was empowered from the outset to lead to the design of the third. To some extent the problems of the second cycle were left behind by moving into different territory. The shift in agenda to that of using Foresight to increase the role of science in policy was probably already expected before the audit got under way.

Table 4.5 describes the evolution of UK Foresight across a number of parameters. This evolution can be accounted for in terms of a number of factors. Some are internal to the Programme – as we have seen, there have been learning processes about what did and did not work in the previous round, even if these are imperfect. In addition the role of individuals should not be underestimated: key figures, especially the Chief Scientific Advisor, have played important roles in the structuring and implementation of the Programme and its projects, and in the positioning of how Foresight could be of value in the UK innovation and policy systems.

Additionally, the role of external factors is significant. The election of the Labour Government in 1997 almost inevitably made for a reorientation of Foresight, as the new administration sought to make its mark. The new government also entered with various ideas in place about evidence-based policy and the knowledge-based economy. These have been reflected in increased interest across Government in futures-oriented thinking (this seems to have reduced OST Foresight's role in 2001, but this role expanded again in 2005 as horizon-scanning activities required coordination). The story is a complex one, however. Shifting problematisations of S&T have occurred over the lifetime of the Programme. In the early 1990s, for example, a shift from a science-led to an innovation-led agenda for public S&T spending required the articulation of S&T priorities to serve predominantly wealth creation. This requirement for priorities underpinned the conduct of the first cycle. By the time of the second cycle in the late 1990s, science and society issues had joined the innovation agenda, as concerns were raised around the social and ethical dimensions of genomics and nano-level technologies. This, too, had an impact on the scale and scope of the Programme. By the turn of the Millennium, a culmination of "shocks" around food and public health and the interrelations between scientific and policy responses to these, led to pressures on a number of ministries (especially Defra)[6] to handle these interrelations better and to deal more effectively with emerging risks by bringing science to bear to inform policy-making. It is clear that this "risk-uncertainty agenda" has strongly influenced the purpose and organisation of the third cycle. So we can conclude that the Programme has certainly been able to adapt to shifting environments. How far Foresight can continue to play a central role in the confluence of these agendas remains to be seen; certainly there could be some turbulent times ahead.

Table 4.5 Schematic picture of the evolution of UK Foresight

	Stylised Particularities of each cycle		
Parameter	Cycle 1	Cycle 2	Cycle 3
Main Rationales	S&T priorities	Business and societal dialogue	Anticipating policy-relevant change and risk
Main Targets	Initially, scientists and research funding agencies; latterly, also the business community	Several actors across government, business (including SMEs), the research world, and society	Predominantly government ministries
Coverage	Mix of sectoral and technological areas spanning most of private sector and some public sector	Mix of sectoral and thematic areas – even wider coverage than the first cycle	Mostly small numbers of focused topic areas of interest to government ministries
Structure	Standing sectoral panels	Standing sectoral and thematic panels with task forces	Rolling projects
Participants	Essentially the same across all three cycles, although fewer industry actors are involved in the third cycle		
Methods	Delphi and workshops used across the Programme, with bespoke methods used by the individual panels	Predominantly scenarios and consultation documents, website for dissemination and interaction	Wide variety of methods (scenarios, simulations and gaming, Delphi, workshops, etc.) used in different projects
Outputs	Panel reports, Delphi results, priorities and recommendations, and various other reports during the implementation phases	Panel & task force reports, many web publications (i.e. scenarios and even videos at one point)	State of science reviews, action plans, scenarios, project reports, academic books, etc.
Reception	Generally positive, though many argued that the Programme failed to realise its full potential, particularly with regards to reaching the business community	Generally negative, with some panel reports dismissed as dull and uninspiring and the Programme being deemed as unfocused	Very positive, with highly regarded outputs that have been taken up in policy formulation and adaptation

To conclude, foresight is now a well-established policy instrument in the UK, and is used across many agencies and departments. The national exercise conducted by the OSI is just one of many such activities, and is a far more modest and focused activity than its predecessors. Notably, there is no attempt any longer to establish priorities across the whole of the science base. Indeed, such attempts were largely dropped after the first cycle. Nonetheless, agencies and ministries (as well as the OSI) continue to use foresight to identify emerging areas offering opportunities and the need for future policy action.

NOTES

1. From its establishment in 1992 until April 2006, the OSI was known as the Office of Science of Science and Technology (OST).
2. For example, see the results of the European Foresight Monitoring Network (EFMN), where a few hundred foresight-type exercises conducted in the UK have been mapped (http://www.efmn.eu).
3. This is the list of panels effective at the end of the first phase of UK Foresight. The Marine Panel was added to the original 15, and in a further reorganisation two panels were created from the original "Agriculture, Natural Resources and the Environment" Panel, while the IT and Communications Panels were merged.
4. The Cabinet Office can be thought of as the ministry for the civil service and is perceived as being at the heart of Government.
5. Work here was funded by the Home Office, the UK version of the ministry of the interior, responsible for policing among other functions.
6. Department for Environment, Food & Rural Affairs.

REFERENCES

Flanagan, K. and Keenan, M. (1998), 'Trends in UK Science Policy', *Science and Technology in the UK* (2nd edn), Cartermill, pp. 21–68.

Georghiou, L. (1996), 'The UK Technology Foresight Programme', *Futures*, **28**(4), pp. 359–377.

House of Lords Select Committee on Science and Technology (1997), *Science and Technology Third Report*.

Irvine, J. and Martin, B. (1984), *Foresight in Science: Picking the Winners*, London: Pinter.

Keenan, M. (2000), *An Evaluation of the Implementation of the UK Technology Foresight Programme*, Doctoral Thesis, Manchester, UK: PREST, The University of Manchester.

Loveridge, D., Georghiou, L. and Nedeva, N. (1995), *Technology Foresight Programme: Delphi Survey*, Manchester, UK: PREST, The University of Manchester.

Miles, I. (2005), 'UK Foresight: Three Cycles on a Highway', *International Journal of Foresight and Innovation Policy*, **2**(1), pp. 1–34.

OST (Office of Science and Technology) (1993), *Realising Our Potential: A Strategy for Science, Engineering and Technology*, Cmnd 2250, London: HMSO.

OST (1996), *Winning Through Foresight: A Strategy Taking the Foresight Programme to the Millennium*, London: Department of Trade and Industry (dti).

OST (1998), *Blueprint for the Next Round of Foresight*, Department of Trade and Industry.

OST (2001), *Foresight Review – A Summary Report*, London: Department of Trade and Industry (dti).

OST (2004), *Foresight Future Flooding – Executive Summary*, Available at http://www.foresight.gov.uk, last accessed 12 July 2007.

POST (United Kingdom Parliamentary Office of Science and Technology) (1997), *Science Shaping the Future: Technology Foresight and its Impacts*, London: POST, June.

PREST (2005), *Evaluation of the United Kingdom Foresight Programme*, Office of Science and Technology, http://www.foresight.gov.uk.

5. Foresight in France

Rémi Barré

HISTORICAL DEVELOPMENTS

The Emergence and Development of *La prospective*

There is in France a long experience of both theory and practice of *La prospective*. Its roots are to be found in the post-war reconstruction movement, with its State coordinated planning of long-term infrastructural investments and its co-management between high level technocrats, industry and trade unions. The framework was the specific economic model set up in the 1950s and 1960s, based on a powerful central State and an important public industrial sector (utilities, but also energy, most of banking, transport and heavy industries). Planning was at the centre of this enterprise of "building and mastering a future" at the national level.

In the search of a blend of national planning, open to social negotiation and openness to technological innovation, the concept of *La prospective* came very early (see also Chapter 2). The philosophers Bertrand de Jouvenel (1954) and Gaston Berger (1967) paved the way in their path-breaking work by relating the notions of collective will, fate and forecasting in novel ways. Thus, since the late 1960s, the concepts and methods of *La prospective* have been used by the General planning Commission (Commissariat Général du Plan – CGP), which became an active centre in the field (Massé, 1973), along with DATAR (the National Land Planning Commission).

In the 1990s: the Rise of Participative and Strategic Prospective

In 1996, the "General Planning Commission" (CGP) celebrated its 50th anniversary with a conference which was attended by most political leaders. It was an occasion to discuss the meaning of long-term approaches and *La prospective* in public policy in general, and research policy in particular. A broad consensus was found on the following points (CGP, 1997):

- It is one of the duties of the political system to produce long-term "visions" through public debate;
- Such long-term "visions" are a mean to re-assess the "social contract" among citizens in a globalising world, where margins of collective choice are increasingly less clear;
- In such long-term "visions" exercises, it is essential to give room for debate among social actors and to identify the role of political decision-makers.

The aim is thus to foster collective understanding of societal problems, to make more explicit what is at stake and to generate public debate.

In line with this, the influential "Energy 2010–2020" panel report on "long-term challenges" (Dessus, 1997) stated that *La prospective* "aims at building coherent and contrasted images of the future (...) allowing for debates on the hypothesis and consequences of various options".

Also in the mid-1990s, the Ministry of Industry launched its 'Key technologies for 2000' study, followed by a similar exercise in 1999 and 2004.[1] These studies aim at identifying those technologies that represent a major stake for French industry, at characterising the international strengths and weaknesses of the major countries and at making policy recommendations. In a similar perspective, the Ministry of Research launched in 1996 a Delphi exercise (Héraud *et al.*, 1997). It appears that *La prospective* evolved in the 1990s towards a more open and participative model and, in so doing, the methods have been used in a more eclectic and ad-hoc way. In parallel, the linkage with strategic processes became increasingly prevalent. Nevertheless, the initiatives reported here *ad hoc* operations, with little connection between them and with real policy-making.

As a conclusion of the experiences of this period, the Social and Economic Council (Conseil Economique et Social – CES)[2] published in 1998 a report (see also Bailly, 1998) under the title "prospective, debate, public decision", which called for "new relations between *La prospective* and public debate". It was a call for better public debate on public policy issues and it suggested that *La prospective* should be the proper way to achieve that.

In the early 1990s, Foresight emerged, firstly in the UK, as a tool for S&T policy-making. This was in view of the changing nature of S&T knowledge generation and exchange, the new dynamics of its interactions with the socio-economic base, and the changing requirements of a democratic society (Martin, 1995). In spite of the differences in context of origin and institutional roots between foresight and *La prospective*, the possible distinctions between the two appear to be elusive (Barré, 2000, 2001), both being concerned with "strategic intelligence and future oriented participative activities".[3]

We are witnessing the emergence of a new policy tool, the characteristics of which are its long-term perspective, the diversity of actors and inputs concerned, the interaction among participants, the concern for alternatives, and the articulation to strategy it offers. In what follows we use the term foresight as an exact translation of what has historically been labelled *La prospective* in France.

The S&T Policy Context and its Linkage to foresight

In order to position properly the foresight exercises which will be described in this chapter, it is important to have an overview of the architecture of the system in terms of its governance structure. The "political level" is the "macro" level where general funding priorities and the orientation of research and innovation policies are defined and debated in the political arena. In France, the Ministry in charge of research articulates the national S&T policy through the Civil R&D budget,[4] which does not include the Defence R&D budget, nor University institutional research funding, while innovation policy is largely dependent on the Ministry of Industry. Without interministerial coordination,[5] or a strong advisory structure[6] together with a very complex programming level (see below), it is hardly surprising that no foresight activity has taken place at this level in France since the late 1990s.[7] The FutuRIS operation, which will be presented as our first case-study, emerged as a reaction to this situation, the initiative having been taken by the national industrial association dealing with research.[8]

The "programmatic level" is the "meso" level in charge of translating the overall policy orientations and funding structure into allocation of resources to specific institutions and projects usually through programmes. It is represented in most countries by a dual structure of institutional support and project funding from one or more agencies through research programmes. In France, this level is traditionally split between:

- The Ministry of Research which uses two instruments, firstly the pluri-annual institutional contracts[9] with the research institutions and the universities, and secondly direct funding of projects through small programmes directly under its control;[10]
- A number of technical Ministries to which parts of the research budget are attributed to develop sectoral programmes (environment, infrastructures – housing and transport, industry, defence, etc.);
- Public research organisations which are more than research institutes since their role includes the national programming responsibility in their area of competence.[11]

This fragmented picture of the programmatic level is the basic reason for the complexity of the system, hindering a transparent relationship between the political and the programmatic levels, hence the possibility of meaningful foresight at national level.[12] But at the level of those institutions having a programmatic role in their own specific area, foresight can and indeed has been used. Our second foresight case study, Agora 2020, is being carried by the research directorate of the Ministry for Infrastructure, Housing and Transport. Our third case, INRA 2020 is also partly at this level. It has been carried out by INRA (Institut National de la Recherche Agronomique), the research organisation in charge of the agronomic sector, and extended to cover aspects of environment and nutrition. INRA 2020 also addresses the "operational level" or "micro" level where research is performed by scientists in laboratories linked to research institutes or to universities.

THE FUTURIS EXERCISE[13]

The FutuRIS exercise[14] took place from February 2003 to June 2005 under the auspices of the Association Nationale de la Recherche Technique (ANRT), being jointly financed by the Ministry of Research and of Industry, firms, and ANRT itself. The FutuRIS exercise was initiated by the chairman of ANRT, M. Francis Mer,[15] who then gained support from the government; his successor, M. Jean-François Dehecq[16] continued the process.

The European Union's Lisbon objectives for growth and jobs from a knowledge-based society, as well as the European Commission's Open Method of Coordination related exercises (benchmarking) of the last three years, provided an intellectual and a political context for raising the question of the unsatisfactory performance of the French National Innovation System (NIS). Concerns were raised on the overall efficiency and on the relevance of the architecture and governance of the NIS, leading to increasing doubts concerning the ability of the French research and innovation system to meet the needs of French society in the future. An aspect of that "archaism" of the NIS is that it is very fragmented, with a lack of interactions between its different parts. Reforming the NIS requires an in-depth review and a long-term perspective, drawn together by the various stakeholders; that is what FutuRIS sought to provide. It considers the NIS as a whole, with a systemic approach, addressing issues related to its governance and organisation, the interactions between players from research, academia, business and society.

The general objectives of FutuRIS were "to bring together leading players of the public and private sectors with the aim of laying the foundations for the future of the National Innovation System"; "to review and then launch the debate on the challenges the NIS is likely to face in the future"; and "to build a shared vision of the future of the NIS between research, academia, business and society".

Such general objectives were translated into the following operational objectives:

- To explore alternative states of the NIS at Horizon 2020 described in terms of financial flows, the human resource situation, and governance structure, in an international and European socio-political context;
- To assess their contribution to national objectives (industrial innovation, contribution to public goods and public policies, defence, security, etc.);
- To outline the policies needed to bring them about (institutional reforms, financial flows, education, etc.);
- To identify the key-issues to be addressed and discussed for reforming the NIS in view of the Lisbon objectives.

It should be noted that FutuRIS had nothing to say on research themes, nor was it concerned with substantive priorities identification, or sectorial analysis, but only "systemic" issues.

A one-year feasibility study was conducted by ANRT in 2002, which led to a formal launch of phase I in February 2003, followed by phase II (May–December 2004) and phase III (January–June 2005). The phases were overseen by a Steering Committee composed of about 25 top-level personalities from research, business, government and society and which met every four months. The project team was led by a director,[17] and involved about ten people, most of them part-time.

In *Phase I* the objectives were to build scenarios in order to raise and document the key questions to be faced by the NIS in the coming 15 years. Four working groups were set up (research excellence; competitiveness through innovation; science and society relationships; and trends in the international, European and national contexts) using a process of identification of members with the co-nomination procedure and selection based on criteria of balance between categories such as: experts; stakeholders (research and higher education, business, government, society); gender; age; sector; professional profile and position; geographical area of origin; etc. Beyond the four working groups, circles of external ("second circle") participants were set up, and interactions by collaborative virtual tools were organised (to react to papers, to complete questionnaires, etc). The four groups met 8–10 times (once a month on average) to identify key factors, to document them, to define hypotheses about the possible ways in which events might evolve. Then the hypotheses were combined to build the scenarios concerning their field. From these, the project team built global scenarios.

Phase II took place at a time when researchers were involved in a series of public protests against proposed reforms. It consisted of disseminating the results of Phase I and preparing recommendations for the Bill on research and innovation prepared by the Government. The dissemination strategy focused on the participation of FutuRIS in events and debates prepared by other institutions. In addition, *ad hoc* working groups addressed points on which additional understanding and insight were needed: the strategic governance of the NIS, the synergies between higher education, research and innovation, and the attractiveness of France and its regions for research and innovation activities.

Phase III was a follow up activity for disseminating the recommendations, as well as working on issues such as the perspectives of employment and careers for PhD graduates, the development of innovative firms and the challenge of dealing with the inter-sectoral problematique for resources allocation. Phase III was also used for preparing and launching an evaluation of the exercise, as well as preparing for the follow up of FutuRIS on a regular basis, a smaller scale.

At the end of Phase I of the exercise, under pressure from the researchers, the Government began to prepare an important bill on research and innovation policy and asked FutuRIS, among others, to contribute to the process. Even though one cannot "prove" impacts, it can be said FutuRIS was influential in the process,[18] all the more since its internet site has been considered to be an independent source of documentation and analytical references.

In a more general sense, FutuRIS revealed the strong, but often latent need of social actors to address and discuss uncertainties concerning the dynamics of the innovation system. It highlighted the lack of "space" to do so, that is, the need of an instrument of dialogue for the coordination of anticipated developments, the joint production of visions, and the integration of partial analyses which gain meaning in a broader context.

FutuRIS proved to be an instrument able to handle the diversity of perception of issues and the uncertainty of a number of trends, through hypothesis, scenarios or questionnaires. It was also able to confront controversial issues through the work on hypotheses and on longer-term horizons; and to address the issue of coherence (how can things fit together?). It has proven to be a workable and acceptable platform for informed debate, analytical work, exploration of hypothesis, and has been an instrument to accompany the process of structural change.

In many respects, the organisational set-up of FutuRIS was a classical one in that the methodological highlights and lessons are to be drawn in terms of confirmation of the central importance of some features:

- A Foresight exercise is a project and has to be managed as a project, hence the key role of the project director.
- A Foresight exercise, especially if taking place at national level, must be seen as an actor in the political arena; the steering committee should handle the interactions in this realm.
- Thematic working groups have their own legitimacy and become quickly quite independent in their way of handling topics. This ensures the relevance and originality of their contribution, but can raise problems of integration of results in a coherent or understandable whole. Thus the specific role the central team is to produce results in parallel to the working groups.
- Participation of a broader public in the context of the production of the foresight results is difficult. Things become easier when people are asked to react to a set of results or proposals.
- Identification of the people to be invited to participate in a working group is key for the success and legitimacy of the results, but it is a difficult exercise. The co-nomination process is a valuable way to go forward in this matter.

FutuRIS made a specific advance in one methodological aspect, this being the translation of scenarios into quantitative indicators. Scenarios allow us to look "far and wide" (longer term, system-wide), and provide an efficient basis for interaction and debate (the tacit–codification loop of collective learning). They allow for linkage to decision-making through backwards analysis from future to present (strategic foresight).

But there are conditions for scenarios to fulfil in practice the promises they hold in principle. Namely, they have to be transparent (result from explicit operations) to allow for participation and be coherent to have the credibility required for debating on strategic options and implications. A characterisation of the scenarios by quantitative indicators is particularly welcome here and has been realised in the case of FutuRIS (Barré and David, 2005).

THE AGORA 2020 EXERCISE

Agora 2020 took place between March 2003 and December 2005, under the responsibility of the Foresight Centre (CPVS)[19] in the Directorate for

Research (DRAST) of the Ministry in charge of public infrastructures, tourism, housing, urbanism, transport and natural risks.[20]

DRAST had the attributes of a sectoral research agency, having several research institutes under its responsibility (a responsibility shared with the Ministry of Research) and funds for allocation to research projects through programmes, in which any laboratory in the country is eligible to be funded. Agora 2020 was a foresight exercise aimed at contributing to the orientation of the research institutes and programmes. The project had three objectives:

- To produce a shared vision of the medium-to-long term challenges and key-issues for the following areas: transport, civil engineering, housing, urbanism, and natural risks management;
- To allow for the expression of the demands of the public at large and of various stakeholders;
- To derive from the above a number of priorities and orientations regarding research programmes for the coming years.

Agora 2020 was headed by a steering committee, with members representing different stakeholders/actors (public and private) meeting twice a year to decide on the overall orientations of the foresight. A management committee representing the funding body, i.e. DRAST, organised meetings every two months for an operational follow up. The 'Forum RST',[21] composed of the heads of the research institutes and programmes within the perimeter of DRAST, met every six months to discuss the intermediary results and their possible meaning in terms of research orientation; it also discusses the next steps and provides advice and suggestions for their implementation. Finally, there was a central team, of three people (half time) from CPVS and a consortium of consultants of three people (half time), headed by the director of CPVS.

The Agora 2020 exercise consists of three phases of work:

1. *Phase I* aimed to shed some light on the social demand of various constituencies through an expression of their representations and expectations regarding the challenges of the future. Five workshops, each corresponding to one type of stakeholder, were organised (national administration, local administration, NGO, public research and industry). This led to the identification and prioritisation of major trends and challenges in the infrastructure, housing, urbanism, transport and natural risks areas, as well as preliminary ideas of research orientation.
2. In *Phase II* the ambition was to identify, characterise and cluster the key issues and avenues for innovation to 2020–2030 for each theme,

trying to understand the relative positions of the stakeholders. This was achieved firstly through working groups, which were organised for each theme (infrastructures, housing, urbanism, transport and natural risks). Each group met for two days, starting with the results of *Phase I*, producing organised lists of uncertainties, possible shifts in trends and key questions, and concluding with the possible role of research in producing responses. Secondly, in parallel to the working groups, three panels of "lay citizens" were set up in various cities, each panel being made up of a specific socio-economic category (middle class, with children, in middle-sized city, aged 25–50; lower middle class, in the suburbs of a large city; higher income in large city). Each panel met for two half days to discuss their assessment and future vision about housing, cities and transport. A two-round Delphi-type questionnaire closed this phase, to prioritise the issues and to categorise them according to the kinds of action, including research, which they call for. From there, the issues which required research were clustered into a dozen areas.

3.   *Phase III* aimed at translating the expression of the requests and needs identified during the first two phases into priorities for future research. It also assessed the research capabilities and capacity needed to carry on such research programmes. In practice, during a two-day seminar, about 60 high-level scientists, research administrators and decision-makers from public and private organisations, produced between one and four "lead-visions"[22] per cluster. Such lead-visions were then assessed in terms of capabilities of research organisations to handle them[23] and finally discussed in plenary session for final validation.

The conclusions of Agora 2020 consist of a set of lead visions with their associated socio-economic and scientific challenges, plus recommendations and open questions,[24] all this forming the basis for research orientation in the field of transport, civil engineering, housing, urbanism and natural risks management. The main lesson from phases I & II of Agora 2020 was that there is a large diversity of opinions and visions among the different actors (national and local administration, industry, NGO, public research, citizens). The question then is how to integrate this diversity of needs and social demands into meaningful research orientations. It is too early at this stage to assess the full effects of the foresight, but we can already point to a "cultural effect" in the demonstration of such a diversity which research cannot ignore any longer. In terms of methodology and organisation, the key characteristic of Agora 2020 was the careful and systematic implication of the relevant actors in the successive steps for identifying the challenges and issues, this being achieved through a great variety of methods, used in a complementary way: seminars with various kinds of constituencies and working techniques, focus groups, a Delphi survey, and so on.

THE INRA 2020 EXERCISE

INRA 2020[25] was a foresight exercise that ran between 2002–2004. It was rooted in the conviction of its Chairman[26] that strategic orientations to be decided today need a long-term perspective and anticipatory action: the forthcoming demographic shock in the research force of the Institute – about 40 per cent are due to leave in the next eight to ten years – constituted a strong incentive. Furthermore, it was intended that the exercise would help INRA contribute to the national debate on research.

INRA is the French public research organisation on agronomic research. Its areas of research are agriculture, food and nutrition, and environment. It has a permanent personnel of 8,600, of which 4,000 are researchers. It is organised in 21 regional centres (institutes), regrouping 260 research units, more than half being associated with universities. Its budget is around 600 million euros. As with all the large public research organisations in France, INRA has functions both as a national funding Agency ("programmatic level") and as a set of research institutes, in charge of "intramural research" ("research organisation level").

INRA 2020 had its *raison d'être* in the felt need for developing a "culture of change" in the Institute both in terms of organisation and in terms of thematic focus in the field of agriculture, nutrition and sustainable development. The operational objectives were:

- To place INRA strategy into a longer term framework, thus identifying the social, economic, technological and scientific tendencies that are relevant in this respect;
- To develop a culture of change through participation;
- To take stock of the changing context of public research in France.

More specifically, the foresight exercise intended to address the question of the relative weight to be given to the four axes of the work programme of the Institute:

1. Basic research, mostly life sciences;
2. Agricultural and agronomic research;
3. Food and nutrition research;
4. Environment and sustainable development.

The programme was defined by the INRA chairman and his close advisers interacting with external foresight experts. The steering group was chaired by one of the consultants.[27] The exercise had the following steps:

- *Phase 1* (pre-foresight) consisted of preparing analysis on specific topics (young researchers, research fronts...) and of a set of one-day debates between the Chairman, the INRA personnel and stakeholders invited for the occasion, in each of the regional centres of the institute. This consisted of personnel and stakeholders consultation, expert panels and vision creating workshops.
- *In Phase 2* (foresight), the steering committee split into several *ad hoc* working groups to work on the various components of the "INRA system". This resulted in a morphological analysis leading to scenarios of the strategic environment of INRA.
- *In Phase 3* (post foresight, strategy), there was a strategic synthesis prepared by the Chairman of INRA for presentation and debate with the board.

The challenges that define the strategic environment of the Institute were identified as follows:

- *Global change*: climate change, access to natural resources, type of development in the world and in Europe, dynamics of rural spaces, agricultural policies;
- *Social demand for life sciences production and expression*: perception of S&T in France and in Europe, values and perception regarding life and life sciences, perception and acceptability of risks, objectives set for life sciences, the role of research policies in France and in Europe;
- *S&T dynamics*: molecules and molecular interactions, from the gene level to the cell and organism, ecosystems and their evolution, international accessibility of knowledge;
- *Organisation and management of public research in France*: governance, public–private partnerships, research as a profession, governance of public research, financing of public research, institutional organisation of public research, status and career of research personnel.

Scenarios concerning the strategic environment of the Institute were built from the combination of hypotheses related to these items; a second set of scenarios for the Institute was then built, keeping the balance between the four areas (basic research, agriculture, food, environment) or else giving a strong priority to one of them. Phase 3 consisted of assessing the scenarios for the Institute in terms of compatibility with the strategic environment scenarios.

INRA 2020 had almost no direct effects, since its sponsor and client, the Chairman of INRA, was not renewed in his position at the end of his term in office, which happened not long after the completion of the exercise. However, significant indirect results were achieved since a four-page summary was sent to the 8,500 employees of INRA and provided an opportunity for real debates in the administrative board and the scientific committee of INRA. The national press gave accounts of the major results. All these provided a context for strategic and forward thinking which has been useful for the next Chairman and her board of directors.

In terms of the methodological and organisational highlights and innovation, the main lessons of this exercise are the following:

- Foresight dealing with the cognitive aspects of research is particularly difficult to carry out.
- The closeness of the INRA chairman to the exercise was both an advantage (legitimation, motivation of experts and stakeholders, etc.) but also a difficulty in the sense that there was a permanent risk that the exercise could turn into a negotiation between the chairman and the stakeholders (internal and external), squeezing out the foresight itself.
- Alignment with the institutional calendar is essential: here, the chairman left his position just at the end of the exercise, thus limiting its impact.

THE FUTURE: NEW FORMS OF FORESIGHT

Foresight is becoming a standard methodology and practice and a central part of the governance of research and innovation policies and, more generally, of national research and innovation systems. This is the result of an evolution which started at least two decades ago. Thus, there is now a critical mass of experiences, which enables us to draw lessons and to identify the challenges ahead and the necessary developments in order to match foresight better to its goals. We address below some of the key issues for the future of foresight.

Handling the Variety of Foresight: Linking Roles and Models

As we have seen earlier, foresight can play a role at all three levels of research policy, namely, the political (level 1), programmatic (level 2) and research organisation (level 3) levels. At the political level, it can help formulate and debate macro-issues; at the programmatic level, it can address research issues, problems and challenges, as well as their social and economic aspects; at research organisation level foresight can contribute to strategy (new institutional context, ERA, etc.), development of competencies, building alliances, and the orientation of infrastructures.

Since foresight has been used at all those levels, i.e. in the context of different functions, it is not surprising that it has taken a wide variety of forms, varying in terms of the object it addresses ("thematic class"), main objective, mode of involvement of the social actors, cognitive nature and institutional architecture (Table 5.1).

Table 5.1   A variety of foresight approaches

Possible options / Dimensions	A	B	C	D
Thematic class	Technology areas	Activity sectors	Public functions	Strategic issues
Main objective	Research priority agenda	Efficiency of the innovation system	Shared awareness for future techno.	
Mode of involvement of the societal actors	No direct implication	*Ad hoc* limited implication	Systematic implication	
Cognitive nature	Intensive analytic work	Significant analytic work	Interactions and consensus building	
Institutional architecture	Sequential panels, institutionally distributed	Sequential panels, institutionally centralised	Parallel panels, institutionally distributed	Parallel panels, institutionally centralised

The possible combination of those features defines a diversity of foresight approaches, characterised not only by their scope, but also by two major models (Barré, 2001):

1. *The analytic model of foresight*: in this model, there is an important knowledge production activity, leading to data, modelling and formalisation challenges. It involves few people directly, having interactions only with technology and a few social sciences experts. The central focus is the internal coherence and plausibility quality of the analysis and, eventually, of the scenarios.
2. *The societal model of foresight*: the central focus here is on who participates and the outcome is expressed in terms of the creation of new networks and circulation of information. The actors themselves participate (not only their representatives) and this participation is meant to be dealing with matters of substance, including the S&T itself. Such a societal model is rooted in the tradition of public participation, and extended technology assessment. The societal exercises are socio-political processes, which make them a new breed of decision-making instruments in tune with "dialogic" democracy.

The challenge at this stage is the validation of the typologies, stabilising the models and developing a common understanding of the related definitions and methodologies. In addition, there is a need to relate explicitly those models to the levels at which foresight is being implemented, in other words, to link functional and operational categories. Furthermore, there is a need to be more specific in the operationalisation of foresight activities, in terms of their linkage or proximity to: scientific expertise, strategic evaluation, science–society questions, and foresight watch and/or communication partnership activities.

Handling the Epistemological Status of Foresight

What is the robustness of the knowledge and results produced by a foresight exercise? In other words, what is the epistemological status of foresight results?

*   *Scientific legitimacy*: it is true that foresight is aimed at taking into account all relevant variables that influence the dynamics of the issue at stake, whilst mobilising the best available knowledge on such variables and their interactions. Nevertheless, one must admit that the parameters introduced are extremely heterogeneous and have to be over-simplistically defined both in their characterisation and interactions. Furthermore, there is no deterministic theory of social

dynamics. In other words, the images of the future produced by foresight methods are not based on an epistemology of modelling, they cannot be reproduced and therefore are not scientific knowledge. Can we say that such knowledge is pure contingency and worthless for guiding reflexion and action? Not this either, since foresight methods lead to the production of *ad hoc* scale models, which capture certain characteristics considered important, enabling the production of images of the future in a systematic and explicit way. In this sense, foresight is a rational process using the logic of modelling in its own way. It does not qualify foresight as a scientific knowledge producing process, but rather as one which produces results useful for strategic thinking and debating.

- *Political legitimacy*: it is true that foresight is concerned with integrating social actors and developing collective learning processes, and also seeking to identify equilibria positions among the stakeholders. Nevertheless, the results of a foresight do not have any political legitimacy, since they do not have any basis for it.

Suffice to say that foresight is a process that allows scientific knowledge to be useful for action and for political decision-makers to address key research policy questions. Thus, the quality of a foresight exercise depends as much on the robustness of the analysis and hypotheses made, as on the soundness and depth of the interaction processes among the social actors.

In any case, the epistemological status of foresight is not a simple one to manage and the risks are high of ambiguity and even misunderstanding, which can dramatically weaken a foresight exercise. Hence the need of reflexivity and evaluation to address the risks of confusion about the positioning of foresight. Reflexivity means getting assessments and criticism from the academic and professional community, applying the social sciences to foresight, allowing also for accumulation of knowledge. Evaluation provides for an enhanced learning process, allowing for transparency and reliability of the foresight process, and also for its improvement.

Handling Activities Complementary to Foresight

Foresight practitioners are often social scientists, while foresight operations result from the complementarity among activities entailing a diversity of skills. The risk therefore is one of working with sub-professional standards in activities falling outside the competency of social scientists. In particular, the case for a more professional handling of communication and stakeholder participation issues must be made.

Handling Linkages Between Multi-level Foresight Activities

A major issue is the capability of foresight exercises to relate to one another, both for benefiting from the background work done by others, and for coordinating processes and results. In this respect, it is worthwhile distinguishing between, on the one hand, relationships and sharing experiences among foresight activities at national level, where coordination across levels (political, programming, research organisations) can be very useful, and, on the other hand, relationships and sharing among foresight activities at EU level, where coordination can take place at a given level among member states. In this sense, the initiative of the IPTS (Institute for Prospective Technological Studies) and DG Research to establish a foresight platform is very welcome, as is the establishment of the ERANET ForSociety (see Chapter 14).

CONCLUSION

After the period of reconstruction following The Second World War, when foresight played an important role in policy-making, it appears it is regaining its status, in particular for science, technology and innovation policies, in the present period of globalisation. The examples presented in this chapter show that initiatives in this respect have already taken place at the different levels of the innovation system, thus providing a critical mass of experience.

It is important to note that the new Law on Research, which took effect in 2006, clarifies the roles of the institutions, establishing in particular at the macro-level a High Committee on Research and Technology, as well as a strategy–oriented ministry of research, plus a programme agency ("National Research Agency") to shape better the meso-level of the innovation system – in line with the Futuris operation recommendations. This results in opening the possibilities for foresight activities to be embedded in the system, each level having an explicit mandate to elaborate and implement a strategy based on foresight and taking into account the realities of the other levels. In this context, Futuris can be considered as a prototype of what will be happening at the level of the Ministry of Research and the High Committee on Research and Technology; Agora 2020 of what will be happening at the level of the National Research Agency; and INRA 2020 of foresight at the research institutions level. The cases presented in this chapter can be seen as experimentations towards the embedding of foresight at the various levels of the French innovation system in transition.

More specifically, we suggest the case studies presented above show that the challenges to foresight can be addressed with reasonable chances of success, under the following conditions:

- The conduct of systematic reflexive and cumulative work on the experiments and exercises that have taken place over these last few years in the EU Member States;
- Addressing the problems of the evaluation of foresight – in terms of the validity, credibility, quality control, criteria and ethical norms of exercises. This would go along with training and research activities in the field;
- The building of the infrastructures needed for access to background studies and previous results of foresight activities, and the mechanisms for the capitalisation of results, methodologies and experiences. This raises the issue of the interoperability of national practices and methods;
- The emergence of an open professional community drawing upon a variety of backgrounds and competencies.

Under such conditions of more rigorous evaluation and reflexivity on practices and impacts, foresight appears to be central to a governance mode which fulfils the requirements of a "knowledge society", since it places science and technology in the democratic arena needed to build our "common world". In this sense, foresight has to play a major role in the shaping of the European Research Area, which goes with the European social model, for which the science–society relationships is a major component.

NOTES

1   See 'Les 100 technologies-clés pour l'industrie française à l'horizon 2000', ministère de l'industrie, Paris, 1995; Technologies-clés 2005, ministère de l'économie, des finances et de l'industrie, Les éditions de l'industrie, Paris, 2000; a new exercise "technologies-clés 2010" has been launched in November 2004; see also *A Trans-national Analysis of the Results and Implications of Industrially-oriented Technology Foresight*, IPTS Technical Report Dec. 2001.
2   CES is an official consultative body whose members are representatives of a large number of "social actors".
3   See for example "S&T policy-making for the future: new rationales, new design tools, in S&T policies in Europe: new challenges and new responses", STRATA (2002) and also Smits, R. and Kuhlmann, S. (2002).
4   Budget civil de recherche – développement (BCRD).
5   Except at very occasional meeting of an *ad hoc* Interministerial Council for S&T (Conseil Interministeriel de la S&T) (CIRST).
6   The Superior Council of S&T has always played a minor role.

7 The Ministry of Research launched a small and uncompleted Delphi survey in 1996; the Ministry of Industry launched "key-technologies" exercises in 1995, 2000 and 2005, with a strictly defined industrial technology focus; the last General Planning commission report on S&T policy was issued in 1998.

8 Association Nationale de la Recherche Technique (ANRT).

9 "Contrats d'Objectifs" with public research organisations and "Contrats quadriennaux" with universities.

10 The Fonds de la Recherche et de la Technologie (FRT) and Fonds National de la Science (FNS).

11 CNRS for basic research, INSERM for medical research, INRA for agronomic research ; these organisations develop research orientations for their own labs but also for university labs which they finance through the mechanism of "associated labs" (laboratoires mixtes).

12 The rationale of FutuRIS is precisely to demonstrate that and propose reforms and simplifications, allowing for better policy-making.

13 I thank Nadège Bouquin, from ANRT, for her contribution to this description.

14 http://www.operation-futuris.org; in the name FutuRIS, RIS stands for Research, Innovation, Society.

15 M. Francis Mer was named Finance Minister shortly after.

16 CEO of Sanofi-Aventis Pharmaceutical.

17 Jacques Bravo for Phase I, then Thierry Weil for Phases II and III.

18 Completed by the end of 2005.

19 Headed by Jacques Theys.

20 Centre de Prospective et de Veille Scientifique, Direction de la Recherche Scientifique et Technique, Ministère de l'Equipement, des Transports, du Logement, du Tourisme et de la Mer.

21 RST: Réseau Scientifique et Technique (science and technology network).

22 "Axes précurseurs de programmes".

23 Based on a strengths/weaknesses/opportunities/threats (SWOT) analysis.

24 The Agora 2020 was completed in December 2005.

25 See http://w3.inra.fr/l_institut/missions_et_strategie; see also "INRA 2020: des scénarios pour la recherche. Alimentation, agriculture, environnement", 260 pp., TRP n°19, Editions Futuribles, January 2004.

26 Bertrand Hervieu.

27 Hugues de Jouvenel.

REFERENCES

Bailly, J.P. (1998), 'Prospective, Débat, Décision Publique', *Futuribles*, **235**(October), pp. 27–51.

Barré, R. (2000), 'Le "Foresight" Britannique: Un Nouvel Instrument de Gouvernance?', *Futuribles*, **249**(January), pp. 5–24.

Barré, R (2001), 'Synthesis on Technology Foresight', in Tubke, A., Ducatel, K. and Gavigan, J. (eds), *Strategic Policy Intelligence: Current Trends, the State of Play and Perspectives*, JRC-ESTO, EUR 20137 EN, pp. 51–64.

Barré, R. and David, B. (2005), 'Participative & Coherent Scenario building: A Quantitative Translation Method based on an Input/Output Balance Model in New Technology Foresight', Forecasting & Assessment Methods Seminar, JRC, EC.

Berger, G. (1967), *Etapes de La prospective*, Paris: PUF.

CES (1998), *Journal Officiel*, collection of CES Reports, 17 July 1998.

CGP (1997), *Globalisation, mondialisation, concurrence: la planification française a-t-elle encore un avenir? Cinquantenaire du Commissariat Général du Plan*, Paris: La Documentation Française.

De Jouvenel, B. (1954), *L'Art de la Conjecture*, Futuribles, Paris: Editions du Rocher.

Dessus, B. (1997), *Energie 2010–2020 – rapport final de l'atelier "les défis du long terme"*, Commissariat Général du Plan, October.

Héraud, J.A., Munier, F. and Nanopoulos, K. (1997) 'Méthode Delphi: Une Étude de Cas Sur les Technologies du Futur', *Futuribles*, **218**(March).

IPTS (2001), *A Trans-national Analysis of the Results and Implications of Industrially-oriented Technology Foresight*, IPTS Technical Report, December, Seville: IPTS.

Martin, B.R. (1995), 'Foresight in Science and Technology', *Technology Analysis and Strategic Management,* 7(2), pp. 139–168.

Massé, P. (1973), 'De Prospective à Prospectives', *Prospectives*, PUF, **1**(June).

Ministère de l'Industrie (1995), *Les 100 Technologies-clés pour l'Industrie Française à l'Horizon 2000*, Paris: MI.

Smits, R. and Kuhlmann, S. (2002), 'Strengthening Interfaces in Innovation Systems: Rationale, Concepts and (New) Instruments', in European Commission (ed.), *Science and Technology Policies in Europe: New Challenges and New Responses*, STRATA Consolidating Workshop, Report EUR 20440, pp. 300–370.

STRATA (2002), 'S&T Policy-making for the Future: New Rationales, New Design Tools', in *S&T policies in Europe: New Challenges and New Responses*, STRATA Consolidating Workshop, Report EUR 20440, pp. 86–134.

6. Foresight in Germany

Kerstin Cuhls

INTRODUCTION

Germany was not very active in foresight until the 1990s (Irvine and Martin, 1989), but economic and other reasons forced those responsible at the German Federal Ministry for Science and Technology (BMFT, later BMBF) to change their minds. Many of Germany's problem areas became apparent at the time of unification. Stringent requirements of the economy were necessary: setting priorities, the allocation of financial resources, and the strategic orientation of research and development are still challenged. Science and technology policy had to adapt to the fact that national research and development (R&D) budgets would never be sufficient to support all suggested projects. There must be a rational process to set priorities and to concentrate the financial support thereon. Non-financial support is becoming increasingly important. Therefore, the desire to identify the technologies and scientific fields which will have the greatest impact on economic competitiveness and social welfare was expressed from various sides. "Emerging technologies" are increasingly science-based (Grupp, 1992) and are likely to need a high intellectual capacity, which must be provided and supported by the education system.

These reasons prompted Germany to start foresight activities on a national level. S&T shifted towards a longer-term future orientation and new policy strategies. New methods should be tested and used to identify "emerging" technologies and developments of S&T, as well as their general impacts. This was regarded as insufficient so the new concepts in German foresight also take the economy, society, the environment and other impacts into account.

This chapter describes a series of national foresight activities, commencing in the 1990s up to the present day. The examples are a foresight exercise called "Technology at the Beginning of the 21st Century", three Delphi processes and the first national "programme" called "FUTUR" which ended in December 2005.

HISTORY OF FORESIGHT IN GERMANY

The 1980s in Germany were predominantly a decade of strong support for basic research, mainly in large facilities, following the recommendations of scientific advisory committees in the 1970s and at the beginning of the 1980s. After years of technology enthusiasm, the federal government switched to a more reluctant policy, formulating technological goals for the S&T system only in those sectors where a key role in world markets was commonly recognised (Cuhls *et al.*, 1996).

Increasing technological change and the globalisation of markets, as well as the special situation after the unification of Germany with its severe budget restraints made the responsible persons at the Federal Ministry for Science and Technology, BMFT, change their minds (Martin, 1995). Longer-term perspectives and strategies to make better use of limited resources were sought. The selection for the support and the more goal-oriented prioritisation of certain technologies seemed to be necessary. On the other hand, the state had to be careful not to intervene too much in the market and its self-regulating forces or in the self-organised science system. There is always the danger of confusing technology policy with technology planning in the sense of socialist planning, a kind of socialism which had just been overturned in Germany with the unification (concerning the difference between foresight and planning, see Cuhls, 2003). The term "foresight" is used in the sense of "outlook" in the German context. This is not the same connotation as a "prediction" which would be closer to "forecast".

Starting as "risky projects" and earning harsh criticism in the beginning, the German foresight studies later on became widely accepted by those who could make use of them. It was considered a political question whether state bodies should give more emphasis to direct intervention in research matters (e.g. by financing specific R&D projects from industry) or to more indirect support (e.g. tax reductions for R&D projects or subsidies to those companies hiring new scientific and technical staff). The BMFT at first decided not to use one single approach but a broader range of studies to have a fundamental basis to make choices and to combine data. As most of the methods are described in Chapter 3 (for further readings on indicators, literature analysis, scenarios, surveys or trend extrapolation, see Cuhls and Kuwahara, 1994, p. 3; Cuhls, 1996), in the following sections new approaches with relevance tree and key technologies, as well as the different Delphi studies are described.

The methods applied in Germany for longer-term foresight all fulfil the following functions, which are defined as the major classification for purposes of foresight by Irvine and Martin (1989, p. 30): (a) Direction-setting, (b) Determining priorities, (c) Anticipatory Intelligence, (d) Consensus-generation, (e) Advocacy, (f) Communication and Education.

Public and private institutions can make use of these foresight studies (Cuhls, 1998; Cuhls *et al.*, 2002), as they are available to the public. The last process mentioned, FUTUR, has its main objectives in working directly for BMBF. The following foresight processes therefore show a line of first experiments and methodological developments in Germany until the year 2005. All of them are "learning processes" and have their pros and cons.

Technology at the Beginning of the 21st Century

Technology at the Beginning of the 21st Century, abridged T 21 (Grupp, 1993, 1994) was a BMFT-financed project and started in 1992. In the Federal Republic, BMFT, since 1994 Federal Ministry for Education and Research (BMBF), is assisted by several so-called "Projektträger" (programme operating agencies), mostly located within the national laboratories. Representatives from these "programme operators" set up a task group and worked face to face on an assessment of critical technologies for the Federal Republic of Germany. The Fraunhofer Institute for Systems and Innovation Research (ISI), which took the overall responsibility for this task, was asked to devise a comparatively new methodology based on technology lists and relevance trees. Scanning all available studies from abroad and making use of the internal expertise of the "programme operators" in panels, a list of about one hundred technologies was established, redefined and regrouped. The list was relatively detailed and contains items like biochips, data network safety, genome analysis, fuzzy logic, flat displays and the like. Then, a common report form was worked out which is filled with information on the technological item, including product visions considered most important by the staff of the programme operators, and criteria assessing the frame conditions. As a result, different tables, descriptions and a two-dimensional representation of technological overlaps were elaborated. Finally, the dynamics during the following ten years were examined. It is well-known that there is no linear progress in science and technology, but rather several feedback and cyclic effects (Grupp, 1992). A standard scheme differentiating eight typical phases in the research, development and innovation process was agreed upon. As this was a new methodology with some traditional elements from the relevance tree approach, the outcome of this study is difficult to summarise briefly. The growing interdisciplinarity in technological development, first discussions of the programme operators in different workshops (now, in 2005, they even have an established network) about who can make use of the new knowledge generated and the establishment of new methodologies may help to make "better" and more effective decisions about the support of R&D projects. The coordination and communication about these new technologies by the programme operators is facilitated.

The First Comprehensive German Study (Delphi 1993)

In the first German Delphi study (Cuhls, 1998), the Fraunhofer ISI collaborated with the Japanese National Institute of Science and Technology Policy (NISTEP), an institute of the Science and Technology Agency (STA, which is nowadays integrated into the Ministry of Education, Culture, Sports, Science and Technology (MEXT), see also the contribution of Kuwahara in this volume). The German Delphi Team took the 1,150 topics in 16 fields prepared for the Japanese fifth survey and translated them into German (BMFT, 1993; Cuhls *et al.*, 1996). The assessment criteria were the same as in Japan. For results see Cuhls and Kuwahara (1994).

The main conclusion was that Delphi inquiries on science and technology should always be undertaken with an international panel of experts, including people from several countries and continents. However, for many topics no extreme discrepancies in the results were found, but rather congruent and diverging results occurred at the same time.

The Mini Delphi(s)

The Mini Delphi approach was a test to develop the Delphi method further, to meet criticism from the first German Delphi and to gain more detailed data about some of the internationally problematic areas. The areas chosen were:

- *Materials and Processing*:
  (a) Photovoltaics
  (b) Superconductivity
- *Microelectronics and Information Society*:
  (a) Cognitive Systems and Artificial Intelligence
  (b) Nanotechnology and Microsystems Technology
- *Life Sciences and the Future of the Health System*:
  (a) Cancer Treatment and Research
  (b) Brain Research
- *Problems of the Environment*:
  (a) Waste Processing and Recycling
  (b) Climate Research and Technology

The Mini Delphi was more oriented towards the technical solutions for current or emerging problem fields which were identified as the most important in the previous Delphi survey. Expert committees in Japan and in Germany selected the major topics jointly.[1] Between the first and the second round, some of the topics had to be reformulated more precisely because of expert suggestions, and some new topics were introduced.

The whole procedure of the survey was conducted parallel to that in Japan. The cooperation partners were again ISI on behalf of the BMBF in Germany, and NISTEP in Japan. Many data were gathered and sent to the participants and were also published as an official BMBF brochure (Cuhls *et al.*, 1995).

The Second Comprehensive German Study (Delphi 1998)

As foresight gained momentum in Germany and most of the restraints mentioned above still remained, it was obvious that Germany needed further concepts to develop the necessary degree of effectiveness to make innovative leaps. Especially for research programmes or companies' strategies, information about the future is required as the basis for general decisions. Therefore, the German foresight activities were supposed to provide more information about things to come, concerning also those actors who are not able to gain this knowledge alone (e.g. small and medium-sized companies, research institutes, "the public"). The BMBF financed and carried out a new foresight activity in 1997. The Fraunhofer ISI was again given the task of managing this project.[2] Federal Minister Dr Jürgen Rüttgers established a steering committee made up of prominent members from science, industry and the media to advise the Ministry in all decisions concerning the establishment of important framework guidelines. The Delphi method was once again applied to gather information. The topics were worked out in expert groups, by desk research and by taking over the Japanese Delphi topics in order to make comparisons later. Criteria, questions and the new category of megatrends were further developments of the methods (Blind *et al.*, 2001). The most important innovation areas for the future were selected, namely:

- Information and Communication;
- Service and Consumption;
- Management and Production;
- Chemistry and Materials;
- Health and Life Processes;
- Agriculture and Nutrition;
- Environment and Nature;
- Energy and Resources;
- Construction and Living;
- Mobility and Transport;
- Space;
- Big Science Experiments.

In Germany, many companies started to analyse the dataset for their own purposes. The great advantage of a Delphi process is that everyone can make his/her own analysis of the data – depending on individual needs and questions about the future. Therefore, the data were provided to everyone who wanted to use them (Cuhls *et al.*, 1998).[3] Some examples for using the data were already introduced in the reports or the newsletter *Zukunft nachgefragt* (The future in question, BMBF (eds)). Thus everyone can make use of the Delphi '98 as working material, not as a picture of the future itself.

FUTUR

In 1999, the BMBF decided to organise a foresight process in order to counter the criticism that only experts were involved in previous foresight activities, and to open up the German national foresight process to a greater variety of participants. This forerunner version of FUTUR put special emphasis on the use of the internet as a platform for discussing the different topics. The kick-off meeting took place at a conference in June 1999. The process started with a focus on two fields, "Mobility and Communication" and "Health and Quality of Life". The Ministry expected that it would be sufficient to provide a platform and some inputs on the themes to encourage any persons interested in the topics to participate in the discussions. This approach failed because too few people knew about the process, and the questions to be discussed were not well defined. Furthermore, the methodology and objectives were unclear. In consequence, BMBF decided to restart the process.

In 2001, "FUTUR – The German Research Dialogue" was launched on behalf of the BMBF. The procedure was more BMBF-oriented and relied on a wider process, using a variety of methods and instruments such as focus groups, conferences, online votes and scenario writing. It was decided that face-to-face meetings in working groups should be the central medium of discussions, and the internet should be used for information purposes, supporting the transparency and communication of the whole process. The first phase of this "new" FUTUR ran until the beginning of 2003. It was evaluated by an international expert panel in the autumn and winter of 2002 (Georghiou, 2004 or Cuhls and Georghiou, 2004). The process itself was organised by a consortium of five (at the beginning six) institutions, each responsible for different tasks: IFOK (Institute for Organisational Communication, Bensheim, Berlin, Brussels) was the leader of the consortium, ISI (Fraunhofer Institute for Systems and Innovation Research, Karlsruhe), IZT (Institute for Futures Studies and Technology Assessment, Berlin), Pixelpark AG, and VDI/VDE Innovation + Technik GmbH (Berlin-

Teltow). A strategy department of BMBF was responsible for FUTUR (Cuhls, 2004).

The results of this first phase of the FUTUR process consisted of so-called "Lead Vision Papers"[4] describing a broader field important for the future, research necessary in this field and including a scenario to illustrate and visualise the things to come (Gaßner and Steinmüller, 2004). The first Lead Visions were presented to the public in summer 2002. The outcomes are being implemented by the BMBF. As FUTUR was regarded as an interesting new tool by the Ministry, the second phase of the FUTUR process was started in early 2003, with slightly changed procedures and methods. This second phase of FUTUR finished by March 2005 and after another international evaluation (Salo *et al.*, 2005), the third phase of FUTUR was started with the same methodology as the second phase (Hafner and Cuhls, 2004 or Cuhls *et al.*, 2004). Only IFOK, IZT and VDI/VDE-IT remained in the consortium. The final results of the second Lead Vision elaboration process were expected to be available in winter 2004. Starting with the second phase of FUTUR, it was decided to develop one or two Lead Visions per phase of FUTUR on a continuous basis. But they were never published officially. For different reasons, especially the costs of the process and acceptance in BMBF, FUTUR ended in December 2005.

RATIONALES AND OBJECTIVES

The rationales and objectives in German foresight activities have in common that they all provide information about things to come in order to have a better base for decision-making and priority-setting. Nevertheless, the rationales and objectives in the different projects changed, and even broadened in FUTUR. In the first national project, the T21, the objectives were rather modest and adequate for a "project". The main motive was to complement economic growth criteria by the idea of growth using intelligent new technologies. Secondly, learning from Japanese and US sources, a stricter and more transparent methodology for listing and assessing technologies was to be tested. The approach also aimed to mobilise the in-house expertise of German research administrators for foresight purposes.

The rationale behind the first German Delphi was to discover more about future science and technology, to determine their time horizon and to test the Delphi methodology for this purpose. In order to keep the project costs modest, the Japanese Delphi topics were translated. If the final results would prove very similar, it was expected that in the future, the German government could adopt the Japanese results. If not, a German Delphi every five years (to update) was intended, also to start a communication process about the future.

A side effect of the project was to gain insights into Japanese visions about the future and scientific solutions or even products developed in Japan.

The results were published as a BMBF report to provide all interested persons and organisations with the information. It was not intended to make use of the data for "strategic planning" in BMBF in the field of science and technology. The German constitution says "Science shall be free". Therefore, planning from the side of the state would not be accepted – and the five-year planning of the GDR which was regarded as exceptionally unsuccessful had just been abandoned.

This was why Japanese–German Mini Delphi studies were conducted only for methodological improvements. New in these studies were the start from the demand side (what is the demand for science and technology reflected by the most important topics in the Delphi '93), the strict cooperation with Japan in keeping the questionnaire equivalent but with new questions and a more complicated survey design, conducting the same study concurrently and analysing the results in a joint workshop.

The second comprehensive German Delphi '98 was started especially at the request of industry because the information about future science and technology was regarded as very valuable for strategic planning. The second aim in 1996 was to make the different experts in the system aware of the future, think long term, take their views for granted and create a certain commitment for actions in the different fields (see the 5 Cs of Martin, 1995).

FUTUR started with a new and ambitious set of objectives, principles and a kind of "hidden agenda" (to have an impact on the internal organisation of BMBF). Officially, FUTUR, as started in 2001, aimed to introduce new perspectives into the existing research agenda of BMBF, by adding to the traditional mechanisms for agenda-setting and prioritisation. The conventional decision-making process is characterised by a closed and rather non-transparent interaction between research institutions, industry, project operating agencies and ministerial bureaucrats in charge of research funding. Strategically oriented officers within the ministry were increasingly concerned about the risk of missing important new issues on the funding agenda, if this were solely based on traditional mechanisms driven by the involved actors. FUTUR[5] therefore is oriented towards the identification and inclusion of societal needs in future research agendas and serves as a means of priority-setting for future innovation-oriented research policies. Interdisciplinary, problem-oriented 'Lead Visions' (Leitvisionen) are the major outcomes of the process. They should reflect the demand for research and be translated into publicly funded research programmes or projects. Participation of a broader audience in various kinds of activities and the combination of different creativity, communication and analytical tools are additional characteristics of the process.

The objectives broadened in the second phase of FUTUR starting in early 2003. In addition, it was intended to start public discussions in so-called "Future Dialogues" (Zukunftsdialoge), but the first tests were not very successful. The following characteristics and principles were to be integrated in the FUTUR process: first and foremost, the process should involve a broad variety of societal actors from different disciplinary, thematic and professional backgrounds to ensure interdisciplinarity. Also, non-thematic experts should be involved at certain stages of the process to open up traditional networks and ways of thinking. This diversity is reflected in the combination of various methods used during the process.

To manage this diverse process, a considerable level of transparency was necessary, not only for the participants, but also for the FUTUR consortium and the Ministry. This helps to ensure the continuity and stability of the whole process. As FUTUR is a pioneer process, reflexive learning is also an important aspect, to be able to modify and adapt the process to new conditions and experiences, so continuous quality management and evaluation of the process by independent experts were provided to attain this end. The process had to be need-oriented, meaning the results should reflect future needs of German society. This does not necessarily mean all topics needed to be completely new. Instead, besides up-and-coming future topics, also inventive and unobserved aspects of existing themes can be part of the outcomes. This is reflected in the fact that the process was open to results, which means that it was independent of current funding programmes or "hot" topics within the Ministry and it had no thematic restrictions. At the same time, even if the process was broad and diverse, it had to be result-oriented towards the generation of a usable output for the Ministry (Cuhls *et al.*, 2002; and Cuhls, 2003). The process was not only supposed to give input to the Ministry, but also to promote awareness-raising and future-oriented thinking in society (Banthien *et al.*, 2004). To achieve this, all FUTUR outcomes were designed to be "understood by everybody".

Although the main objectives were all met during the process, there were too many principles and objectives to be met and communicated, which made the process very complicated – this was one observation of the second FUTUR evaluation performed by an international evaluation panel (Salo *et al.*, 2005). It recommended a concentration on certain objectives and clarification if the process should only be directed towards the BMBF or also towards other actors in the innovation system. In that case, the objectives and methodology used would need to be changed accordingly. A simplification of the complicated process would be helpful.

Concerning the principles of FUTUR, interdisciplinarity was met sufficiently and provoked problems inside BMBF since some of the topics identified were not within the responsibility of BMBF. This led to implementation problems. More problematic was the need-orientation, which for example was not reflected in the methodology. The questions remain: What is the society's demand of the future? And what or who is this society?

Tensions occurred between the openness to any result at the beginning and, on the other hand, focusing on BMBF Lead Visions with filtering out topics BMBF is not responsible for. Additional tension was felt between the declared principle of participation of new actors and lack of knowledge on the part of the participants. The meaning of societal "participation" also had to be communicated, because it was obvious that the general public was not involved and the number of participants was not greater compared with the previous studies (Cuhls *et al.*, 2004).

METHODOLOGICAL HIGHLIGHTS AND INNOVATIONS

Technology at the Beginning of the 21st Century was remarkable for its time because it included the people responsible for new projects at BMBF in workshops and thus provided some stakeholder participation already in this foresight. It utilised the already known technology list approaches, as in the USA. earlier (Wagner and Popper, 2003) or later influenced the ones in Italy (Fondazione Rosselli, 2003), France (Durand, 2004) or indirectly the Czech Republic (Technology Centre AS, 2002). The methodology applied was new, with some traditional elements from the relevance tree approach, which is known as a "normative" method. The time horizon of the study was approximately the year 2000.

By means of multi-dimensional scaling of the technologies (whereby their distances represent the closeness of the technical content as judged by the technical experts), it was shown that the current borderlines between individual technologies will become less distinct in the next decade. The results manifested for the first time the importance of nanotechnology and of cross-disciplinary work, because new disciplines are often being shaped outside the classical research areas. This certainly has dramatic effects on the necessity of technology monitoring, on technology policy implementation of R&D programmes and the adaptation of technological opportunities by firms.

Results are twofold (Grupp, 1993, 1994): the growing interdisciplinarity in technological development is shown and first discussions of the programme operators in different workshops (now, in 2006, they even have an established network) about who can make use of the new knowledge generated, and the establishment of new methodologies may help to make "better" and more effective decisions about the support of R&D projects. The coordination and communication about these new technologies by the programme operators is facilitated.

Parallel to this, the Delphi method was tested for the first time on a national scale. The German-based survey was conducted principally along the same guidelines as the fifth technology forecast survey in Japan, although it took place with a one year delay. The questionnaires were sent out to a group of experts from industry, universities and government over two rounds. In order to make the two investigations independent of each other ("double blind"), it was arranged that because of the time lag, the German experts did not know any results from the Japanese sample because the translation into English was not published until the German survey was already finished. In case of the German inquiry, the compiled data were published in August 1993 (BMBF, 1993). The Mini Delphi studies were conducted to test for methodological improvements and therefore contained many innovations. New in these studies were the start from the demand-side (what is the demand for science and technology reflected by the most important topics in the Delphi '93), the strict cooperation with Japan in a workshop and then via fax and mail in keeping the questionnaire equivalent, but with new questions and a more complicated survey design, conducting the same study at the same time and analysing the results in a joint workshop. The topic generation was performed in German and Japanese language, so as not to lose information. Only the external communication was in English.

An unintended test of the Delphi method was made: when the results of the first round were automatically copied into the second round questionnaire, there was a mistake for the last topics – the results were written into the line below. A cross-check did not detect the mistake. But interestingly, the Delphi method worked. Experts with high expertise noticed the mistake and did not change their previous opinion, but made comments about the strange result. Lower level experts (in their own self-estimation) often did not notice the mistake and many of them changed their minds in the second round towards a "consensus opinion" (Cuhls *et al.*, 1995).

Delphi '98 made use of the experiences: a steering committee was established, expert groups helped to propose and elaborate the topics, and again 30 per cent of the topics remained equivalent to the Japanese topics. As usual, steering committees have their own ideas and therefore influenced the process according to the participants' backgrounds. This is time-consuming, but on the other hand, the "standing" with the public is much better. Expert groups of more than one hundred individuals with specialised knowledge from industry, higher education and other institutions were responsible for gathering the most important information about the above-mentioned fields. All remaining assessments were undertaken by a significantly larger circle of specialists in the various areas of science and development (Cuhls *et al.*, 2002). The definition of exactly who is considered to be an expert was very broad: persons actively carrying out research in a particular field, and those who regularly obtain first-hand information about the field or just read the literature. But also persons from clubs, associations or sociologists and other soft sciences were included to a certain extent. The results were published in the usual way, with new examples of calculations and mini-roadmaps or mini-scenarios. This was appreciated by the media and non-scientific users. The data were also provided on a disk and via internet for all interested. A tool for searching topics was developed to help analyse the data. This made applications more comfortable than printed versions.

FUTUR was planned as an answer to the criticism that only experts were involved in the prior surveys. The first two phases of FUTUR differed, but in general, the instruments were the same,[6] a strict methodology was not applied. The third phase of FUTUR was just a continuation of the second (see Figure 6.1). Facilitated workshops with creativity methods to have brainstorming sessions about "Society 2020" were held in the first phase of FUTUR only. Then a filtering process was started, in which the first instrument was an open space conference, where 25 themes were given and the participants could decide what they were interested in and whether they would like to form a workshop group. The conference had about 400 participants, but the problem was that some participants came from strategic departments or were decision-makers, sometimes even higher ranks, who had problems with this approach, especially as persons with less knowledge about the subject chose the same group and discussed on a "low detail level" with them.

The next new method was an online vote (via the FUTUR internet page) in different variants with different questions. All have in common that the persons actively or inactively involved in FUTUR could vote on certain topics. The aim was to reduce the number of topics for further work. The response rates were in the normal range, but it was regarded as problematic that the sample was unknown and that it was somehow possible to vote twice or more often. Therefore the result had to be scrutinised very carefully. The other problem was that in the end, BMBF had the last word on whether the topic fitted into the BMBF portfolio so that the more participative instrument of the online vote was regarded as "pseudo democratic" (remarks from participants). When starting the third phase of FUTUR, the online vote tool was even used to generate new topics. The participants (all in the database) were asked to mention important topics. The results were not very new, but interesting. Nevertheless, most of the topics did not meet the BMBF criteria, especially the question of BMBF responsibility, and were not taken into account for the next round. In addition to the online vote, topics were generated in different ways. One was desk research by the consortium, screening articles, internet, conferences, international foresight activities, and so on.

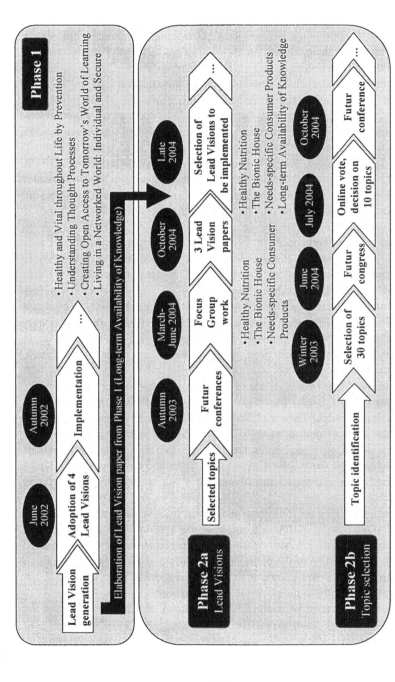

Figure 6.1 FUTUR: main phases and activities

Additionally, in the second phase of FUTUR, a workshop with German futurists was conducted in Berlin and one in London with future experts (futurists, trend researchers) from the UK and from Germany. The results were rather disappointing, because most of the topics mentioned were already known, some of them already dealt with in FUTUR but filtered out at a certain stage of the process.

Conferences were also regarded as a methodological part of FUTUR. Different formats were used, e.g. one at the beginning of the process to outline the topics, another very large one in June 2004 with young persons who could formulate their views about the future in different forms (e.g. SMS to the future, brainstorming sessions, discussions with politicians, etc.). During all conferences, the participants were very enthusiastic about FUTUR and formulating futures. But in the end, many were disappointed, because their ideas were not really integrated into the whole process or could no longer be detected, because they had been integrated into mainstream topics. Follow-up actions to the conferences were often missed. It cannot be expected that in-depth work or structuring of foresight topics should take place at such occasions. Some persons questioned whether the conferences were foresight activities or just public relations conferences to create awareness or public acceptance of science.

In order to foster creativity and introduce the long-term view, an abridged form of Future Workshops (according to Jungk and Müllert, 1996, later only called Creativity Workshops) were held for those fields chosen for in-depth work later. Different persons from the other workshops were involved here, which made knowledge and information transfer rather difficult in the first phase. In the second phase, some of the participants were the same as in the following Focus Groups. The major work in FUTUR was performed in so-called Focus Groups. Their task was not only to focus the area, but also to formulate science- and technology-based solutions in the field. A creative session was held to work out the basis for the scenarios. Between the Focus Group workshops, there were always online votes and decisions, which topics would be further elaborated and which groups had to stop, so that there was competition among the groups. That sometimes caused frustration, because the participants invested their time in formulating new themes for the BMBF but the subjects were not chosen, meaning the investment in influencing the BMBF agenda was in vain. It also transpired that in the first phase of FUTUR, the expertise of the participants was often insufficient. Therefore, in the second phase, more people with scientific expertise were invited, which was actually contrary to the principle of participation.

Conclusion: it must be decided what is wanted, and at which stage of the process – expertise from experts or more participation of non-experts with potentially fresh ideas.

The Focus groups provided the input, but the scenarios were written by IZT and included in the Lead Vision paper, which was edited by a task force and was the major result of FUTUR. It included the topic as such, proposals for research directions and the scenario for illustration. Scenarios were also written in a "scenario contest". One took place after the first phase of FUTUR with limited participants from schools, another one during the topic generation of the second phase. People could send in their illustrations of the Lead Vision topics via e-mail, but the number of participants was limited as the action was not well publicised.

The planning of the second phase of the FUTUR process included the design of a new tool, the so-called "Future Dialogues" (Zukunftsdialoge). They are supposed to shape and discuss the future challenges of relevant topics derived from the topic generation process, which are assessed by the BMBF as relevant and which are not appropriated as Lead Vision topics. Future Dialogues shall involve a wider public than the actual pool of FUTUR participants. One Future Dialogue, undertaken in 2004, was about the issue of demographic change.

In FUTUR, some new tools were tested (e.g. online vote), but in general, existing methods and formats were applied and adapted. The major innovation in this process is the combination of different methods and approaches (Banthien *et al.*, 2002) and the direct link to the BMBF, which on the other hand made things more complicated and time-consuming, because the consortium often had to wait for decisions necessary to proceed. For more information about FUTUR, there are unpublished descriptions (Cuhls *et al.*, 2002a; Cuhls *et al.*, 2004; Hafner and Cuhls, 2004).

POLICY AND OTHER EFFECTS OF FORESIGHT

To open foresight processes to all kinds of experts and the general public is a new feature of foresight and for Germany a change in policy. When the different approaches are regarded in general, it must be stated that no direct use of the data can be noted in the T 21, but BMBF made use of them as an information source for different purposes (e.g., calls, programmes, new projects, etc.). A more direct impact of the workshops was a series of annual meetings of the programme operating agencies, which previously had no direct links or institutionalised contacts. A few years ago, they even established a formal network with regular meetings.

The initial result of the Delphi '93 survey as well as the Mini-Delphi and the Delphi '98 was a large volume of data which is the base for further analysis and discussion. The data do not have one addressee but are provided to all who are interested: companies use them as an input in their strategic planning and as additional information about their future framework conditions, ministries to re-evaluate or pre-evaluate their research agenda, research institutions or associations for strategic thinking or evaluation (e.g. the Fraunhofer Society made use of the data during its systems evaluation, Cuhls *et al.*, 2002), or the general public and the media for information and transparency about what is going on in research and technology.

Some firms have managed their own survey (Reiss *et al.*, 1995; Cuhls *et al.*, 1996). One large chemical company started with topics of the Delphi '93 survey, made their own evaluation of the topics and built up a strategy until 2010. In working groups, the information was discussed and distributed. Some smaller-scale comparisons of the business portfolios to the future-oriented areas are also being done in other companies, sometimes assisted by external consultants or the Delphi Team at ISI.

Industry and industrial associations have their own subject-tuned activities on behalf of their member firms, either in preparation (in case of the industrial association of machinery and apparatus manufacturers VDMA) or completed (in case of the association of electrical instruments ZVEI).

The main "user" of all Delphi studies in Germany was supposed to be the national government (federal level). The results of the surveys already contributed to decisions such as orientation of the education and research system, as well as to strategic talks between industry and large research organisations. But as the timing before general elections was bad in both of the comprehensive studies, there was no strategic use in BMBF. The regional administrations (Länder) are also interested in the results; they tried to analyse and interpret the data from their point of view (Schmoch *et al.*, 1995; Blind *et al.*, 1997; Cuhls *et al.*, 2002). Another follow-up project was a European effort in the field of biotechnology in food and food processing which was conducted on behalf of the European Commission and compared the opinions of producers, consumers and other stakeholders of five European countries in more detail. The results were especially interesting because in this conflicting area, no consensus between the different opinion groups could be observed. This evidently shows that foresight can also be used to identify the cases in which there is consensus and in which conflict potentials are especially high (Menrad *et al.*, 1999).

The impact on German society is also linked to the widely discussed results in the media, leading to interesting debates on the desirability of specific technologies. This was especially vivid in the run-up to the year 2000. There were times when the ISI server nearly broke down because too many persons accessed the Delphi dataset.

FUTUR has a different target audience. Although there seems to be a learning impact on all participants, these are not the major clients. FUTUR is conducted on behalf of the BMBF and directly involves it. Therefore, one should expect an impact mainly here, but a direct impact is difficult to state. In general, interdisciplinary project teams were established within the Ministry to manage the implementation of the Lead Visions from the first phase.

The four Lead Visions "Understanding Thought Processes", "Living in a Networked World", "Healthy and Vital throughout Life by Prevention" and "Creating Open Access to Tomorrow's World of Learning", which were developed and adapted during the first phase of the FUTUR process, have undergone different stages of implementation by the Ministry until now. In detail, the procedure of implementation was different for the four different Lead Visions (Dietz, 2004; Cuhls *et al.*, 2004), e.g. for the Lead Vision "Understanding Thought Processes", the BMBF project team elaborated a detailed concept of implementation together with different BMBF divisions. In autumn 2003, a first call for proposals was launched to establish centres for computational neuroscience. Two centres started their work in autumn 2004, two others followed at the beginning of 2005. The funding will be a total of €5m per year for a period of five years. A brain research network (Bernstein Centres) will be supported with €34m.

The Lead Vision "Living in a Networked World" lead to an implementation strategy being developed by a high-ranking expert group working for the BMBF sub-directorate "Information and Communication Technology". It was decided to give priority to the topic of heightened IT security. Meanwhile, projects derived from topics recommended by the FUTUR Focus Group started or are about to start: applications in cars (network on wheels, embedded systems), mobile internet and next generation internet, human machine interface, especially service robotics, smart web, augmented reality. Funding in 2004 amounted to €7.9m, while €64m are allocated to running projects in 2005–2007.

The implementation of the third Lead Vision "Healthy and Vital throughout Life by Prevention" started with a BMBF project team involving health research, educational aspects, biotechnology and innovative workplace development. A call for proposals was formulated by the division Health Research, based on a further expert workshop. Before the call was launched, it was discussed and accepted in the national Health Research Council. First projects started at the end of 2004. The total budget planned amounts to €5m per year. The recommendations for the fourth Lead Vision "Creating Open Access to Tomorrow's World of Learning" were found to be difficult to implement. Budgetary restrictions lead the BMBF to decide not to implement the Lead Vision now (Cuhls *et al.*, 2004).

In BMBF, the Lead Visions and the whole FUTUR process were not easily accepted. Doubts were also voiced whether the topics of the Lead

Visions were really new to BMBF or would have been supported, anyway, but maybe at a later time.

A priority-setting fund (Priorisierungsfond) was established by the BMBF to implement the Lead Visions that were transferred into research projects in 2004. This fund contains €10m, and is planned to be expanded to €25m per year. This is an innovation. The sum seems to be high at first sight, but it is not very high compared to the large programmes. The divisions are also reluctant to apply for this money (in competition), because the money is only provided for a limited time and when projects are started, in the end, the division would later be responsible for paying the money from their own budget.

To sum up, the effects of the relatively large FUTUR programme are different from the previous foresight exercises because the target audience and the concept are different. Although the figures for implementation sound high, the direct impact can be rated as relatively modest. For the previous Delphi surveys, there is no direct measurable impact, but when the companies' strategic planning and the activities started with the data are regarded, the impact seems to be even higher, as it was a much more resource-saving foresight process (including all publications Delphi '98 cost less than €750,000). Companies are relatively critical of FUTUR, because their own output (learning or influencing research programmes was an expected effect) is modest and no "data" or "facts" about the future are derived from the process.

CONCLUSIONS AND OUTLOOK

Looking at foresight in Germany in the last 15 years, after much initial scepticism, it even became a kind of fashion to apply foresight, but also the tendency emerged to use foresight for everything to do with priorities and to try to meet too many objectives with the foresight approaches. This became especially obvious in the FUTUR programme, in which the input–output relation is more and more imbalanced. With time, even the distinction between foresight activities in support of priority-setting and general public relations activities (i.e., public awareness, acceptance of science) became blurred. Both kinds of activities may be needed; but pursuing them simultaneously through mixed instruments (e.g., organisation of large conferences, participatory generation of new ideas) requires considerable management activities and seems to appear confusing to the participants.

The process of generating new topics, which was in the forefront of the Delphi method, was originally part of FUTUR, but the filtering process let only more general, well-known topics "survive". Generating new themes is necessary, and maybe needs a complementary process, focusing not only on science and technology but society-driven topics, in particular. Even in the international context, there are no straightforward approaches for articulating the implications of "societal demand", which is why such a process may call for new methodologies also. In FUTUR, they were not developed. The methods of FUTUR were not new, but their combination was an innovation. Some instruments served their purpose, others did not or served even different ones (like the conferences) under the heading of FUTUR.

The interlinking of methods and results was often a challenge. By mainly using papers for the transfer, often interesting aspects, perspectives and details were lost. This is still a challenge to be solved in German foresight. What can be unequivocally stated is that applying only "soft" methods such as workshops and focus groups is meanwhile regarded as problematic; industry and BMBF demand "harder" results (facts, figures, and indicators) about the future to work with, although it is known that the future cannot be foreseen.

The audience addressed by foresight in Germany was unclear in the Delphi processes; it emerged that there is a broad range of addressees, which made the use of the results more self-organised and arbitrary than intended. On the other hand, the concentration on the only addressee BMBF means that the Ministry needs to be more directly involved and the process needs to meet its requirements better than at present. The other possibility is to once again broaden a foresight process in Germany, focusing not only on BMBF but also on other ministries or even more on industry. This is a decision that has to be taken by those who may finance the process.

At the time of writing, foresight in Germany is at a crossroads. Depending very much on government policy and the new ruling parties, and a potential shift in the general assumptions on the functions of foresight, it is hard to predict the coming foresight activities in detail.

One possibility is that a similar process to FUTUR continues, with slight modifications. This is rather unrealistic, as the criticism of BMBF itself concerning the usefulness of the process is rather harsh. It is more realistic that the soft discussions about the future will have their (more inexpensive) forum in different dialogue formats, but that the priority-setting based on foresight will be conducted in more formal ways with structured methodologies. People like to see "facts" about the future. That means pictures or scenarios as well as data and indicators. Therefore, surveys, maybe even a Delphi, indicators, foresight processes based on interviews and expert rounds could be the methods of choice, again – as they are in industry, for associations, etc.

Participation and demand-orientation are regarded as important but, until now, no satisfactory methodological solutions have been found to serve a

BMBF or a company, association or others in answering their questions. If participatory methods are applied in future foresights, a definition is needed to what extent participation is meant. To involve more external persons (experts, general public, etc.) needs a definite decision. To involve only the general public or even try representative surveys is possible, but then the standard is different. And participation in these cases cannot mean that "the public" decides or that a majority vote automatically leads to a decision.

When summarising the main elements of the R&D policy that are currently being implemented, there is first of all the goal of achieving awareness for the challenges Germany has to accept. This is true for any country facing a shift in policy. It needs to enter into competition on coming problems and solutions, which require concepts and visions of the social and economic future. There has to be an interdependent process of science-based creativity between the fields of emerging technologies and the problem-generated demand for science and technology.

The future is shaped by looking into the future and asking questions about what to do – or not to do. We all decide about it – and act or desist from action. This is more than self-fulfilling prophecy. We tried to make German policy more proactive to support this aspect of shaping the future. And it is hoped that foresight will help to shape a (little bit) better future. For these reasons, there is a need for foresight and although the methodology may change, foresight in Germany will somehow continue. This is the author's prediction.

NOTES

1. Topics were selected in a conference in Berlin 1994, working in "virtual groups".
2. See Cuhls *et al.*, (2002); Cuhls *et al.*, (1998); and Cuhls and Blind (1999).
3. See also www.isi.fraunhofer.de/P/Projektbeschreibungen/Cu-delphi.html.
4. See www.futur.de.
5. See Cuhls *et al.*, (2004) or Hafner and Cuhls (2004).
6. See Cuhls *et al.*, (2002); Cuhls *et al.*, (2004).

REFERENCES

Banthien, H., Ewen, C., Jaspers, M. and Mayer-Ries, J. (2002), 'Welche Zukunft für Foresight und Forschungspolitik? Futur als methodische, inhaltliche und institutionelle Innovation', *Development & Perspectives*, 1, pp. 24–46.
Banthien, H., Cuhls, K. and Ludewig, N. (2004), 'Introduction to Futur – The German Research Dialogue. About the Futur Process', Cuhls, K. and Jaspers, M. (eds.), *Participatory Priority Setting for Research and Innovation Policy*, Stuttgart: IRB Publishers, pp. 5–23.

Blind, K., Grupp, H. and Schmoch, U. (1997), *Zukunftsorientierung der Wirtschafts-und Innovationsstrukturen Nordrhein-Westfalens*, Karlsruhe.

Blind, K., Cuhls, K. and Grupp, H. (1998), 'The Influence of Personal Attitudes on the Estimation of the Future Development of Science and Technology: A Factor Analysis Approach', *Proceedings of the 42nd Meeting of the International Society for the Systems Sciences (ISSS)*, Atlanta, Georgia, 19–24 July 1998 (CD ROM).

Blind, K., Cuhls, K. and Grupp, H. (1999), 'Current Foresight Activities in Central Europe', *Technological Forecasting and Social Change*, **60**(1), pp. 15–37.

Blind, K, Cuhls, K. and Grupp, H. (2001), 'Personal Attitudes in the Assessment of the Future of Science and Technology: A Factor Analysis Approach', *Technological Forecasting & Social Change*, **68**, pp. 131–149.

BMFT (Federal German Ministry for Research and Technology/Bundesministerium für Forschung und Technologie) (ed.) (1993), *Deutscher Delphi-Bericht zur Entwicklung von Wissenschaft und Technik*, Bonn: BMFT.

BMFT (Federal German Ministry for Research and Technology/Bundesministerium für Forschung und Technologie) (ed.) (1998-2001), *Zukunft nachgefragt*, Newsletter 1–8, Karlsruhe: ISI.

Cuhls, K. (1996), *Foresight in the German Science and Technology System*, Contribution to the Expert Group Meeting on Technology Forecasting and Foresight Activities in Latin America, Beyond Latin America 2000, Santa Cruz, Bolivia, 11–13 December.

Cuhls, K. (1998), *Technikvorausschau in Japan. Ein Rückblick auf 30 Jahre Delphi-Expertenbefragungen* [Foresight in Japan. Hindsight on 30 Years of Delphi Expert Surveys], Heidelberg: Physica.

Cuhls, K. (2003), 'From Forecasting to Foresight Processes – New Participative Foresight Activities in Germany', in Cuhls, K. and Salo, A. (Guest Editors), *Journal of Forecasting*, Special Issue, **22**, 93–111.

Cuhls, K. (2004), 'Futur – Foresight for Priority-setting in Germany', *International Journal of Foresight and Innovation Policy*, **1**(3/4), pp. 183–194.

Cuhls, K. and Kuwahara, T. (1994), *Outlook for Japanese and German Future Technology*, Comparing Technology Forecast Surveys, Heidelberg: Physica.

Cuhls, K. and Blind, K. (1999), 'The German Foresight Study '98 on the Global Development of Science and Technology', in Kocaoglu, D.F. and Anderson, T.R. (eds), *Technology and Innovation Management*, PICMET '99, Portland, pp. 577–582.

Cuhls, K. and Georghiou, L. (2004), 'Evaluating a Participative Foresight Process: "Futur – the German Research Dialogue"', *Research Evaluation*, **13**(3), pp. 143–153.

Cuhls, K., Breiner, S. and Grupp, H. (1995), *Delphi-Bericht 1995 zur Entwicklung von Wissenschaft und Technik – Mini-Delphi*, [Delphi Report 1995 on the Development of Science and Technology] Karlsruhe (same as brochure of BMBF, Bonn 1996).

Cuhls, K., Uhlhorn, C. and Grupp, H. (1996), 'Foresight in science and technology – Future Challenges of the German S&T System', in Meyer-Krahmer, F. and Krull, W. (eds), *Cartermill Guide for Science and Technology*, London, pp. 63–81.

Cuhls, K., Blind, K. and Grupp, H. (eds) (1998a), *Delphi '98 Umfrage. Zukunft nachgefragt. Studie zur globalen Entwicklung von Wissenschaft und Technik* [Delphi '98 Survey. Study on the Global Development of Science and Technology] Karlsruhe: www.isi.fraunhofer.de/P/Projektbeschreibungen/Cu-delphi.htm.

Cuhls, K., Grupp, H. and Blind, K. (eds) (1998a), *Delphi '98 - Neue Chancen durch strategische Vorausschau* [New chances through strategic foresight] Tagungsband der Tagung in der Deutschen Bibliothek in Frankfurt/Main, 1 July 1998, Karlsruhe.

Cuhls, K., Blind, K. and Grupp, H. (2002), *Innovations for our Future. Delphi '98: New Foresight on Science and Technology, Technology, Innovation and Policy,* Series of the Fraunhofer Institute for Systems and Innovation Research (ISI), **13**. Heidelberg: Physica.

Cuhls, K., Ludewig, N. and Kuhlmann, S. (2002a), *Futur. The German Research Dialogue,* Information Document for the Evaluation Panel, Karlsruhe: Fraunhofer ISI.

Cuhls, K., Hafner, S. and Rainfurth, C. (2004), *Futur – The German Research Dialog,* Information Document for the Evaluation Panel, Fraunhofer ISI, Karlsruhe.

Dietz, V. (2004), 'The German Foresight Process Futur: Implementation of Results in Research Funding', in Cuhls, K. and Jaspers, M. (eds), *Participatory Priority Setting for Research and Innovation Policy. Concepts, Tools and Implementation in Foresight Processes,* Proceedings of an International Expert Workshop in Berlin, December 2002, Stuttgart: IRB, pp. 104–110.

Durand, T. (2004), 'Selecting Technologies and Themes in "Technologies Clés 2005" (Key Technologies 2005)', in Cuhls, K. and Jaspers, M. (eds), *Participatory Priority Setting for Research and Innovation Policy. Concepts, Tools and Implementation in Foresight Processes,* Proceedings of an International Expert Workshop in Berlin, December 2002, Stuttgart: IRB, pp. 54–65.

Fondazione Rosselli, CeS&T (2003), *National priorities for industrial R&D in Italy,* http://www.jrc.es/projects/enlargement/FN/ThematicNetworkMeetings/Nicosia-00-03/Positionpapers/Roveda.htm.

Gaßner, R. and Steinmüller, K. (2003), 'Szenarien, die Geschichten erzählen. Narrative normative Szenarien in der Praxis', *Wechselwirkung und Zukünfte,* **2**.

Georghiou, L. (2004), 'Evaluation of Futur: Intermediate Results', in Cuhls, K. and Jaspers, M. (eds), *Participatory Priority Setting for Research and Innovation Policy. Concepts, Tools and Implementation in Foresight Processes,* Proceedings of an International Expert Workshop in Berlin, December 2002, Stuttgart: IRB, pp. 18–23.

Grupp, H. (1992), *Dynamics of Science-Based Innovation,* Heidelberg and New York: Springer.

Grupp, H. (ed.) (1993), *Technologie am Beginn des 21. Jahrhunderts,* Heidelberg: Physica (2nd edn 1995).

Grupp, H. (1994), 'Technology at the Beginning of the 21st Century', *Technology Analysis & Strategic Management,* **6**, pp. 379–409.

Grupp, H. (1996), 'Foresight in Science and Technology: Selected Methodologies and Recent Activities in Germany', *Science Technology Industry (STI) Review,* **17**, Paris: OECD, pp. 71–99.

Hafner, S. and Cuhls, K. (2004), 'Futur – the German Research Dialogue', Paper presented at the EU-US Scientific Seminar on New Technology Foresight, Forecasting & Assessment Methods (Future-oriented technology analysis: FTA), Seville, May 2004.

Irvine, J. and Martin, B.R. (1989), *Creating the Future,* The Netherlands.

Jungk, R. and Müllert, N. (1996), *Future Workshops: How to Create Desirable Futures,* London: Institute for Social Inventions.

Martin, B.R. (1995), 'Foresight in Science and Technology', *Technology Analysis & Strategic Management*, 7(2), pp. 139–168.

Menrad, K., Agrafiotis, D., Enzing, C., Lemkow, L. and Terragni, F. (1999), *Future Impacts of Biotechnology on Agriculture, Food Production and Food Processing*, Heidelberg: Physica (Technology, Innovation, and Policy 10).

Reiss, T., Jaeckel, G., Menrad, K. and Strauss, E. (1995), 'Delphi-Studie zur Zukunft des Gesundheitswesens', *Recht und Politik im Gesundheitswesen*, 1(2), pp. 49–62.

Salo, A. *et al.* (2005), *Evaluation Report on Futur – The German Research Dialogue*; Helsinki/Karlsruhe: BMBF.

Schmoch, U., Laube, T. and Grupp, H. (1995), *Der Wirtschafts- und Forschungsstandort Baden-Württemberg – Potentiale und Perspektiven*, ifo studien zur strukturforschung 19/I, München.

Technology Centre AS CR in cooperation with the Engineering Academy (2002), *Technology Foresight in the Czech Republic 2002*, www.foresight.cz.

Wagner, C. and Popper, S. (2003), 'Identifying Critical Technologies in the United States: A Review of the Federal Effort', *Journal of Forecasting*, Special Issue, **22**, pp. 113–128.

7.   Foresight in the USA

Alan L. Porter and W. Bradford Ashton

BACKGROUND AND CONTEXT

National-level technology foresight studies have not been a prominent activity in the science and technology (S&T) landscape of the United States. In contrast to many other developed economies where foresighting is used to help determine national S&T priorities and set national technology investment strategies, the USA stands virtually alone in specifically avoiding centralised S&T planning. Some American technology foresight studies have been done, but they have been limited in their scope, application and impact. Federal government studies have been restricted to mission agencies with S&T budgets, such as the Department of Energy (DOE), Defence and the Environmental Protection Agency (EPA) and they are rarely prominent above that level.

Many reasons for this state of affairs can be offered, but clearly one of the key drivers is a widespread scepticism of most forms of centralised, formal government planning among USA national leaders. The most prominent central planning at the national level is found in individual government agencies and administrations and even for these economic sector-oriented organisations, attempts to integrate programmes of agencies with similar or overlapping missions are usually limited to inter-agency coordination groups. High level agenda-setting and budgeting in the federal government, even for work that is directed at national goals such as health, defence, energy, education, transportation and environmental protection, is carried out through the dynamic political interactions between the appropriate Executive Branch agency and the USA Congress. Drawing on a long history of tradition since the country was formed in the late 18th century, Americans simply do not like strong central government and government planning.

Despite this experience, the USA has played two important roles in the evolution of technology foresight work. First, a variety of "foresight-like" efforts have been completed in both the public and private sectors. These efforts have some of the key process and method features found in more traditional studies as practiced in foresight-oriented countries such as Japan, the United Kingdom, Canada and Australia. But USA efforts do not have the comprehensive scope and central focus of the national government-sponsored work elsewhere. Government-sponsored foresight-like efforts have shown the following characteristics:

- A focus on anticipating the long-range future, namely the period 20–30 years beyond the study date;
- Participation by numerous stakeholders and subject matter experts;
- Use of process and analytical tools found in non-USA foresight efforts.

Several firms in the USA private sector have also employed "foresight" methods for business strategic planning and to set directions for long-term technology investments.

Second, USA planners, researchers, businessmen and other players have made significant contributions to development or improvement of the methods and tools used in foresight efforts around the globe. Computational, visualisation, information gathering and planning tools along with process innovations to handle groups of experts have all enjoyed active development and application in many USA organisations. These three themes – no comprehensive national foresight studies, important narrowly focused foresight efforts in some federal agencies and a strong track record of methodology development – are described further in this chapter.

HISTORICAL HIGHLIGHTS

We suspect that some of you turn here wondering if it might be a one-sentence chapter (and a short one at that): the USA does not do foresight. When you scan the major national foresight activities of other developed countries, this would not be totally off the mark. However, the USA does perform a range of related studies that contribute to research and development planning. Americans also contribute extensively to pertinent tech foresight methodology. So, we will write a chapter.

We divide this section into three uneven parts: OTA (Congressional Office of Technology Assessment), Critical Technologies, and Other. While we touch on numerous activities under "Other", the first two are, perhaps, the most noteworthy regarding foresight-like activities in the USA.

OTA – The Congressional Office of Technology Assessment

Did OTA conduct foresight studies? No, but it did devise a form and execute a rich set of technology assessments (TAs) to inform national policy deliberations. Each TA was called for by some Congressional client (e.g., a Committee), or by the bi-partisan Technology Advisory Board, or rarely at in-house initiative (distinctions were not always sharp). Over OTA's quarter-century lifetime, it generated some 500 publications, ranging from white papers to comprehensive TAs. Those became a rich resource for public discourse as they were all publicly available.[1]

OTA began operations in 1972. It was created during the same era in which the USA instituted the environmental impact assessment (EIA) procedures under the National Environmental Policy Act of 1969 (NEPA). Concern for the "unintended, indirect, and delayed" impacts (Coates, 1976) of technological and other developments stirred national political action. Another motivation for the creation of OTA was to provide Congress with technical savvy to counterbalance that of the Federal Agencies. These USA innovations in future-oriented analyses, TA and EIA, have spread widely. Over 100 countries mandate some form of environmental impact statements. The USA has not actively implemented the NEPA mandate to assess "proposals for legislation" as well as discrete projects.[2] Other nations, especially throughout Europe, have advanced Strategic Environmental Assessment (SEA). Likewise, Europe continues to use nationally adapted TA offices to inform policy.

OTA's influence on Congressional deliberations has been deemed modest. The extensive formulation and review processes resulted in most major TA's taking a couple of years to deliver. One of the reasons given for doing away with OTA was this poor timing fit with the annual and biannual Congressional legislative cycles. A strong suspicion holds that the right wing, conservative crew who took control of Congress in the mid-1990s felt that OTA was slanted against their interests. OTA also made for a handy scapegoat to show that Congress was cutting its own budget in a time of Federal budget-cutting (1995). Sufficiently few Representatives and Senators had become so engaged with OTA that they would favour saving it at the expense of cutting some personal or committee staff instead.

Critical Technologies

Through the 1990s, the USA undertook a series of exercises to identify critical technologies. Definitions of critical technologies evolved toward specifying those "essential for the long-term national security and economic prosperity of the United States" (Wagner and Popper, 2003). Most prominent was a sequence of four reports in response to Congress' direction that the President's Office of Science and Technology Policy (OSTP) should generate a critical technologies list biennially. The process varied for each such report, involving a National Critical Technologies Panel whose 13 (or so) experts included heads of major Federal agencies and industrial and academic leaders. Staffing moved with the second report to RAND's Critical Technologies Institute, later renamed as RAND's Science & Technology Policy Institute (STPI) (presently STPI is operated by the Institute for Defense Analysis). In addition at least two Federal agencies (Defence and Commerce) issued their own critical technologies lists, as did two industry associations (Aerospace Industries Association and the Computer Systems Policy Project). Comparing the various lists entertained a number of researchers and students – there was much in common among the technologies identified. However, comparison was not trivial as the number, level and details of the technologies varied greatly.

In contrast to most foresight practices, the American critical technologies processes tended to rely almost exclusively on expertise. The Congressionally mandated efforts strove for breadth in terms of including leaders from government, university and industry; broad stakeholder or public involvement was not pursued. Neither were technology foresight (forecasting, assessment) methods used much at all.

The influence of the critical technologies reports was diffuse. The effort stimulated significant "buzz" in the early 1990s, drawing attention to emerging technologies and the relationship between research and economic competitiveness. However, the reports did not directly lead to the many cross-agency technology initiatives that ebb and flow (Wagner and Popper estimate that there tend to be about a dozen such ongoing at a given time through this period). The critical technologies lists tended to be used for justification in support of stakeholders' interests. They seem also to have served as signposts influencing agency, state and university initiatives.

Other Foresight-like Activities

Martin and Irvine (1989), from the perspective of the 1980s, devote considerable attention to an activity known as "The Five-Year Outlook" and attendant efforts to do national research planning. The OSTP was given the responsibility for preparing an annual report; it passed this charge down to the National Science Foundation (NSF). The lack of enthusiasm for an analytical process – by OSTP, by NSF, and by the candidate users – to set R&D priorities led to revisions. The National Academies were then asked to prepare sets of "Research Briefings" which the Committee on Science, Engineering and Public Policy (COSEPUP) undertook. Prominent scientists resisted any cross-disciplinary prioritisation efforts, and these also eventually disappeared.

Various successive efforts involved the National Academies and their National Research Council. These do not much reflect the broad purview and participatory processes associated with Foresight. Yet, the studies of science and technology (S&T) issues, often with significant policy issues, generated by the National Academies continue to provide a major resource for public discourse. Their website highlights a staggering range of S&T topical treatments.[3] The site notes (as of June, 2005), that the National Academies Press offers more than 2,800 reports online, free. Topical headings range from Agriculture to Urban Development. The National Academies conduct a rich stream of Federally mandated studies. They also pursue topics that span disciplines and do not involve Federal support. For instance, the National Academies Keck Futures Initiative is a 15-year programme to foster interdisciplinary research – c.f., Committee on Facilitating Interdisciplinary Research (2004). A "New Releases" and "Best Sellers" sampling shows that these studies can and do address national technology policy concerns:

- Dietary Reference Intakes for Water, Potassium, Sodium, Chloride, and Sulphate;
- The Geological Record of Ecological Dynamics: Understanding the Biotic Effects of Future Environmental Change;
- Memory: The Key to Consciousness;
- How Students Learn: History, Mathematics, and Science in the Classroom.

Martin and Irvine (1989) observed that the National Institutes of Health (NIH) and NSF favour bottom-up "peer pressure" to generate research priorities. This holds true today. The past couple of years, however, have seen the initiation of an NIH Roadmap. This was developed through a broadly participatory process to "roadmap" medical research for the 21st century. The process is outlined in the Overview.[4] Notably it starts with discussions to prioritise research areas. The purpose was to identify major opportunities and gaps in biomedical research that no single institute at NIH could tackle alone but that the agency as a whole must address, to make the biggest impact on the progress of medical research. The Roadmap identifies the most compelling opportunities in three main areas: new pathways to discovery, research teams of the future, and re-engineering the clinical research enterprise. Nine working groups were then charged to devise implementation plans, including timelines, coordination mechanisms and staffing requirements.

The Environmental Protection Agency (EPA) has had an active interest in foresight efforts for many years. Following a 1995 report from the EPA's Scientific Advisory Board ("Beyond the Horizon: Using Foresight to Protect the Environmental Future") EPA has sought on a regular basis to foresee future environmental stressors and reasonably predict their potential impacts. This knowledge helps EPA work to prevent or avoid these impacts rather than just respond to them. Implemented by the Office of Research and Development (ORD), the systematic foresight effort is focused on potential issues 10 to 20 years into the future in contrast to current research planning and budgeting processes which use a time horizon of approximately three to eight years. The "foresight" approach has two parts in which ORD:

1. Searches for detectable early warning signals and extrapolate them into the future; and
2. Identifies new issues for which an early warning signal does not currently exist.

EPA uses a network approach of workshops involving the broader scientific community, its own programmes, regions, futurists, stakeholders and the public. The EPA goal is to identify and understand potential future risks to human health and the environment, recommend new directions for research and programme management decisions, and identify innovative, cost-effective solutions and alternatives through an ongoing futures programme.

The US Department of Energy (DOE) conducts a foresight-like activity through the DOE Office of Industrial Technologies (OIT), Office of Energy Efficiency and Renewable Energy. In the late 1990s, OIT initiated a programme called Industries of the Future to begin joint government–industry goal-setting and programme-planning for its research and development (R&D) activities. As part of this initiative, OIT established partnerships with representatives of the eight main energy-intensive industries[5] to identify and prioritise the major long-term industry needs for advanced energy-efficient technology. Industry needs represented technical barriers that no single firm could afford to address alone and thus were used to establish broad technology goals for all industry firms out to the year 2020 and interim development milestones that OIT would support. These roadmaps and the OIT-industry teams associated with them became the foundation for resource-allocation and project selection for OIT's R&D programme to ensure that the public investment supports realistic industry needs.

The National Reconnaissance Office (NRO) initiated an intriguing project in 1998 called "Proteus". It strove to develop truly fresh perspectives on intelligence needs and technologies to fulfil them. It did so using the scenario planning approach of a commercial facilitator, Deloitte Consulting. Focusing out to the year 2020, the project generated nine insights – i.e., fresh lenses different from the Cold War themes. These metaphors could provide new ways to consider issues in a changing world, and subsequently, new ways to plan to address them. For instance, one insight keyed on "Herds" – people and ideas on the move (Loescher *et al.*, 2000). Three workshops involving a range of intelligence professionals and outsiders helped compose five scenarios – characterisations of the world of 2020 to stimulate consideration of issues and solutions. For instance, one was named "Amazon.plague", wherein mutating viruses wrack the world, shrinking trade and the world's economy, with governments turning authoritarian or chaotic, and reliance on the global information grid in lieu of reduced physical interchanges. Follow-on stages aimed to transfer Proteus thinking, implement gaming environments, and assess the potential of emerging technologies to contribute to multiple future environments (Krause, 2002). In the wake of 9/11, the project and its approaches took on heightened significance to help in dealing with terrorism, including pick-up by other agencies (Waddell *et al.*, 2004).

Nowadays, the Department of Defense (DOD) has a strategic planning process for development of defence S&T. Elements include:[6]

- Defence Science and Technology Strategy, including principles such as dual use (to facilitate a common base with commercial developments);
- Basic Research Plan, providing strategic basic research objectives and outlining planned investments in a dozen technical disciplines;
- Joint Warfighting Science and Technology Plan, to plan for transitioning basic research findings through applied research and development toward joint (Army, Navy, Air Force) and coalition warfighting objectives;
- Defence Technology Area Plan, seeking to chart multi-agency investments relating to given technologies;
- Defence Technology Objectives, setting forth some 300 specific desired technology advances and their timing, along with funding requirements.

These involve the Secretary of Defence, the military services and various defence agencies (e.g., the National Security S&T Council). The plans are shared with defence contractors and allies to help focus collective efforts.

Another agency of note is the Office of Management and Budget (OMB). They have a cross-agency purview. However, they mainly exert pressure to constrain federal expenditures without delving into topical assessments.

Of Non-Governmental Organisations (NGOs) we note "The Foresight Institute", partly because of its name. This small NGO was founded in 1986 by Erik Drexler and colleagues to promote consideration of nanotechnology development, its potential impacts and policy options. It continues to stimulate nano-related research, information exchange and discussion but not to promote formal foresight studies.[7]

The Millennium Project of the American Council for the United Nations University (ACUNU), an NGO that happens to be located in the USA, is conducting foresight activities internationally. Jerome Glenn and Theodore Gordon coordinate an array of activities, including global lookout studies that engage an international set of volunteers connecting through the web. Their website describes their participatory process to collect and assess judgements from several hundred participants to produce an annual "State of the Future" and various special studies.[8] They also compile a fine volume on Futures Research Methodology.

METHODOLOGICAL CONTRIBUTIONS

Foresight processes can involve widely varied use of methodology. For cases where methods and tools are important in the study, USA work in tool development has laid important foundations for foresighting. Tools such as scenario planning, Delphi methods, economic analysis, technology scouting and scanning, technology roadmapping and technology assessment and impact assessment are examples. In this section we spotlight some of the specific methods that can contribute and significant American contributions to their development.

Glenn and Gordon (2002) offer a great resource that provides a rich compilation of short papers covering the essentials of some 25 methods. Each addresses the method's history, description, how to do it, strengths and weaknesses, frontiers and application examples. For instance, Delphi is covered by Theodore Gordon. He notes its invention by RAND in the early 1960s – indeed, most of the technology forecasting and assessment tools were devised by Americans. Futures Research Methods also covers environmental scanning, trend impact analysis, scenarios, participatory methods, simulation, etc. It also compares suitability and integration, and treats "meta-issues," such as Hal Linstone's presentation of multiple perspectives.

Joe Coates, in particular, merits appreciation for efforts to enable technology assessment (TA). His diverse intellectual contributions abound. But additionally, he led the NSF Program Office of Technology Assessment through the early 1970s. They supported at least 24 exploratory TAs that contributed greatly to methodological development – far more than did the ensuing work of OTA.

American texts have helped provide methodological foundations for technology foresight. Without pretence of comprehensive or unbiased coverage, we note:

- Ashton and Klavans (1997) provide the foundation for technical intelligence;
- Martino (1993) and Porter *et al.* (1991) lay out technology forecasting methods, with express treatment of how these pertain to technology management;
- Porter *et al.* (1980) cover technology assessment and other forms of impact assessment, while Porter and Fittipaldi (1998) edited a compilation of environmental methods.

Comments on three important methods developed or improved in the USA are given below (see also Chapter 3):

1. *Scenarios*: Scenario writing is a cornerstone methodology in virtually all major foresight efforts. Today, scenarios are a widely used strategic planning technique in government, private industry and other sectors throughout the world. One basic reason for their use is the ability of scenarios to deal with a variety of intersecting social, political, economic and technological forces that form the basis for what the future will bring. To deal with complex, uncertain, dynamic and risky matters facing strategic planners in the public or private sectors, scenarios have proven to be an enormously valuable tool. As a planning tool, scenarios were popularised by Royal Dutch Shell well over 20 years ago (Schoemaker and Corelius, 1992), but earlier techniques for using scenarios were developed and applied by a number of USA organisations, including Battelle Memorial Institute. Strategic planners from the business and government sectors use some variation of scenario writing as standard practice.

2. *Technology Intelligence* (Scanning and Scouting): Although originally developed in the USA national security arena, forms of technology intelligence (TI) have been practiced in the private sector for many years (Ashton and Klavans, 1997). Technology intelligence refers to the process of gathering information from a variety of "open sources" to develop an understanding of current activities, emerging trends and future directions of key companies, technologies and other players of interest. Environmental and, more specifically, technology scanning is one form of TI that is important in foresight studies. Scanning involves systematic surveying of a broad landscape to

identify activities, organisations and events that warrant further study as inputs to strategic planning processes.

In the USA, scanning has been a component of some strategic planning process in a variety of companies for many years. The term "environmental scanning" was coined by strategic planners about 40 years ago (Aguilar, 1967) and has grown in use with the advent of computer-based literature and web site sources and automated search tools. A review of relevant media covering the social, technological, economic, ecological and political (STEEP) environments – sometimes called "360 scanning" – is a foundation practice in futures research.

An emerging TI practice called technology scouting also has the potential to become a more prominent element of foresight studies. Scouting refers to the purposeful searching for specific technology-related entities of high interest such as products, companies or individuals. Both technology scanning and scouting provide essential foundation data into the foresight process for further analysis or consideration by the working teams evaluating potential directions for an entity.

3. *Technology Roadmapping*: Technology roadmapping has become a widely used technique during the past few decades, primarily in business but increasingly in government. The main purpose of a technology roadmap is to chart an overall direction for technology development or usage for a long planning horizon, typically 5–20 years. The concepts were initially popularised by Motorola, Inc in the USA to help decrease development cycle time of new products and decrease time to market for new products (Willyard and McClees, 1987). This methodology later turned in to a process to tie strategic planning with technology roadmapping and even evolved into corporate vision management. Use of roadmapping has been a recent development in foresight studies compared with techniques like scenarios or Delphi surveys. But its use is expected to grow given the benefits that roadmaps provide. Once created, a technology roadmap serves as a tool to provide essential understanding, orientation, context, direction and some degree of consensus in planning technology developments and implementations.

A variety of USA Federal agencies have used roadmapping and the practice is increasing. Examples include NASA, the National Science Foundation, the Department of Energy. Private sector R&D organisations, such as the Electric Power Research Institute (electricity technology roadmap) and The Santa Fe Institute (a Novel Computational Roadmap to synthesize and guide the research needed now to create the computing technologies), have both developed science roadmaps to help guide their research for the next 10–20 years.

The symposium on EU–US Future-oriented Technology Analyses provides a nice perspective on recent methods development (IPTS – Institute for Prospective Technological Studies, 2004). Background papers reviewed the state of the art of technology forecasting, assessment, intelligence and related methods *vis-à-vis* emerging societal foresight challenges (Coates *et al.*, 2001; Technology Futures Analysis Methods Working Group, 2004). We note eight challenges of note to those involved with foresight (drawing heavily, but not exclusively on these sources):

1. Changes in the nature of "technological change" with increasingly science-based innovation;
2. Shift in the prime drivers of technological innovation from the more narrowly technical concerns of Soviet–American Cold War military systems to industrial competitiveness concerns requiring inclusion of contextual influences;
3. Renewed attention to societal outcomes (impact assessment and sustainability);
4. Opportunities to exploit electronic information resources to enrich tech intelligence and foresight;
5. Better capabilities to address systems complexity in technological innovation;
6. Gauging irreducible uncertainties to devise adaptive foresight and technology management processes;
7. Interactive, participatory methods (scenario planning, multi-actor gaming, simulations);
8. Suitable approaches to anticipate potential discontinuous advances and radical innovation.

An interesting facet of this joint EU–US symposium was the limited American participation, even in this workshop devoted to methods. The Synthesis paper (IPTS, 2004) offers challenges in terms of foresight utilisation, credibility, intellectual interchange (academics, practitioners, users; public and private sectors), and how best to address such challenges.

CONCLUSIONS

The Foreword of an OECD report (1999) asserts:

There is a growing interest in technology foresight in the OECD Member countries because of the need to set priority in research and development in the context of the increasing cost of research and the tightening public budget for research. R&D efforts also need to be directed towards fulfilling social needs at the same time as providing sources of innovations that contribute to sustainable growth.

These premises do not hold for the USA. There is no growing interest in foresight. There is no momentum, still less consensus, that R&D prioritisation should be set by spelling out goals and ascertaining relative priorities for various research domains. Indeed, the USA seeks to be a predominant player in all R&D domains (National Academy of Sciences, 1993). And there are no strong indications of commitment to link R&D priorities to societal goals.

What are the consequences of this "anti-foresight" situation? Can a nation fare well scientifically and economically trusting to a marketplace of ideas and pressures to establish priorities and allocate resources? If so, why are other nations investing in extensive foresight activities?

The American political process essentially treats science and technology via pluralistic processes. Voices cry out (for resources) from scientists, industrial interests and activists (pro- and anti-science). Multiple agencies support and sometimes conduct R&D in somewhat overlapping spheres of interest. Multiple Congressional committees exert power over legislative, budgetary and regulatory aspects. Two political parties strive to effect agendas. The Executive Branch balances against the Congressional Branch, where the House and the Senate jockey between themselves. Annual Federal budget cycles predominate. In recent years, pork-barrelling has arrived in R&D, with earmarking of pet projects for institutions in particular political constituencies. However, most R&D funding continues to rely on some variation of scientific peer review to assure quality. This is not a milieu conducive to far-reaching national foresight. But it lends a multi-voice robustness to science and technology development.

Wagner and Popper (2003), reflecting particularly on the decade of USA critical technologies activities, identify the poor fit between foresight-like processes and the American S&T system. While some of us lament this in whole or in part, Americans and others should also reflect on its strengths. The lack of an established priority-informing process allows, and may facilitate, adaptive response to changing concerns. The lack of clear agency R&D responsibilities enables alternative routes to pursue new initiatives. Examples abound, but consider how many USA Federal agencies have a hand in "nano" R&D. This risks redundancy and waste of precious resources. However, the increasing availability of accessible S&T information resources on research programmes, projects and outputs (papers, patents) offer the potential to coordinate (not that this is well done now).

To the extent that technology foresight is able to foresee emerging technologies and shifting needs, those who use foresight may gain a significant advantage over those who do not. Wagner and Popper (2003) assert that "Foresight cannot provide predictions or even leading indicators." We disagree, yet must admit that it is not easy to present clear evidence of foresight providing effective "early warning" that made a difference.

Another argument in favour of the USA non-foresight stance is that foresight is usually open information. Hence, American agencies, companies and researchers can, in principle, make use of foresight knowledge generated by others. Albeit, we do not have evidence that much attention is being paid to such knowledge in any explicit way.

To what extent does a "triple-helix" innovation system involving academic, industry and government contributors to and users of R&D attend to foresight? In countries that strive to do so, national governmental roles take a strong leadership position – i.e., set priorities. But the pluralistic American innovation system may make up for such focused effort through more multi-path explorations – or as Wagner and Popper (2003) call it, "self-assembly". The more the innovation system is subject to unpredictable, rapid changes, the more advantage to the pluralistic approach. Good technological intelligence to pick up quickly on emergent opportunities may outweigh careful foresight. Such a messy, opportunistic system may especially do better at "Radical Innovation" (Dismukes *et al.*, 2005).

That said, how about the value in spelling out R&D fit to societal goals? As we write the USA is fighting about stem-cell research but has opted in to genetically engineered foods without much of a murmur. The American scientific community does not seem overly anxious to engage a broad public in deliberating about research agendas. While agenda-setting ought to be inclusive in a democratic society, do we want a populace whose majority (+/-) disbelieves evolution directly involved? As foresight proponents, we might well counter that well-orchestrated processes have a strong educational component preceding judgemental elements. Were foresight able to widen popular interest in science, this might yield significant impetus to fixing an educational system that fares badly in teaching science, math and engineering compared to other nations.

FUTURE DIRECTIONS FOR USA FORESIGHT EFFORTS

The information regarding USA foresight activities presented in this paper suggests that "the future of USA foresight efforts will probably be very much like the past". This means that decentralised foresight studies will continue to be undertaken. Agencies such as those of the EPA, DOE and DoD, will probably continue foresight work, but it is not likely that national level efforts encompassing a broad slate of national goals with national leadership will occur. The pluralistic nature of the USA R&D system, the diverse, dynamic nature of USA national political bodies and the limitations of USA foresight history makes centralised goal-setting across many national issues in the USA very unlikely.

However, in line with other past foresight contributions, the strong interest in methods, techniques and tools exhibited by USA researchers and practitioners is also very likely to continue. This means that new methods and tools that can be applied to foresight studies around the world will probably continue to emerge from USA developers. Tool development will also probably be diffuse and diverse with inputs coming from a wide range of strategic planners, policy analysts, business leaders, the academic community and other political players.

Three themes surrounding tool development seem likely as we move forward into the future. First, the increasing widespread availability of data of all sorts is not likely to abate, making advanced tools that help process, search, mine, organise, display and interpret the many forms of data we expect to become available a probable USA contribution. Second, the need for better methods of extracting, organising, comparing and combining a wide variety of human judgments will also continue to prompt method development in the USA. Taking a vast array of expressed interests and opinions into account seems to be a continuing driver to improve foresight studies. Third, the increasing proliferation of rapid and comprehensive communication tools, such as those building on progress with the internet and other electronic networks, will permit vast numbers of individuals around the world to participate in decision-making situations like foresight studies. For instance, it is not difficult to imagine that future foresight and other large-scale analysis efforts will one day incorporate inputs from experts and stakeholders using some form of election-style electronic voting processes. Networking and collaboration tools seem to be attractive solutions to dealing with large diverse contributors and stakeholders and these tools will undoubtedly improve over time.

In short, it appears that contributions to foresighting from USA players are likely to continue as they have in the past. While not fully embracing centralised national-level foresight studies, the USA does place a high value on development and use of advanced strategic planning and decision-making methods and tools.

NOTES

1 The "OTA Legacy" [available at http://www.wws.princeton.edu/~ota/] tells about OTA and provides the collection of OTA publications in searchable form.
2 The vision of the National Environmental Policy Act (NEPA) is remarkable – see http://ceq.eh.doe.gov/nepa/regs/nepa/nepaeqia.htm.
3 See http://www.nationalacademies.org.
4 See http://nihroadmap.nih.gov.
5 The OIT Industries of the Future are: Aluminum, Chemicals, Forest products (pulp and paper, wood manufacturing), Glass, Metal Casting, Mining, Petroleum Refining, Steel.
6 c.f., www.milnet.com/pentagon/dto/intro.htm.
7 www.foresight.org.
8 www.acunu.org/millennium.

REFERENCES

Aguilar, F.J. (1967), *Scanning the Business Environment*, New York, NY: McMillan.
Ashton, W.B. and Klavens, R.A. (eds) (1997), *Keeping Abreast of Science and Technology*, Columbus, OH: Battelle Press.
Coates, J.F. (1976), 'Technology Assessment: A Tool Kit', *Chemtech*, **6**, pp. 372–383.
Coates, V., Faroque, M., Klavins, R., Lapid, K., Linstone, H.A., Pistorius, C. and Porter, A.L. (2001), 'On the Future of Technological Forecasting', *Technological Forecasting and Social Change*, **67**(1), pp. 1–17.
Committee on Facilitating Interdisciplinary Research, National Academy of Sciences, National Academy of Engineering, Institute of Medicine (2004), *Facilitating Interdisciplinary Research*, Washington, DC: National Academies Press.
Correia, Z. and Wilson, T.D. (2001), 'Factors Influencing Environmental Scanning in the Organizational Context', *Information Research*, **7**(1), Available at http://InformationR.net/ir/7-1/paper121.html.
Delbecq, A.L. and VandeVen, A.H. (1971), 'A Group Process Model for Problem Identification and Program Planning', *Journal of Applied Behavioral Science*, VII (July/August), pp. 466–491, reprinted in Del becq, A.L., VandeVen, A.H., and Gustafson, D.H. (1975), *Group Techniques for Program Planners*, Glenview, Illinois: Scott Foresman and Company.
Dismukes, J.P., Miller, L.K., Bers, J.A. and McCreary, W.N. (2005), 'Technologies of Thinking. Seen Key to Accelerated Radical Innovation,' *Research Technology Management*, July–August.
Glenn, J. and Gordon, T. (2002), *Futures Research Methods: Version 2.0*, AC/UNU Millenium Project, Washington, DC: http://www.acunu.org/millennium/FRM-v2.html.
IPTS – Institute for Prospective Technological Studies, [European Commission, Directorate-General], (2004), 'New Horizons and Challenges for Future-Oriented Technology Analysis', Proceedings of the EU-US Scientific Seminar: New Technology Foresight, Forecasting & Assessment Methods [EUR 21473 EN; www.jrc.es/projects/fta/].

Irvine, J. and Martin, B.R. (1984), *Foresight in Science: Picking the Winners*, London: Pinter Publishers.

Krause, P.H. (2002), 'The Proteus Project – Scenario-based Planning in a Unique Organization', *Technological Forecasting & Social Change*, **69**(5), pp. 479–484.

Linstone, H.A. and Turoff, M. (eds) (2002), *The Delphi Method: Techniques and Applications*, (http://www.is.njit.edu/pubs/delphibook/).

Loescher, M.S., Schroeder, C. and Thomas, C.W. (authors); Krause, P.H. (compiler), (2000), *Proteus: Insights from 2020*, The Copernicus Institute Press.

Martin, B.R. and Irvine, J. (1989), *Research Foresight: Priority-Setting in Science*, London: Pinter Publishers.

Martino, J.P. (1993), *Technological Forecasting for Decision Making*, 3rd edn, New York: McGraw-Hill.

National Academy of Sciences (1993), *Science, Technology and the Federal Government: National Goals for a New Era*, Washington, DC: National Academy Press.

OECD – Organization for Economic Co-operation and Development, Working Group on Innovation and Technology Policy, Directorate for Science, Technology and Industry, Committee for Scientific and Technological Policy (1999), 'Technology Foresight and Sustainable Development', Proceedings of the Budapest Workshop (11 December 1998).

Osborn, A.F. (1948), *Your Creative Power*, New York: Charles Scribner's Sons.

Porter, A.L. and Fittipaldi, J.J. (eds) (1998), *Environmental Methods Review: Retooling Impact Assessment for the New Century*, Atlanta, GA: Army Environmental Policy Institute.

Porter, A.L., Rossini, F.A., Carpenter, S.R. and Roper, A.T. (1980), *A Guidebook for Technology Assessment and Impact Analysis*, New York: North-Holland.

Porter, A.L., Roper, A.T., Mason, T.W., Rossini, F.A. and Banks, J. (1991), *Forecasting and Management of Technology*, New York: John Wiley.

Schoemaker, P.J.H. and Van der Heijden, C. (1992), *Integrating Scenarios into Strategic Planning at Royal Dutch/Shell Case Study*, Planning Review, pp. 41–46.

Skumanich, L. and Silbernagel, K. (1997), *Foresighting Around the World*, Battelle Memorial Institute, Seattle, WA.

Technology Futures Analysis Methods Working Group (2004), [Alan L. Porter, W. Bradford Ashton, Guenter Clar, Joseph F. Coates, Kerstin Cuhls, Scott W. Cunningham, Ken Ducatel, Patrick van der Duin, Luke Georghiou, Theodore Gordon, Hal Linstone, Vincent Marchau, Gilda Massari, Ian Miles, Mary Mogee, Ahti Salo, Fabiana Scapolo, Ruud Smits and Wil Thissen], 'Technology Futures Analysis: Toward Integration of the Field and New Methods', *Technological Forecasting and Social Change*, **71**, pp. 287–303.

Waddell, W., Kim, J. and Smith, J. (2004), 'Proteus, New Insights for a New Age', *2004 Command and Control Research and Technology Symposium: The Power of Information Age Concepts and Technologies* [www.dodccrp.org/events/2004/CCRTS_San_Diego/CD/papers/011.pdf].

Wagner, C.S. and Popper, S.W. (2003), 'Identifying Critical Technologies in the United States: A Review of the Federal Effort', *Journal of Forecasting*, **22**(2/3), pp. 113–128.

Willyard, C.H. and McClees, C.W. (1987), 'Motorola's Technology Roadmap Process', *Research Management*, pp. 13–19.

8. Foresight in Japan

Terutaka Kuwahara, Kerstin Cuhls and Luke Georghiou

INTRODUCTION

Japan has gone through many phases of development on the way to becoming an economic superpower, with the industry-driven post-War economic miracle ending in the 1990s when the speculative wave of the "bubble economy" burst and a decade of deflation and stagnant GDP began. More recent recovery has followed substantial reforms (Lee, 2006). Similar changes may be perceived in Japan's self and external images with a wave of commentators exhorting imitation of Japan's success being superseded by a critical reaction. Against this background some concerns have remained constant. The geography of Japan, which sees most of the population densely packed into the habitable coastal zones, is combined with high tectonic activity and a lack of natural resources, especially energy. Industrial success has rested very much upon the skills of the population, government–industry cooperation and technology-based industries.

Science and technology have also gone through phases. During the catch-up period Japan was frequently accused of making use of the scientific advances made and paid for by Western nations, succeeding in markets through superior manufacturing ability built on an R&D base dominated by industry. By the late 1980s it was recognised that this strategy had reached its limits and that Japan would have to take its place at the scientific frontiers if its industry was to be able to remain competitive in an increasingly high-tech era. Relatively low public expenditure on science would need to be rapidly increased but in parallel a rigid and over-regulated university system would need to be galvanised. The key legislative step came with the Basic Law on Science and Technology in 1995[1] which laid the ground for a succession of five-year plans which set budgetary and reform targets across government and the "privatisation" of the universities (MEXT, 2003).

This chapter tracks the development of foresight in Japan as applied to the science and technology system. It should be stressed that Japan has a more general foresight culture with future-looking activity frequently undertaken by most public and indeed private organisations. Here the focus is on those activities undertaken by the State with the specific purpose of orienting and understanding scientific and technological developments at a holistic level. An unprecedented example of institutional continuity combined with gradual evolution and adaptation will be seen.

DECADES OF DELPHI SURVEYS

Since activity began in 1969 national technology foresight in Japan has been based on a large-scale Delphi survey addressed to experts in a wide range of fields. The technology forecast has been repeated approximately every five years and is now in its 8th iteration – a wholly unprecedented level of continuity. In the meantime the scope of work and the range of methods applied have also expanded. The stability of practice in foresight has masked a gradual evolution and adaptation to the position of Japan in the world. In Japan the perception is that the process has moved through the first three generations of foresight as described in Chapter 1 of this volume. This chapter sets out to track those changes and to present the most recent work in greater detail.

Kuwahara (1999) has presented the Delphi survey as the underpinning element of a four-layer model of foresight in Japan, providing a holistic backdrop upon which the other activities depended. The second level was that of macro-level surveys, which are carried out by many government ministries and agencies. For example the former Ministry of International Trade and Industry (MITI) released mid-to-long-term visions regarding the direction of Japan's industrial technology development every two to three years. Another example mentioned was that of the Economic Planning Agency, with an economic and social outlook. Survey fields in such studies are limited within the mandate of ministries or agencies, and time horizons are usually 10 or 15 years. The third level is done by groups of private firms or semi-public organisations. The fourth level is the forecasting activities of private firms to help their own business decisions. Usually, the survey areas are limited, and the time range is short (Kondo, 1993).[2]

When foresight was introduced, essentially as an import from the USA and long before the term "foresight" was in common usage, the environment was one where Japan was in a catch-up and growing process from an economic point of view. Industry was the major and most active R&D player. Government perceived a lack of strategic vision in the area of science and technology and the initial motivations for foresight were to form common vision/consensus on future priorities and perspective and through this to lead industry through "long-term visions". There was no explicit public policy role but nonetheless a moderate link to government's S&T policy existed and there were indirect effects on R&D resource allocation.

Hence it may be seen that the Delphi did not target a single group or policy. NISTEP (National Institute of Science and Technology Policy), the institute which for many years has operated the survey together with the Institute for Future Technologies (IFTECH), believes the Delphi process provides several advantages (Cuhls and Kuwahara, 1994):

- The S&T community must periodically think seriously and in detail about significant science and technology trends relative to important socio-economic priorities and obstacles.
- Participation of science experts outside of government helps maintain information flow into the government and improves the ability to assess future demands on national infrastructure.
- The Delphi provides a disciplined way to handle a broad range of topics, including new and/or cross-cutting areas of science.

A further factor in favour of this particular approach in the Japanese context is that it provides a vehicle for developing consensus while at the same avoiding any direct confrontation of views between participants, any conflicts in judgement of topics being resolved "on paper".

The full set of surveys is set out in Table 8.1. It can be seen that the range of fields and topics increased through the first few surveys and then stabilised until the most recent where there was a reduction, while other approaches were added.

Table 8.1   Japanese Delphi surveys

No.	Survey Year	Fields	Topics	Experts
1	1970–1971	5	644	2482
2	1976	7	656	1316
3	1981–1982	13	800	1727
4	1986	17	1071	2007
5	1991	16	1149	2385
6	1996	14	1072	3586
7	2000	16	1065	3106
8*	2004	13	858	2239

* *Note*: Other methods also used in the 8th foresight

The first Delphi Report looked at areas such as the development of society, information, medicine and health, nutrition and agriculture as well as industry and resources. In all these fields, the issues for consideration were formulated by experts. The assessment made was about the timescale for the topics to be realised, the technological potential and where catch-up with other countries would be necessary. There were three rounds in the survey and 2,482 experts participated in all of them.

The second Delphi report in 1976 covered seven fields with 20 sub-fields, and 1,317 experts were involved throughout the study. More criteria were specified, including the importance of the topic and measures to be taken. From the second report on, only two rounds were conducted. In the third study, 1,727 experts participated, and the number of areas as well as the number of topics increased further. The fourth survey involved 2,002 experts and 1,071 topics, with the criteria remaining nearly unchanged. The same was true for the fifth study, which had 16 areas, 1,150 topics and 2,385 experts. By this time the methodology was well-established, but the organisers at NISTEP were of the opinion that improvements were necessary. It was at this time, therefore, that the cooperation with the German Fraunhofer Institute for Systems and Innovation Research was initiated (see Chapter 6). Over the following years, through Mini-Delphi studies, the cooperation was enhanced and the Delphi methodology improved. The sixth Delphi study was also performed in cooperation with Germany, about 30 per cent of the topics and some criteria being the same. Nevertheless, there were separate German and Japanese reports.[3] For the public and for companies, from the fourth Delphi survey on an easy to read publication was produced. Later on it even included manga (comics). The sixth study in 1997 was set in the context of economic stagnation.

An interesting question is that of why the activity has persisted over such a long period. A very early challenge came with the 1970s "oil shock" which was felt acutely in Japan, a country without its own resources. Although in most other countries forecasting activities fell into oblivion in the 1970s because they had not foreseen the oil shock and the kind of "limits to growth" that it presented, the Japanese Delphi process continued. In Japan, it was observed that it was more important to make the future happen, and to shape it actively by using the information gained in foresight activities, setting stable framework conditions for development in certain fields, and making use of foresight procedures to update the information. Given the unknowability of the future, it was considered important to update the knowledge and the information that was available about it.

A consequence of the long-standing continuity achieved in the Delphi survey has been the possibility of assessing whether the predictions of the early exercises have been realised by means of asking the experts in later cycles to make that judgement. This is discussed further in Chapter 16 but here we may note that the percentage of topics fully or partially realised has been calculated for the first four surveys (up to 1986). The picture is mixed, reflecting perhaps both the relative pace of advance between fields and the knowledge base of Japanese experts. Taking the more generous measure of fully or partially realised, the scores for the surveys in chronological order are 69 per cent, 68 per cent, 73 per cent and 66 cent. These are very consistent around the two-thirds mark. Fields with high realisation percentages include life sciences, health and medical care, agriculture, forestry and fisheries, environment and safety, and cities, civil engineering and construction. Low realisation percentages were obtained for traffic and transportation and energy and resources. On the other hand, this measure is not the right one because actively stopping to support subjects is also an important "success" of self-destroying prophecies, but makes the calculation of "predictions" a wrong one.

CONSIDERING SOCIO-ECONOMIC NEEDS

Science policy has been undergoing fundamental change in Japan from the mid 1990s when the first Science and Technology Basic Law was introduced in 1995 and implemented through the first Basic Plan which ran from 1996–2000. Among the many changes this embodied was a growing emphasis upon socio-economic dimensions to S&T. The fifth and sixth surveys had of course embodied assumptions about socio-economic needs but these were framed by the technological experts responsible for Delphi topic selection. Hence in the 6th survey (1997) topics relating to four areas were extracted:

1. Counter measures for ageing society (creating a barrier-free environment, maintaining quality of life, assisting aged people to be independent, etc.);
2. Maintaining safety (prevention of natural disasters, reducing crime including computer crime, etc.);
3. Environmental preservation and recycling (developing new energy sources, low energy consumption initiatives, recycling, etc.);
4. Shared fundamental technologies (design techniques, processing technologies, handling systems and techniques for observations and measurements).

Technology sub-committees were asked to include considerations of these areas when they set the topics and subsequently when they reported back on findings. The Sixth Delphi was then followed up with a separate study, published in 1999 as a report, "The Analysis of Future Needs for Science and Technology based on National Lifestyle in 2010s". An analysis was made of Government White Papers to extract factors impacting deeply on human lifestyle, housing and diet. The resulting "Citizens' Lifestyle" had 12 categories including, for example, education and social insurance. Further inputs were collected from public opinion surveys and overseas comparisons with maximum consideration given to the views of ordinary citizens.

Against this issue list comparisons were made with the technological topics and their assessed importance and expected time of realisation. Seven aspects of lifestyle were identified as fields closely related to science and technology, covering 326 technological topics in the Delphi. The seven were Health, Dietary, Housing, Water, Information, Safety and Infrastructure. While most were well-covered by topics, some were not, indicating either a need for new technologies or an area where the solution was non-technological (for example a lifestyle change).

In the Seventh Delphi (NISTEP, 2001) a different approach was adopted, building in the consideration of needs from the start. Again, 14 "classical" technological fields such as ICT, Electronics, Life Sciences, Health and Medical Care, Agriculture, forestry, fisheries and food, Distribution, Transportation or Services were in the centre of the Delphi survey. But this time, three sub-committees discussed future "needs". The committee comprised experts from the cultural and social sciences who were asked to identify possible future trends in socio-economic needs over the coming 30 years. The fields they selected were: (a) New socio-economic systems; (b) Aging Society; and (c) Safety and Security. The committee handed in three reports about the perspectives in these fields and the results of the technological fields in the light of these perspectives. The design of the Eighth Japanese Foresight gave even more weight to socio-economic aspects; below it is described in detail.

GOVERNMENT REFORM: NEW INTEGRATION OF TF

In 2001 a major reorganisation of central government ministries took place in Japan. For foresight the most significant development was the establishment of the Council for Science and Technology Policy within the Cabinet Office. This Council, chaired by the Prime Minister and with a Minister of State for Science and Technology in the Cabinet Office, discusses comprehensive national measures and other issues concerning science and technology. It compiles the Basic Plans which structure science and technology spending in Japan. At ministry level, the Science and Technology Agency in which NISTEP previously sat was integrated with the Ministry of Education,

Science, Sports and Culture, in a rationalisation which reduced the cabinet from 22 to 13 ministers.

The Second Science and Technology Basic Plan was approved on March 2001. This built upon the First Plan which had doubled government R&D expenditure but had concentrated on S&T fundamentals with priority-setting being non-explicit, though technological fields were favoured. By the time of the Second Plan the budget had again increased and four very broad areas were presented as priorities: Life Sciences, IT, Environment and Nanotechnology/Materials. From a foresight point of view, eyes were already on the content of the Third Plan[4] due to begin in 2006, and with the new structures the possibility of a stronger top-down influence was in place. In order to deliver necessary data, the eighth Japanese Foresight was performed earlier than the usual five-year interval would have suggested.

NISTEP adapted its structures to meet the enhanced need for future-oriented policy guidance. The group which had produced the Delphi surveys was reconstituted and strengthened to form the Science and Technology Foresight Centre (STFC). This brought together researchers from mixed government, university and industry backgrounds and associated some 2,800 experts in a wider network. A key mission was to support the development of the Third Plan. Figure 8.1 shows the linkage between foresight and policy-making under the new structures. NISTEP as a whole (including the STFC) was also engaged in an evaluation of the First and Second Plans known formally as the "Study for Evaluating the Achievements of the S&T Basic Plan in Japan"[5] or "Basic Plan Review" in short. This was a comprehensive exercise in benchmarking the Japanese S&T system in an international context, and identifying changes and impacts deriving from S&T activities. However, here we will focus upon the parallel exercise of the Foresight Survey. Both exercises were supported by Special Coordinating Funds for Promoting Science and Technology.

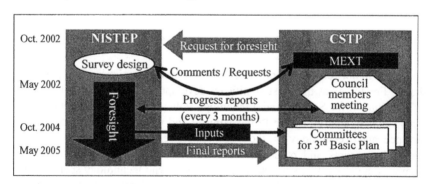

Figure 8.1   Linkage between foresight and policy-making in Japan

The inputs to the top-down prioritisation of the Third Plan came from four distinct elements of the foresight programme. Their interrelation is shown in Figure 8.2. The aim was to get a spread of approaches which would cover the spectrum from basic research, through application to broader societal issues (reflecting a continuation of the earlier exercises concern with societal needs).

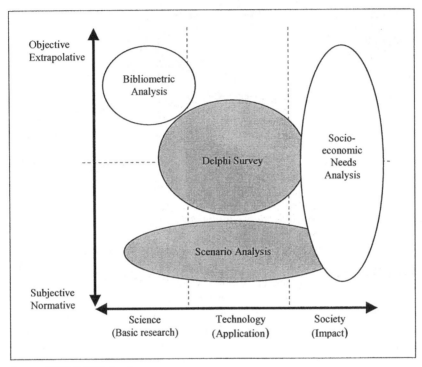

Source: NISTEP 2005a,b,c,d

Figure 8.2   Design of the 8th Foresight Programme

SPECIFIC ACTIVITIES IN THE EIGHTH PROGRAMME

Study on rapidly developing research areas (bibliometrics) [6]

This study aimed to identify rapidly developing research areas through the use of citation databases, and to examine the presence of Japanese papers in these areas. As indicated in Figure 8.2 the focus was on science and in particular upon basic research. The methodology is summarised in Figure 8.3 (below).

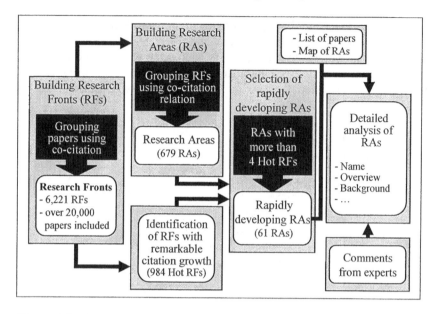

Figure 8.3   Process used to study rapidly developing research areas

The basic dataset for the analysis was highly cited papers (the top 1 per cent) in 22 research fields (for example clinical medicine, plant and animal sciences, chemistry, physics, etc.). There were about 45,000 papers for each year between 1997 and 2002. These papers were clustered using co-citation to identify larger research areas ("research fronts"). The basic assumption of co-citation is that papers which are frequently co-cited by third parties are assumed to have similarities in their content. This produces a structure of "core papers" which constitute the cluster, together with "citing papers" – those that cite the core. From an initial 5,221 research fronts, sub-clustering produced 679 research areas. Among the 679, 153 were identified as including more than one research front where the number of citing papers is rapidly growing. These "hot fronts" constitute the developing areas sought by the study. The process was completed by a qualitative expert interpretation, first by Centre staff and subsequently by field experts.

The results were analysed in terms of the ratio of Japanese papers to the total number of papers in a research area with 7 per cent being taken as a benchmark. Areas in excess of 7 per cent included physics, chemistry and plant and animal sciences, while underperformance was seen in clinical medicine, environmental sciences and ecology and, perhaps surprisingly, engineering. Dynamic patterns were also analysed, for example, tracking Japan's declining share of citing papers in the area of carbon nanotubes – initially a Japanese discovery.

Study on Social and Economic Needs [7]

The aim of this exercise was to collect information on the needs of society and the economy and to link them with specific areas of science and technology, and then to assess the potential contribution of science and technology to satisfying those needs. The time-frame covered the next 10–30 years. Building upon the need categories identified in the Seventh Delphi exercise, a detailed draft list of needs from the citizen's perspective was compiled. This also drew upon needs identified in other documents such as government white papers. A literature survey was used to identify industrial needs and the list was completed by consultation of intellectuals.

The next step was to structure the draft needs list using cluster analysis. This involved a web questionnaire survey of 4,000 people with use of the Analytical Hierarchy Process to weight the need categories. The resulting refined needs list contained 12 clusters of similar content. Examples of main headings were:

- Society is peaceful, safe and provides peace of mind (preventing traffic accidents, crime and terrorism);
- Actively contribute to solving global problems.

The resulting list was put to three panels, consisting of intellectuals, the public, and business executives respectively who were asked to summarise needs over the identified period. In addition a trial survey of 109 experts was made on how much science and technology could contribute to the listed needs. For two of the clusters, concerned with personal and social fulfilment and with education and learning, initiatives well beyond science and technology were seen to be required. The survey was carried out by setting three future scenarios and giving different priorities to the needs categories within each one. In addition 130 Delphi survey topics were related to the needs categories and the 170 members of the technical subcommittees were asked to assess the level of the contribution of each Delphi topic to the needs, the type of contribution it might make (direct or indirect), and any concrete examples of areas of S&T requiring attention or raising concern.

Delphi Survey [8]

The Eighth Delphi Survey centred on applied science and technology but as in the previous surveys also contained topics relating to basic science and societal impacts. It addressed the 30-year period between 2006 and 2035. As noted above 13 fields were covered. While these appeared superficially similar to previous exercises, there were significant differences. They were mainly concerned with taking over information from the other three "pillars" of the Japanese foresight already when formulating the topics. Of course, some topics from the previous surveys were included again in order to be able to follow the expert judgements over time, and the classical definition of areas which occupied a middle layer in a hierarchical structure consisting of field, area and topic layers was also kept. For example, a field was "Information and communication", areas within this field "very large-scale information processing" or "information security", and a topic within the information security area "a highly reliable network system capable of protecting the privacy and secrecy of individuals and groups from intrusion by malicious hackers".

The new feature was that this time, respondents were asked questions at multiple levels. At a general level they were asked to identify fields where fusion and collaboration should advance, at an area level the focus was on expected impacts (now and in the medium term), and at topic level the more traditional questions of importance, time of realisation, leading countries and necessity of government involvement and measures. The time of realisation question also broke with the past by separating judgements on when technological realisation would be achieved and judgements on when the technology would become available as products and services (social realisation). This gives some measure of the experts' assessment of the time taken to commercialise technologies. Early applications were most frequent in ICT and industrial infrastructure. Despite the different methodologies, an effort was made to compare the nature of the 100 topics receiving the highest importance rating. Between the Seventh and Eighth surveys the biggest increase was in "disaster-related" issues, with over half related to earthquakes, a natural Japanese preoccupation.

Scenario Analysis [9]

In response to concerns about the consensual nature of Delphi there was a desire to have an element in the foresight exercise that highlighted subjective and normative future visions for wide areas of basic science, technology and societal impacts. It was decided to pursue this by inviting distinguished individuals who were outstanding in a research area to write a scenario on a related theme. Before engaging the writers, "progressive scenario themes" were identified. A committee used the interim results of the Study of Rapidly Developing Research Areas, the work of the Delphi analysis sub-committees and external suggestions to find draft themes. These were science and technology areas with the potential to make major social and economic contributions or bring ground-breaking knowledge 10–30 years into the future. Forty-eight themes were eventually developed from these inputs and these were divided into two rounds.

The next step was to select scenario writers. The two rounds of themes were sent to a large number of academic societies and other organisations and nominations were requested. Emerging candidates were assigned to a theme on the list and the result was sent to electors in agencies, societies, etc. These were invited to vote for one or two candidates from up to five themes in the first round (four in the second). This process resulted in 96 scenario writers being selected. The eventual result was that 85 scenarios were written for 47 progressive scenario themes. Two writers were sought for each theme though this was not achieved in every case.

Examples of themes are "regenerative medicine for a long-lived society", "reconstruction of science and technology evaluation models" and "energy conservation". Scenarios included an analysis of the current situation, the progressive element which indicates key developments and dates, and a list of actions that Japan should take in the area.

CONCLUSIONS

As noted at the beginning of this chapter, the history of Japanese foresight has seen remarkable continuity in its setting and in its adherence to the Delphi methodology. This must be interpreted as a success indicator – no administration would have maintained the activity unless it was demonstrably useful. Of particular significance was the early period through the 1970s and 1980s when Japan stood almost alone in maintaining this kind of effort. Mirroring the story of the Japanese economy, the Delphi methodology was of course imported but then adapted and improved to suit local circumstances. Certain features of Japanese society were propitious, notably the willingness of experts to make serious time and effort available for the collective good, demonstrated for example in the high response rates always achieved.

When Europeans began to take an interest in Japanese approaches in the early 1990s through studies such as that of Irvine and Martin (1984) and the later link with Germany, Japan not only influenced Western practice but also began an interactive relationship which saw concepts and details of technique flowing in both directions. Again, the Japanese concept of acquiring knowledge is successful. At a point in time, when foresight in Japan seemed to get "boring" and repetitive, they invited foreigners to bring in fresh ideas (Cuhls and Kuwahara, 1994). The same occurred before the 8th Foresight started: NISTEP organised a conference to learn from others, especially from the German Futur process. Thus, they avoided reinventing the wheel and a lot of mistakes were not made – by adapting some elements of a foreign concept to their own situation but not by just copying. Currently, a mutual learning round with Asian countries is being prepared to extend this learning circle to Asia.

The most recent phase of foresight in Japan has been characterised by two major linked changes – one being a much closer engagement with policy-making and the other an expansion of the toolbox and the broader concept so as to able to deliver on this. With a more stable science policy framework for the next five years it will be interesting to see the next steps in Japanese foresight.

NOTES

1　The Science and Technology Basic Law (1995), Law No. 130, unofficial translation, http//:www8.cao.go.jp/cstp/english/law.html, accessed 31/1/2006.
2　Kondô, Satoru: R&D Senryaku Ritsuan no tame no "Gijutsu Yosoku" Katsuyô Gaidobukku (Handbuch zu Technikvorausschau-Aktivitäten für die strategische FuE-Planung), Tôkyô, 1993.
3　See Cuhls and Kuwahara (1994).
4　Sôgô Kagakugijutsu Kaigi (Council of Science and Technology, CSTP) (2005): Heisei 18nendo no kagakugijutsu ni kansuru yosan, jinsai nado no shigen haibun no hôshin (Directions for the resource allocation of the budget, human resources etc. concerning science and technology from the year 2006 on), draft version, 16/6/2005, long version 18/10/2005, www8.cstp.go.jp, accessed 31/1/2006.
5　Kondo, Masayuki (2005): Comprehensive Review of the Achievements of Japanese Science and Technology Basic Plans, Presentation at the PRIME Conference 'Evaluation – connecting Research and Innovation', 17/11/2005, Manchester.
6　See NISTEP (2005a).
7　See NISTEP (2005b).
8　See NISTEP (2005c).
9　See NISTEP (2005d).

REFERENCES

Cuhls, K. (1998), *Technikvorausschau in Japan. Ein Rückblick auf 30 Jahre Delphi-Expertenbefragungen*, Heidelberg: Physica Verlag (Dissertation, ursprünglicher).
Cuhls, K. and Kuwahara, T. (1994), *Outlook for Japanese and German Future Technology, Comparing Technology Forecast Surveys*, Heidelberg: Physica-Verlag.
Cuhls, K., Breiner, S. and Grupp, H. (1995), *Delphi-Bericht 1995 zur Entwicklung von Wissenschaft und Technik – Mini-Delphi* , Karlsruhe: Druck des BMBF, Bonn 1996.
Cuhls, K., Blind, K. and Grupp, H. (ed.) (1998), *Delphi '98 Umfrage. Zukunft nachgefragt, Studie zur globalen Entwicklung von Wissenschaft und Technik*, Karlsruhe.

Cuhls, K., Blind, K. and Grupp, H. (2002), 'Innovations for our Future. Delphi '98: New Foresight on Science and Technology', *Technology, Innovation and Policy*, Series of the Fraunhofer Institute for Systems and Innovation Research ISI, **13**, Heidelberg: Physica.

Irvine, J. and Martin, B.R. (1984), *Foresight in Science*, London: Pinter Publishers.

Kuwahara, T. (1999), 'Technology Forecasting Activities in Japan – Hindsight on 30 years of Delphi expert surveys', *Technological Forecasting and Social Change*, **60**(10), pp. 5–14.

Lee, C.H. (2006), 'Institutional Reform in Japan and Korea. Why the Difference?', in Blomström, M. and La Croix, S. (eds), *Institutional Change in Japan*, London and New York: Routledge.

MEXT (2003), *White Paper on Science and Technology 2003*, Tokyo, http://wwwwp.mext.go.jp/hakusyo/book/hpag200301/hpag200301_2_009.html, accessed 10/3/2006.

National Institute of Science and Technology Policy (NISTEP) (2001), *The Seventh Technology Foresight – Future Technology in Japan toward the Year 2030*, Ministry of Education, Culture, Sports, Science and Technology (MEXT), NISTEP Report 71, Tokyo: NISTEP.

National Institute of Science and Technology Policy (NISTEP) (2005a), *Kyûsoku ni hattenshitsutsu aru kenkyû ryûiki chôsa (The 8th Science and Technology Foresight Survey – Study on Rapidly-developing Research Area)*, Ministry of Education, Culture, Sports, Science and Technology (MEXT), Report 95, Tokyo: NISTEP.

National Institute of Science and Technology Policy (NISTEP) (2005b), *Kagakugijutsu no chûchôki hatten ni kakawaru fukanteki yosoku chôsa (The 8th Science and Technology Foresight Survey, Needs Survey)*, Ministry of Education, Culture, Sports, Science and Technology (MEXT), Report 94, Tokyo: NISTEP.

National Institute of Science and Technology Policy (NISTEP) (2005c), *Kagakugijutsu no chûchôki hatten ni kakawaru fukanteki yosoku chôsa (The 8th Science and Technology Foresight Survey – Future Science and Technology in Japan, Delphi Report)*, Ministry of Education, Culture, Sports, Science and Technology (MEXT), Report 97, Tokyo: NISTEP.

National Institute of Science and Technology Policy (NISTEP) (2005d), *Kagakugijutsu no chûchôki hatten ni kakawaru fukanteki yosoku chôsa (The 8th Science and Technology Foresight Survey – Scenarios)*, Ministry of Education, Culture, Sports, Science and Technology (MEXT), Report 96, Tokyo: NISTEP.

9. Foresight in Nordic Countries[1]

Annele Eerola and Birte Holst Jørgensen

INTRODUCTION

This chapter offers a summary and analyses of technology foresight activities in the Nordic countries. It investigates the potential of joint Nordic foresight efforts in strengthening an integrated Nordic knowledge region. The chapter focuses on the foresight activities of, and scope for cooperation within, five North-European countries: Denmark, Finland, Iceland, Norway and Sweden. By describing and analysing the Nordic joint efforts in technology foresight, together with national and regional foresight efforts in some Nordic countries, the chapter contributes to the understanding of the potentials and challenges of cross-border foresight exercises in their real-world political and economic context. Some ideas for theoretical framing of cross-border foresight exercises, with implications to process design, are also presented.

The Nordic countries have a long tradition of cooperation within research, education and innovation. Although the Nordic Council of Ministers aims at "developing the Nordic region as the most attractive region in terms of education, research and industry", it is considered that there are still insufficient effective mechanisms capable of embracing the various activities and initiatives at Nordic level. Nordic-level technology foresight projects and effective exchange of experiences between the foresight actors and users of foresight knowledge are, however, considered as promising tools in this respect.

The chapter is based on a feasibility study that the authors conducted for the Nordic Industrial Fund (Eerola and Joergensen, 2002) and follows subsequent Nordic-level developments. In the feasibility study the prospects of joint Nordic efforts in technology foresight were examined. As a response to the recommendations of the study, the Nordic Innovation Centre (NICe)[2] has now funded three Nordic level foresight projects: Nordic Hydrogen Energy Foresight,[3] Nordic Biomedical Sensor Foresight[4] and Nordic ICT Foresight.[5] The first Nordic joint effort was co-financed by Nordic Energy

Research in 2003–2005. The two subsequent NICe projects were started in mid-2005 and will be completed in 2006/2007. In addition, NICe has funded a Nordic Foresight Forum facilitating the exchange of experiences between the Nordic foresight actors and the users of foresight knowledge.[6]

By increasing Nordic competence and shared knowledge in technology foresight, and by creating critical mass behind specific long-term proposals, Nordic foresight exercises have made it possible to allocate Nordic-level research funds more effectively. In this way, Nordic joint efforts have also some potential to move R&D and innovation policies into new directions at national and EU levels. Furthermore, the Nordic foresight cooperation gives a good ground for cooperation with some nearby areas such as Baltic and Arctic regions, and other Northern regions in the world with similar types of climate conditions. The transnational foresight approach, in this case exercises which cluster together small countries in a benchmarking mutual learning mode, can prove highly effective, if well-designed and implemented using appropriately-adapted foresight approaches. In such cases, the transnational framework can act as an important policy lever, in redefining policy-making structures and networks and opening up the space for creative, forward-looking approaches to innovation and governance.

POLITICAL AND ECONOMIC CONTEXT

The Nordic countries have in total 24.7 million inhabitants.[7] The average GDP per inhabitant was €27,700 in 2004, demonstrating considerable wealth in the region. The Nordic countries have also eagerly invested in research and development: the average R&D/GDP of the five countries was 2.95 per cent in 2003 (led by Sweden and Finland with 4.3 per cent and 3.4 per cent respectively) (OECD, 2004; Nordic Council of Ministers, 2005).

Nordic cooperation rests on a long and shared history, which for centuries has influenced the political, economic and cultural ties among the Nordic countries. These ties foster shared values – values that are inherent in the Nordic welfare states – with their stable and well-functioning democratic institutions, highly developed economic sectors and safe communities. According to a newly published report on the Nordic region as a global winner region, these Nordic values are associated with some fundamental conditions such as social systems, and similar levels of self-realisation in terms of life style. Shared values include, for example, equality, trust, proximity to power, inclusion, flexibility, respect for nature, the protestant work ethic and aesthetics (House of Monday Morning and Nordic Council of Ministers, 2005).

Despite their high taxation, large public sectors and comprehensive welfare systems – and perhaps also partly due to these conditions – the Nordic countries are among the most competitive countries in the world and come top in a large number of league tables of the world's most competitive nations (see e.g. Table 9.1). The five countries, however, differ in their industrial structure and to some extent also language,[8] which complicate the process of finding common priority areas in R&D and innovation policies.

Table 9.1   Position of Nordic countries in competitiveness measurements

Competitiveness	Technology	Creativity
1. **Finland**	1. Singapore	1. **Finland**
2. USA	2. **Iceland**	2. **Norway**
3. **Sweden**	3. **Finland**	3. **Sweden**
4. **Denmark**	4. **Denmark**	4. **Denmark**
5. Taiwan	5. USA	5. Holland
6. Singapore	6. **Sweden**	6. Switzerland
7. Switzerland	7. Hong Kong	7. Germany
8. **Iceland**	8. Japan	8. France
9. **Norway**	9. Switzerland	9. UK
10. Australia	10. Canada	10. Luxembourg

Source: World Competitiveness Index 2004 and Networked Readiness Index, 2005, cited by
          House of Monday Morning and Nordic Council of Ministers, 2005

In an increasingly globalised world, in which economic and political ties span the globe, a key question is what in particular distinguishes Nordic cooperation from other international cooperations in which the Nordic countries participate. How do spatial proximity and cultural and historical ties contribute to the new learning economy?

The future global economy may also represent a historic challenge and opportunity for the Nordic countries. Competition in the future will be increasingly about the ability to innovate and produce value-creating solutions that are difficult to imitate. This requires a strong innovation culture and the ability to use one's own unique core skills. Recent international assessments suggest that the Nordic welfare model holds partially untapped competitive potential and that this represents a culturally based strength (see e.g. World Economic Forum, 2004, 2005; Campbell and Pedersen, 2005).

The Nordic region has consolidated its participation in European co-operation. Denmark, Finland and Sweden are members of the European Union while Iceland and Norway are associated members. Since 2001, the Nordic countries have been seeking to strengthen cooperation between the Nordic EU members and at the same time involve the two Nordic associated countries in consultation and knowledge-sharing in order to signal a more proactive joint EU-line. The aim is to maintain and develop Nordic influence on European cooperation at a time of EU expansion from 15 to 27 member states (Nordic Council of Ministers, 2003a, p. 23). A joint Nordic regional EU approach is not meant to be a closed Nordic bloc policy, but rather to bring Nordic cooperation closer to the EU agenda and to organise supplementary cooperation structures that fit in with EU cooperation (ibid, p. 24).

The Nordic countries have built their research cooperation on strong national priorities as well as the EU research system. According to a recent White Paper published by the Nordic Council of Ministers, the weaknesses are the lack of sufficient critical mass, visibility and attractiveness, and groundbreaking innovations (2003b, p. 5). The White Paper recommends the establishment of NORIA (Nordic Research and Innovation Area) with well-established and leading networks and partnerships, high R&D investments, high mobility, and improved international higher education systems and structures (ibid, p. 5). A more detailed description of the history and context of Nordic cooperation in the areas of research, education and innovation can be found in Eerola and Joergensen (2002).

FORESIGHT ACTIVITIES IN THE NORDIC COUNTRIES

Governmental institutions, academia and private institutions in the five Nordic countries are increasingly using foresight approaches as tools for creating commitment and shared understandings on longer-term R&D and societal developments. The emphasis, methodologies and focus areas of the foresight exercises, however, vary by country and the ultimate purpose of the foresight effort. This section briefly describes the recent developments of activities in five Nordic countries: Denmark, Finland, Sweden, Norway and Iceland. Joint efforts – in the form of Nordic foresight cooperation – are discussed in the subsequent sections.

Foresight in Denmark

In Denmark, until very recently, technology foresight activities were not considered an appropriate tool for strategic decision-making, connectivity and efficiency, and awareness-raising about future technologies (Dannemand Andersen and Borup, 2006). Instead, forward-thinking activities were performed in various sectors, in particular in the energy sector, and technology assessments were performed with an eye to the problems and challenges within controversial technologies – for example, biotechnology and surveillance. However, in recent years, more and more technology foresight activities have been performed by different actors – government, technology boards and centres, associations and regions. These activities have had diverse objectives, used a range of methods and tools, and generated a variety of results. The Danish activities can be clustered in three groups (Andersen and Joergensen, 2004):

- Top-down government-initiated technology foresight or RTD strategies – central actors include the Ministry of Science, Technology and Innovation and the Danish Energy Agency;
- Intermediate non-governmental initiated technology assessment and foresight – central actors include the Danish Board of Technology and the Danish Society of Engineers (IDA);
- Bottom-up sectorial or regional initiated foresight activities – examples include the County of Ringkoebing.

Dealing with each of these three groups of activities in turn, the *raison d'être* of Danish technology activities depends very much on who is the owner of the foresight. Among the top-down government actors, it is interesting to observe that it is not the Ministry of Science, Technology and Innovation that use this new tool for policy-making, being in the field of prioritisation, better framework conditions, and bringing science and technology closer to society. In 2000, the Danish government earmarked 24 million DKK (€3.2M) to a programme for a national technology foresight project for the period 2001–2004. After the change of government in 2001 this was cut back to 18 million DKK (€2.4M) The policy initiative was started in the Ministry of Trade and Industry, but with the new Government's reshuffle of responsibilities it was moved to the Ministry of Science, Technology and Innovation. Exercises have been performed in the field of Bio- and Healthcare, Green Technologies, Hygiene, Nanotechnologies, Pervasive Computing, Ageing Society 2030, ICT – from Farm to Table, Cognition and Robotics, and Mobile and Wireless.[9]

The experience so far is that the use or implementation of the results takes quite some time – several years in many cases. In some cases there has been a direct implementation of the results. The first round of Green Technology Foresight was followed up with three more targeted foresights (Green Technology Foresight about environmentally friendly products and materials, Sustainable agriculture, and ICT from farm to table). The Nanotechnology Foresight contributed to the Government's political decisions on the focus areas of the newly established High Technology Foundation. Furthermore, at least one university has used some of the recommendations in its contract with the Ministry (Dannemand Andersen *et al.*, 2005). The Council for

Technology and Innovation has launched tenders for a number of High Technology Networks within the areas of the finalised foresight projects. Other impacts of the foresight exercises are of a more indirect nature: creating debate, raising awareness, etc. Based on recommendations from foresight exercises the Danish Board of Technology has initiated political debates in projects such as Pervasive Healthcare in the Danish Healthcare Services and Toxicology and Nanotechnology.

Among the governmental foresight-like activities the Globalisation Council must also be mentioned. The council was set up in the spring of 2005 with the Prime Minister as chairman and with the task of advising the Government on a strategy for Denmark in the global economy. Over a year the Council had 14 meetings, received input from 48 international and Danish speakers and held discussions with 111 representatives of organisations and individuals. All ministries contributed with a range of background reports. The work concluded in a main report with the title: "Progress, Innovation and Cohesion Strategy for Denmark in the Global Economy". Among the many recommendations, under the title "Better basis for prioritising", was one that the basis of political prioritisation of strategic research should be strengthened. Furthermore, "a broad-based survey should be regularly carried out to identify the research needs that society and business developments create as well as the capabilities of Danish research institutions to meet these needs". With the reorganising of the Ministry of Science, Technology and Innovation in 2006 foresight-like activities were moved from the Ministry itself to the reinforced Danish Agency for Science, Technology and Innovation (DASTI). DASTI has among its responsibilities "dialogue on priorities in research and technology initiatives". Also, the Danish Energy Authority has conducted strategy processes with all key stakeholders in selected fields of new energy technologies so that research and innovation funds from different sources may be aligned and optimised.

Among the intermediate group of actors highlighted above, the Danish Board of Technology is an internationally highly-respected institution carrying out technology assessments of controversial technologies in open, participatory processes. It is accountable to the Danish Parliament on these activities and hence provides decision support to the political process. The board do not carry out Technology Foresight in a narrow meaning, but foresight elements are very much present in projects such as "RFID – Risks and Opportunities" and "The Future Danish Energy System". Participatory elements and the close interaction with members of parliament are central traits of the Board's activities.

The Danish Society of Engineers (IDA) has given priority to using technology foresight to mobilise its own members in a discussion of future technologies. Simultaneously they also proactively use results from these foresight processes to influence political discussions and agendas to focus on technology opportunities in strengthening Danish competitiveness in the knowledge society. Especially in the area of energy, IDA has carried out a very large foresight effort. IDA named 2006 as its energy year with the headline "Energy for the Future". Along with political pressure from other groupings, the efforts of IDA probably contributed to a reformulation of the Government's energy policy and its expenditures on energy-related R&D. The process consisted of several rounds of foresight projects with elements such as: Brainstorms, Workshops, Internet-based dialogue with members, Scenarios, and Roadmap seminars on six areas of energy technology.

Finally, the regional hydrogen foresight in the County of Ringkoebing, was tied to the overall aim of contributing to the coherence and efficiency of the regional innovation system and at enhancing the anticipatory intelligence of the developers, users and producers of science and technology. This also included a prioritisation of the application areas upon which the region should focus. The results of the foresight have been actively promoted to influence national decision-making on taking into account regional clusters and innovation centres. As an outcome of the process, a Hydrogen Innovation and Research Centre has been established by regional and national stakeholders from the innovation system.[10] On a regional and municipal level foresight activities are likely to take place over the years to come (See Table 9.2).

Foresight in Finland

In Finland too, the foresight landscape is quite diverse and fragmented. On the governmental side, the Ministry of Trade and Industry initiated a pre-study "On the Road to Technology Vision", which it coordinated together with Tekes (The National Technology Agency) in 1996–1997 (see KTM, 1997). The pre-study was not, however, followed by a more comprehensive national foresight project. Instead, the Ministry of Trade and Industry commissioned an assessment report on the present state and development needs of Finnish foresight activities. The report recommended a clear institutional frame for foresight work and better coordination of the diverse activities that contribute to forward thinking (Salo, 2001). As a result, the Ministry launched a four-year foresight development and coordination project in 2001, though with a relatively modest budget. Within the framework of the project a number of minor-scale foresight exercises/pre-studies were carried out. A small amount of the budget was also allocated to piloting a Finnish Foresight Forum that organises future-oriented dialogues around selected topics on a "voluntary basis" (the costs of the organising consultant and the travelling cost of the experts participating in the dialogues are covered by the Forum budget). The Finnish Foresight Forum also maintains a website[11] that informs Finnish actors of the various foresight organisations and activities, and of relevant news in the field.

In addition, a number of foresight studies have been carried out under the auspices of the Ministry of Labour and the Ministry of Education, often co-financed by European Structural Funds. The Finnish National Board of Education also administrates a foresight data base and Internet-based foresight knowledge service (ENSTI) with the focus on future education and labour demand. The broad-scope future outlooks and the more focused future-oriented technology assessments of the Committee for the Future of the Finnish Parliament play, in turn, an important role in raising the awareness and the level of knowledge of the Members of the Parliament.[12] The Committee for the Future also prepares the Parliament's response to Government's Report on the Future during each electoral period.

Furthermore, public sector organisations, such as The Academy of Finland, Tekes, SITRA (Finnish National Fund for Research and Development), and VTT (Technical Research Centre of Finland), have been active in trying to make the Finnish innovation system more effective and transparent, in part by learning from foresight experiences of other countries (Hjelt *et al.*, 2001; Eerola and Väyrynen, 2002). In the recent FinnSight2015 exercise,[13] the Academy of Finland and Tekes joined their efforts in supporting the priority-setting of basic and applied research. In particular, the change factors that will have impact on Finnish business, industry and society were examined, with the aim of identifying the most important challenges for innovation and research activity. The focus of the panel-based exercise was on social and global issues.[14] The results were published in 2006.[15]

In parallel with the FinnSight2015, SITRA started its own foresight exercise in the form of a national foresight network in late 2005. The aim of this foresight work was – instead of concentrating on in-depth analyses – to recognise the changing trends and crucial future challenges to which decision-makers should already pay serious attention. The five topic areas included welfare and everyday living, work life, public sector, multiculturalism, and environmental technology. The topic areas complemented the thematic work of the Finnish foresight forum. The first results were published in 2006.[16] The continuation of SITRA's foresight network is still under consideration at the time of writing.

The Employment and Economic Development Centres (TE-Centres) have, in turn, played an active role in regional foresight, together with research centres such as the Lahti Centre of Helsinki University of Technology, Lappeenranta Technical University and Finland Futures Research Centre. In addition to regional foresight exercises, various types of strategy workshops have been organised (partly also for educational purposes). Considerable efforts have been put to relevant data gathering and societal embedding of innovations and new technologies (Kivisaari *et al.*, 2004). Furthermore, there are various foresight studies focusing on some specific topics. These foresight studies are typically carried out in collaborative work undertaken by sectoral research institutes, academic researchers and private consultants, supported by the National Technology Agency. On the educational side, a Graduate School in Future Business Competencies (TULIO) has also been established in 2005/2006. The network programme is coordinated by the Finland Futures Research Centre and Finland Futures Academy.

In the private sector, technological developments have been anticipated by companies and industrial federations as part of their strategy processes. Industry is also increasingly interested in "technology roadmapping" as a tool for managing technological change. For instance, the Finnish Technology Industries federation has actively promoted this type of work. Finnish companies and VTT have also been actively involved in technology roadmapping processes carried out by international industrial federations and international consortia. A number of sectoral and regional foresight studies have also been carried out by VTT, Finland Futures Research Centre, other university research groups and private consultants. For example, studies that combine regional cluster strategies with regionally applied technology foresight approaches have been carried out (see e.g. Ahlqvist *et al.*, 2007).

The Finnish Association of Graduate Engineers has, in turn, actively promoted a technology barometer approach with the aim of alerting attention to the views of their members, industrial decision-makers and younger generations on issues related to the functioning of the innovation system (see TEK, 2004/2005). Furthermore, the Confederation of Finnish Industries is active in foresight processes focusing on future jobs, skills and competencies. The activities include a repetitive Future Radar, development of an Education Intelligence System (Ståhle and Ståhle, 2006) and foresight on the future of services (See Table 9.3).

Foresight in Sweden

Sweden, in turn, is the only Nordic country in which a wide-ranging national-level foresight approach, similar to those in countries like the UK and Germany, has been adopted. The first Swedish Technology Foresight was carried out by eight panels in 1999–2000, followed by a series of more focused regional dissemination events. The second round of the Swedish Technology Foresight in 2003–2004 took, however, a somewhat different approach. It consisted of five sub-projects that analysed the foresight studies of other countries, updated the outcomes of the previous round by involving even younger researchers and entrepreneurs, examined the changes in the global environment, assessed the potential breakthrough areas in technology and knowledge, and compiled a synthesis of the results of the other sub-projects (STF, 2004a, 2004b).[17] The idea was to bring together people with varying backgrounds to discuss the driving forces, preconditions and possible strategies for the Swedish innovation landscape (including not just technology, economy and education but also society at large).

The first Swedish Technology Foresight was initiated by four organisations: the Royal Swedish Academy of Engineering Sciences (IVA), the Swedish National Board for Industrial and Technical Development (NUTEK), the Swedish Foundation for Strategic Research, and the Federation of Swedish Industries. It was co-financed by the Swedish Foundation for Strategic Research, NUTEK (in its old form), and the Swedish Government Offices. The organisations behind the second round were partly the same: IVA, the Swedish Business Development Agency (NUTEK, in its reorganised form), the Swedish Governmental Agency for Innovation Systems (VINNOVA, founded in 2001), the Swedish Research Council, the Knowledge Foundation, the Swedish Trade Union Confederation, and the Confederation of Swedish Enterprise. The study was conducted in close cooperation with the Swedish Government, companies, public agencies and other interested parties. In addition to the national-level foresight exercises, some sector-specific and territorial foresight projects have been carried out in Sweden (including the energy sector foresight by IVA and the territorial foresight exercise in western Sweden).

IVA has had the leading role in Swedish foresight activities since the mid 1990s, inspiring and/or initiating most of the prominent foresight exercises. In addition to the two broad-scope national-level foresight projects, an Energy Foresight was completed by IVA in early 2004. The Energy Agency (STEM) played an important role in its funding. The Energy Foresight affected public energy discussions and also facilitated more specific governmental proposals on energy research, as well as proposals in some other fields. The most recent foresight exercises by IVA have focused on environmental issues and crises management. Another consistent foresight actor is FOI (Swedish Defence Agency), although the old Environmental Strategies Research Group (ESRG) of FOI was moved to the Royal University of Technology. The FOI/ESRG grouping has provided methodology experts for Nordic foresight projects, as well as European and national foresight projects, especially in the energy, environment and defence fields. The Foresight Laboratory at Örebro University, in turn, mainly focuses on regional foresight. In addition, there are individual consultants and academics who at times operate under the foresight label. Their role in the overall picture is, however, relatively marginal. On the methodological side, there seems to be a gradual change towards softer, smoother and more autonomous processes when the constitution and design of Swedish foresight activities is considered (See Table 9.4).

Foresight in Norway

In Norway, foresight approaches have been actively implemented by the Research Council of Norway (RCN). Following an evaluation of the Council's activities in 2000 (Kuhlman and Arnold, 2001), a foresight project group was established to design and implement foresight exercises as a strategic input to the Council's funding activities. The following foresight exercises have been conducted: Advanced Materials Norway 2020, Biotech Norway 2020, Energy Norway 2020+, Fish-farming 2020, UTSIKT (ICT) and Roadmap 21.[18] The SURPRISE project at The Norwegian Computing Centre and Oslo Teknopol in 2002–2003 was the first project that addressed both regional innovation and foresight methodology. Some years ago (1998–2001) the Norwegian Government conducted a comprehensive scenario-project, Norway2030, which aimed at identifying roadmaps for the Future of the Public Sector.

In 2006 the Governmental Innovation Policy Agency, Innovation Norway, established a project called INfuture, which aimed at developing and adapting foresight methodology to its strategic agenda. In spring 2006 The Royal Ministry of Education & Research established a new Unit within the Department for Universities and University Colleges addressing Research and Innovation Policy issues and the use and development of foresight methodology. A mapping study that was published in 2003 documented more than 80 foresight projects in both public and private sectors since 1996. During the last three years, foresight has had its breakthrough in several different areas and sectors in Norway. Therefore, the number of foresight projects has increased substantially during this period (See Table 9.5).

Foresight in Iceland

In Iceland, too, foresight approaches have been actively promoted by private organisations such as Icelandic New Energy and the Icelandic Confederation of Industries. Recently, RANNIS (The Icelandic Centre for Research) and Iceland's Science and Technology Policy Council have expressed their interest in developing the national foresight activities of the country and they are also proceeding in this direction. Recently, the Science and Technology Policy Council has announced that foresight approaches will be used for future strategy formulation and programme preparation for the next four years. The first attempt to work on a national foresight project got underway in 2006. A foresight project has also been launched in conjunction with the evaluation of the Icelandic health and medicine field. This will be performed by RANNIS and a Belgian consulting firm (See Table 9.6).

Table 9.2   Overview of foresight actors and activities in Denmark

Public Sector	*Ministry of Science, Technology and Innovation*: 4-year pilot project with sectorial TF 2001–2004
	Danish Agency for Science, Technology and Innovation (DASTI): foresight for supporting RTD priority-setting from 2006, after reorganising the Ministry of STI
	Globalisation Council : foresight elements included in innovation and globalisation strategy advice from 2005
	Danish Energy Authority: Strategic energy analysis and action plans
	Danish Board of Technology: foresight elements included in technology assessment activities, embedded in broad consultations
	County of Ringkoebing: regional hydrogen foresight
Universities and Research Institutes	*Risoe*: Research and development in the field of technology foresight, sectoral and topic-specific TF projects at National, Nordic and European level
	DTU: Foresight network and advisory tasks
	Danish Technology Institute (DTI): involvement in foresight activities focusing on business development and information technology
Private Sector	*Danish Society of Engineers*: TF processes to facilitate discussion of future technologies among the members and to influence political discussions and agendas (main focus on energy)
	Copenhagen Institute for Futures Studies: Future seminars and consultancy

Table 9.3   Overview of foresight actors and activities in Finland

Public Sector	*Ministry of Trade and Industry*: pilot study 'On the Road to Technology Vision' (1996–97), TF needs assessment (2000–2001), TF development project (2001–2005), including a pilot of a Finnish Foresight Forum and mapping of governmental TF activities
	Ministry of Education/Finnish National Board of Education: foresight database and Internet-based service
	Tekes, Academy of Finland, and Sitra: co-funding of TF studies and outlooks, internal vision and strategy processes
	Parliament of Finland, Committee for the Future: TA studies on various topics (come close to TF studies)
	TE-Centres: series of TF seminars, regional foresight
Universities and Research Institutes	*VTT*: sectoral studies, technology roadmaps, research on TF activities and methodologies, international TF cooperation at Nordic and EU European level
	Systems Analysis Laboratorio of Helsinki University of Technology: research on TF methods and practices, expert assignments at national and EU-levels, methodological support to public/private processes
	Finland Futures Research Centre/ Turku School of Economics: broad scope and sectoral foresight studies, scenario and vision seminars, methodological R&D. Coordinating a Graduate School in Future Business Competencies TULIO, together with Finland Futures Academy
	IAMSR/Abo Academy: scenario/vision work in cooperation with companies, industrial federations and Tekes programmes
Private Sector	*Industrial Federations*: sector- and tech.- specific TF studies, roadmaps, strategy outlines, etc.
	ETLA: Cluster studies, including TF
	Finnish Association of Graduate Engineers (TEK): Round-table TF exercises, "Future Engineer" project, repetitive Technology Barometer study since 2004, etc.
	Confederation of Finnish Industries: foresight focusing on future jobs, skills and competencies (incl. a Future Radar, development of an Education Intelligence System and a Services2020 project)
	Private consultants: technology- and sector-specific TF studies, Delphi-surveys, scenario processes for companies and industrial federations, comparative studies of national level, etc.

Table 9.4   Overview of foresight actors and activities in Sweden

Public Sector	*Swedish National Board of Industrial and Technical Development (former NUTEK) and the Swedish Government Offices* co-financed the broad scope TF exercise "Teknisk framsyn" in 1998–2000 NUTEK was also one of the key-actors initiating and carrying out the TF project
	The Swedish Governmental Agency for Innovation Systems (VINNOVA) and the Swedish Business Development Agency (NUTEK, in its reorganised form) were also behind the second broad-scope TF in 2003–2004, together with the Swedish Research Council and some other actors
	Swedish Energy Agency financed the energy foresight in 2002
	Public service organisations: involvement in regional foresight exercises
Universities and Research Institutes	*Swedish Defence Agency (FOI)*: methodological expertise for national, Nordic and European foresight projects, esp. in energy, environment and defence fields, together with the Env. Strategies Research Group of the Royal Institute of Technology
	Swedish Institute of Futures Studies: demographic foresight studies
	Foresight Laboratory of the Örebro University: regional foresight
Private Sector	*The Royal Swedish Academy of Sciences (IVA)*: one of the key actors initiating and carrying out the broad-scope TF exercises in 1998–2000 and 2003–2004. In addition, sector- and topic-specific foresight projects (incl. energy and environment foresight projects)
	Federation of Swedish Industries: key actor in the broad-scope TF exercises in 1998–2000 and 2003–2004, industry representatives participated in the TF process
	Companies and industrial associations: involvement in national and regional foresight exercises

Table 9.5   Overview of foresight actors and activities in Iceland

Public Sector	*Ministry of Industry and Commerce*: short-term development plans within innovation, regional development, etc.
	Science and Technology Policy Council: interest in developing national foresight activities for future strategy formulation and preparation for the next four years. A first attempt to work on a national foresight project started in 2006
	IceTec: pragmatic prioritisation
Universities and Research Institutes	*University of Iceland/spin-off companies*: future hydrogen society
	RANNIS (The Icelandic Centre for Research) and S&T Policy Council: interested in developing the national TF activities
Private Sector	*Icelandic Confederation of Industries*: Vision Seminars and Forums of Technology
	Icelandic New Energy: TF methodologies in energy field

Table 9.6   Overview of foresight actors and activities in Norway

Public Sector	*Ministry of Labour/Government adm.*: Norway 2030 (roadmaps for the Future of Public Sector)
	Research Council of Norway: short-term R&D prioritisation. Strong or weak arena for R&D? A number of exercises on different topics (incl. materials, biotech, energy, fish farming, ICT + roadmap21)
	Norwegian Defence Research: strategic analysis and foresights
	Board of Technology: TA challenges and possibilities
	Innovation Norway (Gov. Innovation Policy Agency): foresight methodology applied to its strategic agenda in 2006 (INfuture)
	The Royal Ministry of Education & Research, Dept. for Universities and University Colleges: addressing research and innovation policy issues with the help of TF methodology
Universities and Research Institutes	*SINTEF group*: sectorial projects (national, Nordic and EU level)
	STEP group: some elements of TF in future-oriented activities
	Stavanger University: energy foresight, EU expert assignments
	Norwegian School of Management, BI: expertise in business devel.
	Norwegian Computing Centre and Oslo Teknopol: regional innovation and TF methodology (SURPRISE project in 2002–2003)
Private Sector	*ECON center*: Horisont 21

NORDIC-LEVEL JOINT EFFORTS

Feasibility Study on the Prospects of Nordic Foresight Cooperation

In the beginning of the new millennium there was not a clear picture of the differences, and similar and complementary features, of the foresight approaches taken in the Nordic countries. Neither was there a clear picture of the potential of intensifying the efforts in strengthening the Nordic knowledge region with the help of Nordic foresight cooperation. A feasibility study on the prospects of Nordic foresight cooperation was thus conducted in 2001–2002 (Eerola and Joergensen, 2002). The study was funded by the Centre for Innovation and Commercial Development of the Nordic Industrial Fund, Risoe National Laboratory in Denmark and VTT Technical Research Centre of Finland. The final report of the feasibility study on the prospects of Nordic foresight cooperation recommended (ibid.):

1. The establishment of a Nordic forum for technology foresight practitioners and researchers;
2. The creation of a common follow-up system – scanning and reporting – for relevant international technology foresight exercises (including monitoring of selected trends and early signals);
3. The realisation of technology foresight exercises that involve participants from various Nordic countries (e.g. focusing on specific technologies or problem areas or in selected cross-border regions).

The regional cross-border perspective was also highlighted in the feasibility study as a promising field for foresight collaboration on a more limited scale (the Øresund Region, the North Calotte Region, and the ScanBalt Region were given as examples). In particular, it was suggested that instead of focusing upon specific technologies, cross-border foresight exercises could aim at developing the framework conditions, e.g. by strengthening possible knowledge synergies between universities and public and private institutions. Cross-border foresight exercises were also seen as tools for developing competitive advantage for local companies and attracting foreign investments into the region.

Nordic Foresight Projects and the Nordic Foresight Forum

The Nordic Innovation Centre (NICe) paid serious attention to the recommendations of the feasibility study, and responded by funding three Nordic level foresight projects by 2006: Nordic Hydrogen Energy Foresight,[19] Nordic Biomedical Sensor Foresight and Nordic ICT Foresight.

The Nordic Hydrogen Energy Foresight project was carried out in 2003–2005 as a pilot project in the area of Nordic foresight cooperation. The subsequent NICe projects – Nordic Biomedical Sensor Foresight and Nordic ICT Foresight – started in mid-2005 and will be completed in 2006/2007. In addition, NICe has launched a Nordic Foresight Forum that aims at facilitating the exchange of experiences between the Nordic foresight actors, and also between the producers and users of foresight knowledge.[20] A pilot phase of this forum began running in 2005–2006. A common scanning and reporting system for international technology foresight exercises has, however, not been realised at Nordic level so far. The joint efforts by 2006 are here briefly described, followed by a discussion of the methodological and organisational aspects in the next section.[21]

The Nordic Hydrogen Energy Foresight
This project was carried out in 2003–2005, co-funded by the Nordic Innovation Centre, the Nordic Energy Research programme, and 16 Nordic partner organisations from the five Nordic countries. It had a total budget of 730,000€. The project examined the prospects of hydrogen energy technologies and their implications in the Nordic context. The entire value chain surrounding hydrogen energy technologies – production, distribution, storage, and utilisation – and the required infrastructure were covered. The specific focus area was motivated by widespread expression of interests by Nordic stakeholders as well as international activities that pointed to hydrogen energy as a potential new technology worth considering while there was not yet a clear picture of its potential and implications in the Nordic context.

Interaction between research, industry and government, and using a combination of judgmental and formal procedures were essential features of the Nordic H2 Energy Foresight as well. A series of pre-structured interactive workshops (Scenario Workshop, Vision Workshop, Technology Roadmap Workshop and Action Workshop) formed the backbone of the exercise. The interactive workshops were supported by quantitative modelling and calculations (systems analysis and assessment of technical developments). Document analysis and expert interviews were among other important ingredients (Eerola *et al.*, 2004, 2005). All project reports and conference papers produced during the project are available online.[22] The website also formed an important means for external and internal communication during the project.

The Nordic Biomedical Sensor Foresight
This is examining the implications of the applications of biomedical sensors on cost and quality of health care 10–15 years into the future. The important

questions of this foresight exercise concern the type of applications to be developed and introduced to the market, the issues that prevent a full-scale adoption of these applications, and overall solutions that would have the greatest potential. The focus of the project is on three important aspects of health care: home-based care, emergency medicine and chemical hazards. The project has rather ambitious objectives in spite of a relatively small budget. For this reason, the project is focusing upon a small number of crucial tasks in order to realise its goals, i.e. mobilising key players throughout the value chain within the Nordic biomedical sensor arena, and connecting the project work to similar ongoing European and international activity. Facilitation of high-quality workshops and work processes, and dissemination of the project results to decision-makers within all the important sectors (government, health care, research and business) are considered important in this respect. The project consortium consists of six partner organisations and a number of invited experts participating and contributing to the project workshops. The project began in June 2005 and was completed in 2006.[23]

The Nordic ICT Foresight
This was started in parallel with the Nordic Biomedical Sensor Foresight in June 2005, and has been completed by early 2007. The aim of the project is to contribute to the strategic intelligence of the Nordic knowledge region so that the full potential of information and communication technology can be exploited to increase welfare in the Nordic countries and also in other parts of the world. The primary target groups for the results of this project are the Nordic companies and research communities that develop IC technologies, applications and systems, as well as public and private actors making use of them and/or providing framework conditions for their development and implementation. The focus areas of the project are: (a) experience economy (media and communication), (b) health care, (c) production economy (industrial automation, production systems), and (d) security, with the emphasis on information security. The project design resembles that of the Nordic Hydrogen Energy Foresight, with the exception that quantitative systems analysis has been replaced by national-level workshops carrying out SWOT analyses, and discussing and further elaborating the outcomes of the Nordic workshops.[24] The project consortium consists of 22 organisations, including research institutes, private firms, associations and funding organisations in four Nordic countries (Denmark, Finland, Norway and Sweden). Cross-border communication between the Nordic countries and between the various actor groups (industry, research, policy-makers and users) is considered important to increase the mutual understanding of the dynamics of innovation and to find the appropriate ways of exploiting the potential of IC technology.[25]

The Nordic Foresight Forum (NFF)

Launched in 2005 to facilitate the exchange of experiences between Nordic foresight actors and the users of foresight knowledge.[26] It is intended to be a meeting place where Nordic foresight practitioners and the users of foresight knowledge can exchange, learn and identify "good practices" for prioritising in science and technology. In order to facilitate collaboration between the Nordic countries, NFF also seeks to identify complementary elements of Nordic innovation systems. The pilot phase in 2005–2006 included:

- Networking and exchange of experiences through Nordic workshops and meetings;
- Mapping of Nordic foresight actors and an analysis of recent Nordic foresight activities;
- Mapping of the Nordic and national-level research and innovation council systems and analysis of their needs for foresight and similar strategic intelligence;
- Identifying fields of science and technology for possible future Nordic (NICe) initiatives;
- A Nordic Foresight Conference and final reporting.

The main objectives of the Nordic Foresight Forum are to create and operate a Nordic Foresight Forum for technology foresight practitioners and researchers, and to identify "best practices" in the Nordic countries for foresight and similar methodologies for prioritising in science and technology.

Methodological and Organisational Highlights

When foresight exercises are defined as useful tools in shared knowledge creation (see earlier chapters in this volume), reasonable frameworks for examining and designing such processes can be found in organisational studies focusing on innovation processes and the dynamics of shared knowledge creation. In particular, the SECI model,[27] originally developed and applied in innovation research (see Nonaka, 1994; Nonaka and Takeuchi, 1995), has been referred to in the context of the Nordic-level foresight studies funded by NICe (see Eerola and Joergensen, 2002; Eerola *et al.*, 2004, 2005). It was also used as a theoretical reference frame in a study focusing on the development of the Finnish technology foresight and technology assessment activities (Eerola and Väyrynen, 2002). Although this model is just one of the conceptual tools in use in the Nordic countries, it is interesting to have a closer look into its essential features and applicability in the present context.

In the SECI model, shared knowledge creation is envisaged as a spiral process in which tacit and explicit knowledge, as well as the different modes of knowledge conversion – i.e. socialisation, externalisation, combination and internalisation – play a central role. "Shared knowledge creation" refers here to generation of new knowledge that people and communities can share, without necessarily agreeing about the exact meaning when applying it to specific problems and goals collectively or individually. It is thus an essentially different concept than "knowledge sharing", "consensus building" or "learning", although these concepts partly overlap. The spiral nature of shared knowledge creation also means an endless process that builds on knowledge created during the previous steps and times (although project deadlines may, in practice, give an impression that the process has a clear start and end).

Figure 9.1 shows the SECI model applied to foresight exercises and also lists typical tools and practices for each knowledge conversion mode (the tools and practices reported here come from the Nordic Hydrogen Energy Foresight project).

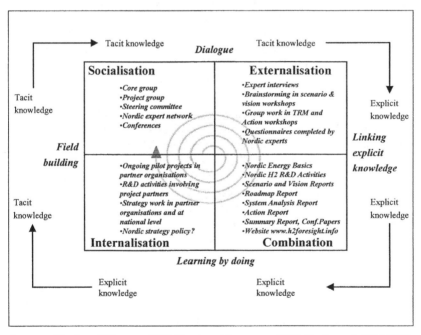

Sources:　Eerola and Väyrynen, 2002; Eerola and Joergensen, 2002; Eerola *et al*., 2004, 2005; SECI model adopted from Nonaka and Takeuchi (1995)

Figure 9.1　Foresight as a dynamic process of shared knowledge creation

According to the SECI framework, our knowledge of future technological developments is a result of a dynamic interaction process where not only facts, but also well-grounded views and opinions, should be treated as important ingredients. As pointed out by researchers already some decades ago, it is not always easy to explicitly express all relevant knowledge (see e.g. Polanyi, 1966). Still, externalisation of valuable knowledge can be facilitated with the help of purposefully assisted dialogues and formal procedures. In the Nordic H2 Energy Foresight, semi-structured expert interviews, organised brain-storming and pre-structured questionnaires were used for this purpose (see the "Externalisation block" of Figure 9.1). On the other hand, the various pieces of explicit information had to be meaningfully linked to render the resulting messages interesting. In the Nordic H2 Energy Foresight exercise the linking procedures resulted in a number of intermediate reports and working papers that included summaries of relevant Nordic facts and R&D activities, external scenarios, Nordic technology visions, maps of potential innovation areas, technology roadmaps and action recommendations (see the "Combination block" of Figure 9.1). The project partners of the Nordic H2 Energy Foresight were also simultaneously involved in various pilot projects, R&D activities and strategy work of their home organisations ("Internalisation block" in Figure 9.1). Involvement in the project and steering groups, and in the wider Nordic hydrogen network, provided, in turn, a convenient forum for reflections and cultivation of knowledge among those sharing overlapping and complementary interests and competencies ("Socialisation block" in Figure 9.1).

A series of interactive workshops formed the backbone of the Nordic H2 Energy Foresight. In parallel, a tentative model of the potential hydrogen energy system in the Nordic area was constructed. Integration of a wide range of information was needed to construct the model and to do the interesting calculations. Finding the right balance between the qualitative and quantitative elements was a key challenge in the project: avoiding model dominance while taking modelling work seriously was not easy. Including quantitative analyses was, however, perceived advantageous for consistent reasoning and credibility. Intensive back-office work, including analyses and consistent and compact reporting of workshop results, served the same purpose. Significant efforts were put in defining the role and nature of the model calculations and stressing the need for open dialogues on the other hand.[28] The SECI model provided a theoretical framework that made it easier to understand to complementary nature of the various tools. Although the multidisciplinary project group needed some time to develop a common language, both qualitative and quantitative analyses were perceived useful in the end.[29]

When the people participating in the foresight process have heterogeneous backgrounds, there are special challenges in shared knowledge creation. In the case of a Nordic joint exercise, the participants represented not just various interest groups (industry, academia, government, NGOs, etc.), but also different countries – each with its own language, cultural character, industrial structure and management style, although they also shared some Nordic heritage. This means that special attention must be paid to the organisation of the process and to appropriate use of formal tools and procedures.

Based upon the Nordic experience, the framework of shared knowledge creation not only helps to understand the dynamics of the knowledge conversion, but also provides a tool for designing well-functioning foresight exercises. Furthermore, it is helpful in communicating the learning process idea of the foresight exercise to relevant key actors. The authors also believe that it provides a useful framework for comparative analyses of various foresight processes, being a complementary perspective to the innovation system framework. By explicating the dynamics of shared knowledge creation the model also complements the "multiple perspectives" and "knowledge-people-system-organisation" frameworks of foresight management (see e.g. Linstone, 1999; Alsan and Oner, 2003).

RATIONALES AND OBJECTIVES

The objectives of various technology foresight activities in the Nordic countries have traditionally been tied to the overall aim of contributing to the coherence and efficiency of innovation systems and at enhancing the anticipatory intelligence of the developers, users and producers of science and technology. Priority-setting has been explicitly included in the objectives of S&T foresight only recently. In Norway, the evaluation of the Research Council of Norway gave rise to reflection on the question of how technology foresight might help the Council to develop into an arena for exploration and have a much more proactive role in prescribing actions. The same is the case for policy development and priority-settings within the innovation policy by Innovation Norway and The Ministry of Education and Research. With the Danish Ministry of Science, Technology and Innovation, the coupling between foresight and STI was established through moving the foresight unit to the agency that is the operative entity and holds secretariat function for research councils.

In Finland, too, priority-setting has been on the discussion agenda, although no direct links between foresight activities and decision-making processes have been established in a broader sense. There are, however, no general guidelines for utilisation of the foresight knowledge generated by the national or regional level exercises. For instance, the FinnSight2015 was intended to "lay the foundation for new internationally competitive centres of excellence and expertise, and reinforce strategy work at the Academy of Finland and Tekes".[30] The foresight exercise was to encourage the organisations to create their own future views rather than take the foresight results by others for granted. The process was expected to create useful background information for defining the scope and priority areas of the new centres of excellence, but the way of utilising the foresight information in these and other possible contexts was not defined more specifically.

The identification of complementary elements in the innovation systems and R&D programmes of the Nordic countries is a challenge to Nordic innovation policy. This is outlined in the recent recommendations for a Nordic cooperation programme in innovation policy (Nordic Council of Ministers, 2004). This is also one of the main motivations behind the exchange of experiences and initiation of joint foresight exercises.

The overall objectives of Nordic-level foresight exercises fall into four categories:

1. Increasing a shared understanding of future developments;
2. Supporting decision-making in industry, research, government and funding organisations;
3. Facilitating useful networking among experts and various interest groups;
4. Gaining critical mass at a Nordic regional level for wider international contexts.

The rationale behind the objectives is that by increasing Nordic competence and shared knowledge in technology foresight, and by creating critical mass behind specific long-term proposals, the Nordic-level research funds can be allocated more effectively. In this way, the Nordic joint efforts may also move innovation policies into new directions at national and EU level.

There is, however, some project-specific variation in the objectives of the three joint foresight exercises funded by NICe so far. All four categories of objectives were present in the Nordic H2 Energy Foresight. As stated in the final documents, the aim of the project was to provide decision support for companies and research institutes in defining their R&D priorities, and to assist governmental decision-makers in making effective framework policies. It was also considered important to highlight new business opportunities and niche areas where the Nordic countries could be front-runners in technological solutions and their application, in order to be able to successfully introduce hydrogen energy to the Nordic energy market more easily. On the other hand, providing a means for developing Nordic networks that would possess the required critical mass in wider international contexts was considered important too. The overall objective was to contribute to the strategic intelligence of the Nordic knowledge region so that technological developments can be harnessed to increase welfare in the Nordic countries and in other parts of the world (Dannemand Andersen *et al.*, 2005; Eerola *et al.*, 2005).

In the case of the Nordic Biomedical Sensor Foresight the objectives are not so clearly connected to policy and decision-making. Instead, creation of shared knowledge and useful networks are emphasised. The aim of the project is to increase understanding of the possibilities and implications of biomedical sensors in healthcare, and to provide a framework for their commercially viable exploitation in the Nordic region by enhancing a network of competencies relevant to technology and applications. By comparison, all four categories of objectives are again included in the objectives stated in the project proposal of the Nordic ICT Foresight.

The positive potential of Nordic foresight cooperation relates to geographical proximity and shared values, exchange of experience and learning, and economies of scale and scope. On the other hand, it has been claimed that foresight exercises can be better conducted at national or European level: the Nordic level is said to be too weak and without political influence, and the research and development funds available for Nordic cooperation are quite modest. Compared to national level efforts there would also be some loss of cultural and local specificity. The potential benefits of Nordic cooperation are, however, considered worth the effort, as a bottom-up contribution of smaller Member States to the wider European Research Area is considered important. In the many ERAnets still under development, the well-developed and sophisticated Nordic cooperation in research and innovation is considered a model for trans-national research and innovation cooperation (Joergensen, 2006). When such trans-national cooperation is further sustained and supported by common foresight activities, as has been the case with the Nordic Hydrogen Energy Foresight, this contributes to strengthening the internal coherence of the innovation systems as well as external positioning in the wider international context.

EFFECTS ON POLICY AND INNOVATION ACTIVITIES

Although foresight exercises may affect the participants' views and actions already during the process, the overall effects and impacts can be assessed only by systematic analyses over a longer time period. Consequently, the focus of this section is on the effects of the Nordic Hydrogen Energy foresight, which, at the time of writing, is the only Nordic-level foresight project that has been completed.[31] Even in this case it is a bit too early to detect the real influence of the joint Nordic foresight effort. But it is still possible to point to some interesting influence paths, including even some impacts that were not considered beforehand.

As reported in Figure 9.1, the project partners of the Nordic H2 Energy Foresight were simultaneously involved in various pilot projects, R&D activities and strategy work of their home organisations. According to the intermediate feedback from the participants, involvement in the foresight process contributed to these tasks by providing an opportunity to reflect on their own ideas in a relatively safe discussion forum within an atmosphere of trust. In addition, the foresight process also extended the participants' existing networks and the established interest areas. A cross-border exercise also turned out to be a good setting for overcoming "black-outs" due to national-level taboos (even sensitive issues were likely to be discussed when the taboos varied by country). The interaction among the core group, the entire project group, as well as the wider network of Nordic experts provided useful opportunities to create shared understandings and to cultivate the tacit knowledge.

Involvement of policy-makers was encouraged throughout the process, but it turned out that attracting their attention was not so easy due to timing and competing agendas. By presenting the final action recommendations as a compact three-page Policy Summary, the project group, however, proceeded in this direction towards the end of the project (the aim was to provide easily digestible ingredients to Nordic and national-level policy-making contexts). Since then, the policy-makers have also adapted some important parts of the recommendations. For instance, during the Norwegian presidency in 2006, the Nordic Council of Energy Ministers discussed how the Nordic countries can develop into a region for demonstration of new energy technologies. An international seminar focusing on this issue was organised by Nordic Energy Research,[32] in cooperation with the Norwegian Ministry of Petroleum and Energy, to coincide with the annual meeting of the Nordic Council of Ministries.

An important aspect of the Nordic H2 Energy Foresight was to bring hydrogen energy developments closer to the various Nordic actors. Not only the key players already involved in the discussions but also a wider range of stakeholders, including the citizens of the Nordic countries, as well as the wider international community were considered as relevant audiences. This implied that the outcomes should be available and comprehensible for those interested in the topic which, in turn, imposed specific challenges for communication channels and presentational forms. Because policy-makers and the wider public do not generally have a basic understanding of the physical laws underlying hydrogen energy technologies, particular attention had to be paid to the way in which the potential of hydrogen energy technologies was described (for instance, it turned out to be very important to explicitly mention in each paper and presentation that hydrogen is not a primary energy source but an energy carrier).

On the other hand, it was noted that Nordic contributions could add some value to hydrogen energy developments in the European Union and enrich the European Research Area. Nordic Energy Research is as an institution the only regional member of the Mirror Group of the European Technology Platform for Hydrogen and Fuel Cells and holds the vice-chair of the Group. Further, Nordic Energy Research is a core partner in the ERAnet on hydrogen and fuel cells, the so-called HY-CO. In the HyWays project, which aims at producing a European roadmap for a hydrogen society up to 2050, there are now also two Nordic member states (Norway and Finland).

The Nordic dimension of foresight has also attracted the attention of hydrogen communities from other Northern regions in the world: the core partners were, for instance, invited as keynote speakers in Hydrogen and Fuel Cells seminars in Canada (Joergensen, 2005) and in Hokkaido, Japan (Eerola, 2006). One of the reasons behind this is apparently the similar type of climate conditions that provide good conditions for testing and developing hydrogen energy technologies. Because English was chosen as the official working language of the Nordic foresight cooperation – in order to guarantee a balanced contribution of each Nordic country – communication in wider international circles was quite straightforward in this respect. The project website has also been actively used throughout the world (most visits somewhat surprisingly from North America). New circles of international visitors may appear after completing the project as the Summary Report of the project was made available on the website also in Japanese in 2006.[33]

Although the participants utilised the resulting knowledge already during the process, utilisation of the results by a wider group of stakeholders has started just recently. In both groups the explicit and implicit uses may still continue for a relatively long period. Profound analyses – paying attention to the indirect and diffuse information flows and impacts of the Nordic Hydrogen Energy Foresight – would be needed in order to be able to adequately assess the real impacts.

CONCLUDING REMARKS

The Nordic countries have a long tradition of cooperation within research, education and innovation. There is, however, still room for improving the mechanisms capable of embracing the various activities and initiatives at Nordic level. Nordic-level technology foresight projects and effective exchange of experiences between the foresight actors and users of foresight knowledge are considered as promising tools in this respect.

In many Nordic countries, increasing attention has been paid to the evaluation and assessment of foresight processes and their impacts since the beginning of the 2000s. The evaluation of major foresight projects is typically commissioned to external consultants that report to those responsible for the project. For example, the broad-scope Swedish and Finnish foresight exercises were evaluated in this way.

It is also worth noticing that foresight components are increasingly also embedded in the evaluation and impact assessment of R&D activities and innovation policy (see e.g. Salo and Salmenkaita, 2002; Salo *et al.*, 2004; Lähteenmäki-Smith *et al.*, 2006). This emphasises the strategic and future-oriented nature of these activities. There, thus, seems to be some convergence between foresight, evaluation and impact assessment activities (Loikkanen *et al.*, 2006). How far this can be brought in the Nordic context remains to be seen, however.

The overall picture of the foresight activities in the Nordic countries is, however, still quite fragmented. Technology foresight approaches have been introduced and used in specific areas, sectors and regions, and also broad-scope national-level foresight exercises have been carried out in some of the Nordic countries, but there are no general guidelines for using foresight knowledge in decision-making in RDT policies at governmental level or in the context of national research councils. Consistent use of foresight knowledge in R&D priority-setting at Nordic, national and EU-levels has not yet been developed, although this kind of need has been raised in some recent discussions.

Instead, the Nordic countries have relied on more indirect and diffuse influence paths, building on the ability of foresight processes to facilitate future-oriented thinking, purposeful networking and dialogues between the various stakeholders. Technology foresight exercises in the Nordic countries have also been successful in facilitating dialogue, communication and consensus on new technologies. When the impacts of new technologies are discussed between the stakeholders in a balanced way already in an early phase, there are better conditions to create shared visions for utilising new opportunities and taking the required action for overcoming the perceived barriers and threats. It is also easier to mobilise joint actions. The experiences from Nordic cross-border foresight activities confirm the potential of foresight exercises in these respects.

NOTES

1   In addition, the following people contributed to completion of the country-specific descriptions: Per Dannemand Andersen, Risoe National Laboratory in Denmark; Bengt Mölleryd and Thomas Malmer, The Royal Swedish Academy of Sciences (IVA); E. Anders Eriksson, Swedish Defence Agency (FOI); Erik F. Øverland, Subito!, Norway; Thorvald Finnbjörnsson, RANNIS Consulting, Iceland; Jari Kaivo-Oja (Finland Futures Research Centre) ; Toni Ahlqvist, Sirkka Heinonen, Sirkku Kivisaari, Totti Könnölä and Torsti Loikkanen, VTT Technical Research Centre of Finland.

2   The Nordic Industrial Fund was reorganised under the label Nordic Innovation Centre, NICe, (www.nordicinnovation.no) from the beginning of 2004.

3   www.h2foresight.info.

4   www.nordic-fobis.net.

5   www.nordic-ictfore.vtt.fi.

6   See www.nff.risoe.dk.

7   Including Denmark, Finland, Iceland, Norway, Sweden and the three autonomous territories of Faroe Islands, Greenland and Åland Islands.

8   Although Danish, Swedish and Norwegian languages are closely related to each other they are still clearly different languages. Following and actively participating in the "multi-dialect Scandinavian discussions" is somewhat problematic at least for the Icelandics and most of the Finns who have Finnish (and not Swedish) as their mother tongue. The same holds for the industrial landscapes of the Nordic countries where a common overall industrial framework does not exist, although there are some sectors – such as forest and electronic industries – that are important for several Nordic countries.

9   See www.teknologiskfremsyn.dk.

10 See www.hirc.dk.

11 See www.ennakointifoorumi.fi.

12 See www.eduskunta.fi.

13 See www.finnsight2015.fi.

14 The ten expert panels focus on the following issues: (a) learning and learning society, (b) services and service innovations, (c) well-being and health, (d) environment and energy, (e) infrastructures and security, (f) bio-expertise and bio-society, (g) information and communications, (h) understanding and human interaction, (i) materials, and (j) global economy.

15 For the report in Finnish, see www.tekes.fi/julkaisut/FinnSight_2015 laaja.pdf.

16 See www.sitra.fi.

17 See www.tekniskframsyn.nu.

18 See www.forskningsradet.no.

19 See www.h2foresight.info.

20 See www.nff.dk.

21 This chapter focuses on the Nordic foresight cooperation funded by NICe. In parallel to these developments, there has been sector-specific foresight cooperation between some Nordic countries and Nordic actors have also been actively involved in European foresight cooperation (see e.g. Könnölä *et al.*, 2007; CMI, 2006).

22 See www.h2foresight.info.

23 For further information see www.nordic-fobis.net.

24   Although the quantitative analyses of the Nordic hydrogen energy system were considered useful and important, modelling of the "Nordic ICT system" was not considered reasonable. Instead, it was important to pay specific attention to national-level involvement and dissemination of information.
25   For further information see www.nordic-ictfore.vtt.fi.
26   See www.nff.dk.
27   The abbreviation SECI comes from the words "Socialisation", "Externalisation", "Combination" and "Internalisation" that refer to the four basic knowledge conversion modes of shared knowledge creation.
28   In particular, it was explained that the "what-if" calculations of the model were needed to concretely illustrate the implications of vision statements and assumptions that were outcomes of the interactive workshops.
29   The project group consisted of experts in innovation and foresight studies, model building, energy systems, and technical fields contributing to the development of hydrogen energy technology and the required infrastructure.
30   See www.finnsight2015.fi.
31   Similar types of influence paths – especially those related to facilitation of fruitful dialogues and creation of networks and joint actions – have, however, been recognised in other Nordic and cross-border foresight exercises too, for example in the STRING project across three national borders – Sweden, Denmark and Germany – in the South-Western area of the Baltic Sea (see Joergensen, 2001; Eerola and Joergensen, 2002).
32   See www.nordicenergy.net.
33   The Japanese version was provided by the Hokkaido Regional Development Centre that prepared a full translation of the Summary Report (Dannemand Andersen *et al.*, 2005) on its own initiative in the context of the Hokkaido Hydrogen Society Seminar in March 2006. The Japanese version can be downloaded from the project home page www.h2foresight.info.

REFERENCES

Ahlqvist, T., Uotila, T. and Harmaakorpi, V. (2007), 'Regional Technology Foresight: Technology Signals and Cluster Strategy in the Region of Päijät-Häme', *VTT Research Notes*, [forthcoming] 2007 (in Finnish, abstract and summary in English).

Alsan, A. and Oner, M.A. (2003), 'An Integrative view of Foresight – Integrated Foresight Management Model', *Foresight*, 5(2), pp. 33–45.

Andersen, M.M. and Joergensen, B.H. (2004), 'Technology Foresight – Wiring up the National Innovation Center', *IAMOT Conference Proceedings*, Washington, April 2004.

Campbell, J. and Pedersen, O.K. (2005), 'Danish institutional competitiveness in the global economy', Paper prepared for the Danish Funktionærernes og Tjenestemændenes Fællesråd.

CMI (2006), *Emerging Priorities in Public S&T Policy in the EU, the US and Japan*, Final report of a Platform Foresight project coordinated by CMI.

Dannemand Andersen, P. and Borup, M. (2006), 'Strategy Processes and Foresight in Research Councils and National Research Programmes', Paper presented at the

Second International Seville Seminar on Future-Oriented Technology Analysis: Impact of FTA Approaches on Policy and Decision-Making, 28–29 September.

Dannemand Andersen, P., Joergensen, B.H., Eerola, A., Koljonen, T., Loikkanen, T. and E., Eriksson., E.A. (2005a), *Building the Nordic Research and Innovation Area in Hydrogen*, Summary Report of the Nordic H2 Energy Foresight, January, available on website www.h2foresight.info (ISBN 87-550-3402-0).

Dannemand Andersen, P., Rasmussen, B., Strange, M. and Haisler, J. (2005b), 'Technology Foresight on Danish Nanoscience and Nanotechnology', *Foresight*, 7(6).

Eerola, A. (2006), 'Nordic H2 Energy Foresight – Approaching a Hydrogen Society', Key note presentation at the Hydrogen & Fuel Cell Seminar, Hokkaido Hydrogen Society, Sapporo, Japan, 23 March.

Eerola, A. and Joergensen, B.H. (2002), *Technology Foresight in the Nordic Countries*, Report to the Nordic Industrial Fund, Center for Innovation and Commercial Development, Risoe-R-1362 (EN), ISBN: 87-550-310992 & ISBN 87-550-3110-2, available on website http://www.risoe.dk/rispubl/SYS/ris-r-1362.htm.

Eerola, A. and Väyrynen, E. (2002), *Developing Technology Forecasting and Technology Assessment Practices on the basis of European experience*, VTT Research Notes 2174, 151 pages (in Finnish, abstract in English; available on website http://www.inf.vtt.fi/pdf/tiedotteet/2002/T2174.pdf).

Eerola, A., Loikkanen, T., Joergensen, B.H., Dannemand Andersen, P. and Eriksson, E.A. (2004), 'Nordic H2 Energy Foresight – Complementary Contribution of Expert Views and Formal Analyses', Paper presented in the EU-US Scientific Seminar "New Technology Foresight, Forecasting and Assessment Methods", Seville, May 13–14; available on websites http://www.jrc.es/projects/fta/index.htm.

Eerola, A., Loikkanen, T., Koljonen, T., Joergensen, B.H., Dannemand Andersen, P. and Eriksson, E.A. (2005), 'Nordic H2 Energy Foresight – Managerial Challengers of the Interactive Process', Paper presented at the 17th Annual International Conference "Foresight Management in Corporations and Public Organisations – New Visions for Sustainability", Helsinki, 9–10 June, http://www.tukkk.fi.

Hjelt, M., Luoma, P., van de Linde, E. Ligtvoet, A, Vader, J. and Kahan, J., (2001), 'Experiences with National Technology Foresight Studies', *Sitra Reports 4/2001*, Helsinki.

House of Monday Morning and Nordic Council of Ministers (2005), *The Nordic region as a Global Winner Region. Tracing the Nordic Competitiveness model*, available at www.norden.org.

Joergensen, B.H. (2001), *Foresight in Cross-border Cooperation*, IPTS Report, **59**, November.

Joergensen, B.H. (2005), 'Building a Nordic Research Area in Hydrogen', Presentation in the Nordic-Canadian Hydrogen and Fuel Cell Partnership Workshop, NRC Institute for Fuel Cell Innovation, Vancouver, Canada, 23–24 August.

Joergensen, B.H. (2006), 'Transnationalt energiforskningssamarbejde – hvor går grænsen?', in Hanne Foss Hansen (ed.), *Den organiserede forvaltning. Politik, viden og værdier i samspil*, Copenhagen: Forlaget Politiske Studier.

Kivisaari, S, Lovio, R. and Väyrynen E. (2004), 'Managing Experiments for Transition: Examples of Societal Embedding in Energy and Health Care Sectors', in Elzen, B., Gells, F.W. and Green K. (eds), *System Innovation and the Transition to Sustainability. Theory, Evidence and Policy*, Cheltenham, UK and Northampton, MA, USA: Edward Elgar.

Könnölä, T., Salo, A. and Brummer, V. (2007), 'Foresight for European Coordination: Developing National Priorities for the Forest Based Sector Technology Platform', *International Journal of Technology Management*, special issue on technology foresight.

KTM (1997), *On the Road to Technology Vision – Technological Needs and Opportunities in Finland*, Helsinki, Ministry of Trade and Industry (in Finnish).

Kuhlman, S. and Arnold, E. (2001), *RCN in the Norwegian Research and Innovation System*, Background Report No. 12, evaluation of the Research Council of Norway.

Linstone, H.A. (1999), *Decision Making for Technology Executives – Using Multiple Perspectives to Improve Performance*, Boston: Artech House.

Lähteenmäki-Smith, K., Hyytinen, K. Kuitinlahti, P. and Konttinen, J. (2006), *Research with Impact. Evaluation Practices in Public Research Organisations*, VTT Research Notes 2336, Espoo.

Loikkanen, T., Kutinlahti, P. and Eerola, A. (2006), 'Towards an Integrated Framework for Impact Assessment and Foresight Studies in Innovation Policy Analysis', Paper presented at the Second International Seville Seminar on Future-Oriented Technology Analysis: Impact of FTA Approaches on Policy and Decision-Making, 28–29 September.

Neumann, Iver B. and Øverland, E.F. (2004), 'International Relations and Policy Planning: the Method of Perspectivist Scenario Building', *International Studies Perspectives*, **5**, pp. 258–277.

Nonaka, I. (1994), 'A Dynamic Theory of Organizational Knowledge Creation', *Organization Science*, **5**(1), pp. 14–37.

Nonaka, I. and Takeuchi, H. (1995), *The Knowledge-Creating Company*, New York: Oxford University Press.

Nordic Council of Ministers (2003a), *The International Nordic Region*, Nordic Co-operation in a European Framework, ANP.

Nordic Council of Ministers (2003b), *NORIA, White paper on Nordic Research and Innovation*, October.

Nordic Council of Ministers (2004), *Innovationsboken - Nordisk styrka, nationell nytta och global excellence: Förslag till nordiskt innovationspolitiskt samarbetsprogram 2005–2010*, Huvudrapport, ANP: 748, ISBN 92-893-1017-0.

Nordic Council of Ministers (2005), *Nordic Countries in Figures*, ANP 2005:757. ISBN 0908-4398.

OECD (2004), *Main Science and Technology Indicators*, Paris: OECD Publication Service.

Øverland, E.F. (ed.) (2000), *Norge2030. Fem scenarier om offentlig sektors framtid, Cappelen Akademiske forlag*, Oslo, book in Norwegian (Norway2030. Five scenarios about the Future of Public Sector).

Øverland, E.F. (ed.) (2003), *SURPRISE. Scenario Use and Research for Planning Regional Innovation Systems*, Report 10/2003 from the Norwegian Computing Centre 2003, Oslo.

Polanyi, M. (1966), 'The Tacit Dimension', reprinted in Prusak, L., *Knowledge in Organisations*, Boston: Butterworth-Heinemann, 1997.

Salo, A. (2001), *A Needs Assessment of Technology Foresight*, Ministry of Trade and Industry, Studies and Reports 2/2001, 76 pp. (in Finnish, abstract in Swedish and English).

Salo, A. and Salmenkaita, J.P. (2002), 'Embedded Foresight in RTD Programmes', *International Journal of Technology, Policy and Management*, **2**(2), pp. 167–193.

Salo, A., Gustaffson, T. and Mild, P. (2004), 'Prospective Evaluation of a Cluster Programme for Finnish Forestry and Forest Industries', *International Transactions in Operations Research*, **11**(2), pp. 139–154.

Ståhle, P. and Ståhle, S. (2006), *Tulevaisuusluotain – Future Radar: Skills needed in the Future. An Education Intelligence System (EIS)*, Report of the Cofederation of Finnish Industries.

STF (2004a), *Choosing Strategies for Sweden*, A synthesis report from Swedish Technology Foresight 2004, available on website www.tekniskframsyn.nu.

STF (2004b), *Inspiration for Innovation*, Swedish Technology Foresight 2004, available at www.tekniskframsyn.nu.

TEK (2005), *Technology Barometer*, The Finnish Association of Graduate Engineers, http://www.tek.fi/teknologia/tulevaisuus/barometer_eng.pdf.

World Economic Forum (2005), *Nordic Countries Lead the Way in the World in the World Economic Forum's Competitiveness Rankings 2004*, October 2004 and September 2005.

10. Foresight in Smaller Countries [1]

Patrick Crehan and
Jennifer Cassingena Harper

INTRODUCTION

In their efforts to benchmark with their larger neighbours, small countries[2] often latch on to foresight as the means for escaping from the constraints imposed by size and related resource constraints, in a "we-too" mode. In practice, small countries tend to have an ambivalent attitude towards foresight, which ranges from high expectations that it can deliver some form of fast-track transition to the knowledge economy to more realistic concerns over the perceived luxury of investment of scarce resources for research in what is often considered a mere form of contemplative, exploratory activity. Thus the start-up of visionary, foresight-type approaches in small countries is often constrained by a number of factors: the concern and pressures of delivering tangible results and value for money; the limited number of small country success stories, a set of methods which have been developed for a large country context, and the limited resource inputs in terms of available funds, foresight and other expertise and background documentation and intelligence on the sector or theme under consideration. This chapter focuses on the sub-set of European smaller countries, Cyprus, Estonia and Malta, with a population ranging from 0.4–1.5 million, where foresight was a relatively new activity until recently and mindsets, competencies and resources for such approaches were highly limited. Common features and needs related to their small size, including governance and resources deficits and lack of critical mass together with the common challenge of being in good shape to join the EU by 2004 provided the rationale for a joint foresight project in 2002–2003.

This is a case study of the first formal national foresight experiences of three small European countries, Cyprus, Estonia and Malta, as part of eFORESEE, a transnational EU-funded project. The project which was implemented in 2001 and 2002, aimed at promoting the Exchange of Foresight Relevant Experiences among Small European Economies (hence eFORESEE). A key connecting factor providing the rationale for this joint activity was a broadly common set of socio-economic challenges facing these countries in the lead-up period to accession to the EU. In particular the eFORESEE project sought to examine the potential role of foresight in dealing with the structural changes to the economy that accompany the accession process, as well as the integration of accession states into the European Research Area (ERA).[3] It also aimed to explore the decision-making processes involved in setting up foresight activities, as well as the challenge of managing and implementing specific foresight actions and creating a community of foresight practice. The qualitative objectives of the eFORESEE project included exploration of a knowledge management approach to foresight, foresight embedding, evaluation and benchmarking, as well as linking accession-level foresight to the development of the ERA and the development of peer networks of professionals from the foresight and policy development communities in EU and other imminent accession states (Pace and Cassingena Harper, 2005). Thus this chapter highlights the need and potential for creative, smart approaches to the implementation of foresight activity in small countries in a critical transition phase. The keywords of eFORESEE emerged as follows: *mutual policy learning, knowledge management approaches, realignment of networks, strategic conversations, common ownership of visions, and common spaces for open thinking on the future.*[4] One of the main insights that can be drawn from the project is that foresight in small countries faces certain constraints in policy-making: resources are limited, the socio-cultural backdrop far from supportive, the scientific and technical pool over-stretched and the political will to take consequent actions has a short time-span and needs to be nurtured through fast, tangible and high profile results together with a visibly effective process of interaction, both formal and informal.

The transnational foresight approach, in this case exercises which cluster together small countries in a benchmarking mutual learning mode, can prove highly effective if well designed and implemented using appropriately-adapted foresight approaches. In such cases, the transnational framework can act as an important policy lever, in redefining policy-making structures and networks and opening up the space for creative, forward-looking approaches to innovation and governance.

HISTORICAL DEVELOPMENTS

Within the context of the European Commission's interest in exploring the potential of foresight in Europe's catch-up regions (sub-national level) and small transition states, the rationale for the eFORESEE project developed in

2001, with Malta and Cyprus being initially involved in the proposal submissions to the EU. There was particular interest in bringing these countries together in a joint foresight activity. Two Mediterranean island states with a recent colonial past, and currently both facing similar concerns over the pressures of fast change imposed by impending accession to the EU. With populations of less than a million (Malta: 380,000; and Cyprus: 700,000), they are the two smallest EU member states, sharing stable economies and a comparatively high GDP per capita (Malta approximately €15,000 and Cyprus: €13,500 in 2004) as compared to other accession countries. It was anticipated that these countries would have a high potential of mutual learning in developing foresight approaches relevant to a Mediterranean small island context, with common themes of interest ranging from information technologies to agriculture, aquaculture and biotechnology. The insights and experiences generated through the joint development and adaptation of appropriate foresight methods could prove useful for other small islands, countries and regions in Europe. Estonia was a latecomer to the project and was invited to join after the project was approved for funding, being regarded as a highly relevant partner to join the team both in terms of contributing its experiences and the learning to be generated from the diversity of bringing together countries at the opposite peripheries of Europe. Estonia is a relatively larger country with a population close to 1.5 million but also with a stable fast-growing economy, though lower GDP per capita as compared to Cyprus and Malta.

Accession to the EU by mid-2004 was perceived as presenting Cyprus, Malta and Estonia with a balance of opportunities and threats. The opportunities related mainly to the economic, political and military security afforded through membership coupled with improved quality of life. The chance however is seen in the possibility to escape from the constraints of a small country and its path-dependencies thereby achieving a faster transition to the knowledge society. Threats identified included potential negative societal impacts, including possible loss of socio-cultural identity and values, the challenge of coping with an increasingly diverse society, the brain drain, and negative effects on marginalised groups. Other identified threats were the heavy burden and competing pressures related to compliance with the EU Acquis Communautaire (all the legislation and case law adopted since the creation of the European Community), the poor/slow visibility of benefits of membership and the capture by vested interests of the opportunities opened up through membership. The drive to launch the eFORESEE initiative emerged from a growing political interest in Malta in 2000–2001 in the introduction of more systematic approaches to Research, Technological Development and Innovation (RTDI) policy formulation and implementation. In 2001, the newly-appointed Parliamentary Secretary, responsible for science and technology within the Ministry of Education, instigated the drive for RTDI capacity-building through the introduction of more systematic approaches to policy development and implementation. Two Cabinet-approved projects were to be given priority by the Malta Council for Science and Technology, namely the National RTDI Audit and the Foresight Exercise. This drive partly related to Malta's commitments under Chapter 17 of the EU Acquis to strengthen Malta's capacity in research and development through the development of an updated strategy. Malta's participation in the EU's Framework Programme was also flagging the need for local capacity-building to support better the participation of Maltese researchers in the programme. The Foresight exercise was given particular importance in view of the fact that Malta's RTDI strategy had to be more forward-looking, since there was a history of very limited RTDI activity (due more to low prioritisation of RTDI and problems of critical mass than to lack of competence) and a poor research and innovation culture.

Foresight-type activity is not new to Malta since the Malta Council for Science and Technology had engaged in this type of vision-building, future-oriented, wide participatory process of strategy development in 1992–1993 with the National Strategy for Information Technology. These kinds of approaches had also been adopted in other sectors such as the Malta Structure Plan which focused on long-term policies for sustainable land planning and development (a critical policy concern in a highly densely populated island). The foresight philosophy is also reflected in Malta's work on future generations and the common heritage principle within the Law of the Sea and Climate Change. Certainly the understanding was present of how foresight could inform and improve the policy development process by encouraging the ownership of visions and commitment to future-oriented strategies. However, the foresight exercise itself helped to clarify better the benefits and ways of maximising the impact of such forward-looking activity in the minds of the sponsors and the community involved. As always in a small country context, the influence of external drivers plays a critical role in the launch of new initiatives and in this case the financial support made available for transnational foresight activity through FP5[5] (under the direction of Unit K2 DG Research) together with the technical support provided by the project coordinator, CKA Consulting Ltd, and PREST, proved instrumental.

For Cyprus, the eFORESEE Project also presented the first opportunity to carry out a national-level foresight exercise and the transnational framework was also seen as providing an essential lever in terms of financial and technical resources and influences which could impact more directly on more long-term strategic policy development. Foresight was earmarked as a strategic policy intelligence activity that could help in illuminating the range of alternative futures (Damvakeraki, 2006). The Cyprus Agricultural

Research Institute (ARI) saw the opportunity to apply foresight to a particular sector in Cyprus where forward-looking, consensus-building approaches were needed as the basis for bringing the national agricultural policy in line with the EU's Common Agricultural Policy (CAP),[6] itself in a transition phase.

Estonia joined the eFORESEE Project with a stronger interest in using foresight as a stimulus for moving towards the knowledge-based competitive economy by capitalising on the national strengths in ICT and biotechnology sectors. As in Malta, considerable forward-looking activity had already been underway, particularly in the ICT sector through the ongoing EU-funded "Vikings" project which provided important inputs to the eFORESEE pilot activity. In general Estonia and Malta had the most potential for sharing know-how and learning more from the eFORESEE project as they shared common broad themes for their pilots. In practice, instituting ongoing learning processes proved difficult to implement, since each partner faced sufficiently tough challenges in handling the national activity. Transnational interactions and learning created additional responsibilities. The Project Coordinator's role in this context was critical, providing the vital drive together with the expertise in terms of knowledge management methods and connecting insights which helped to bring the partners together in a mutual learning mode.

POLICY CONTEXT, RATIONALE AND OBJECTIVES

Among the preliminary tasks faced in designing the eFORESEE project was the need to define the common policy concerns and objectives of the three accession countries in the area of research, technological development and innovation (RTDI). The broad rationale for the foresight exercise related to addressing the challenges faced by policy-makers in implementing foresight activities for smaller economies and regions. This entailed examining the potential role of foresight in dealing with the structural changes to the economy that accompany the accession process as well as the integration of accession states into the European Research Area. The project also explored the decision-making processes involved in setting up foresight activities as well as the challenge of managing and implementing specific foresight actions. The key objectives included building policy formulation and implementation capacities among policy-makers, to further extend and develop the European networks of policy and foresight experts, and also supporting the development of the ERA by contributing to the coordination of policy and policy-related research at the European level, with particular attention to the needs of regions, small economies and accession states.

More specifically, a number of common socio-economic challenges emerged from the first interactions between the partners at the kick-off meeting in February 2001 in Malta which called for:

- Smart strategies to increase the level of public and private investments in RTDI within the context and constraints of ongoing government efforts to reduce the budget deficit;
- Improving the profile and prominence of research and innovation among politicians and policy-makers and on the national agenda;
- Introducing the strategic use of foresight as an input into RTDI Policy, by setting national RTDI priorities and strategies, and more coherent and clear targets for research and innovation in the countries' National Development Plans;
- More systematic approaches to ensure optimal and rational use of scarce resources (particular emphasis on structural funds);
- Efforts to promote an RTDI culture (especially in Cyprus and Malta) through greater public and societal engagement in science and technology;
- More forward-looking mindsets and planning in political and government circles in particular among politicians and strategic policy-makers by encouraging consensus-building across party divides;
- Benchmarking for mutual learning as a tool for promoting competitiveness and innovation-driven growth;
- Tangible strategies for attracting knowledge-intensive and high-tech FDI whilst preventing the growth of the dual economy;
- Knowledge management approaches as the means for investing in the learning economy.

THE FORESIGHT PILOTS

Each of the partners was committed to carrying out two pilots focused on particular sectors or themes of national priority. This allowed the partners some flexibility in their scoping activity and indeed whilst there was some convergence in choice of sectors, the themes finally focused on did vary from one country to the other.

In an effort to allow and encourage some level of benchmarking and learning, the partners with the support of the Coordinator were able to adopt a common set of targets and a similar approach for the pilots which did indeed subsequently facilitate the process of learning through comparing and sharing knowledge.

The common targets and success criteria set were:

- To elaborate a vision for the country as an advanced knowledge economy in 2020 whose main resource is its ability to develop human capital in new economy skills all round the world from a Mediterranean base (adapted for the other pilots);
- To guide the decision-making and inputs into the countries' National Development Plans (2003–2006);
- To mobilise public–private sector partnerships to take action on business opportunities;
- To revitalise old networks and stimulate the formation of new networks (cross-disciplinary/sectoral, involving new players, e.g. management/human resource consultants, researchers and educators);
- To explore methodology/approaches and record the process.

A brief overview of the foresight pilot themes and focus of activity together with the results[7] generated is presented below.

MALTA [8]

The First Pilot was entitled "Exploring Knowledge Futures in Information and Communications Technologies and Education in 2020". The orientation of the exercise was to build on the work which had been developed through the National Strategy for Information Technology with a view to taking the vision-building process forward into the emerging realities of the knowledge society. The emphasis was on the challenge of identifying the drivers of the transition from the information society to the knowledge society and the related knowledge management and transfer aspects. This pilot intended to and succeeded in moving away from a simple concept of a knowledge society being an information technology society. The pilot sparked off a wide ranging debate on alternative future visions in ICT which extended to the broader issues of innovation and creativity. A key output of the pilot was the development of four scenarios, culminating in consensus on a success scenario entitled Knowledge Community 2020 Vision. The major challenges identified included:

- Ensuring a more holistic and integrated approach to strategic policy-making in ICT and knowledge management;
- Embedding foresight in key policy areas for the knowledge-based economy;
- The allocation of appropriate levels of resources.

Given constraints of critical mass, the exercise focused on the need to target resources to a defined set of niche areas with existing strengths and/or with potential for developing and test-bedding new ICT applications and approaches. These included tourism, e-health and local knowledge strengths (e.g. crafts, cultural heritage, fisheries). Among the recommended actions of the Knowledge Community 2020 Vision (Cassingena Harper and Pace, 2005) is the setting up of a National Knowledge Platform for facilitating "joined-up" open governance (considered as a major driver of the Vision). The Knowledge Platform is designed to activate a complementary value-added layer to the current portfolio of information services provided to the community and citizen. The Platform's services and facilities extend beyond information-sharing to knowledge-sharing. The Platform can serve as the main portal for pooling inter-sectoral knowledge, e-learning facilities and knowledge resources, on-line policy discussions and debates and fast-track learning. It is still being discussed and is an excellent example of continuous (maybe perpetual is a better word) foresight as new stakeholders come along, identify themselves as being part of the picture and complete it with their own initiative. In this way it has lead to a number of spin-off foresight activities such as "Futurechild", "Futurefest" and a foresight exercise on the "Future of Drama" in Malta.

The Second Pilot dealt with the theme "Biotechnology" and enjoyed good support from academia, industry and relevant government ministries. The emphasis of the Biotech Pilot emerged as the drive to promote capacity-building in the fledgling local biotechnology sector through efforts to improve science and technology education at all levels and to encourage the introduction of RTDI-related incentives to attract foreign direct investment. It considered both health- and non-health-related biotechnology, identifying issues such as a lack of appropriately trained scientists, a lack of entrepreneurial scientists and a lack of cooperation from government authorities as the most important weaknesses in the sector. The core objectives of this pilot were:

- To produce a plan to develop the fledgling Maltese biotechnology industry into a core sector of the Maltese economy by 2015 through a collaborative venture between academic institutions, the public sector and private enterprises;
- To map biotechnology-related activity and resources in Malta current and as projected by 2015;
- To identify developments in biotechnology that will impact on the Maltese economy and society by 2015;

- To develop a basis for a national biotechnology strategy that will provide the basis for the national investment of resources in this area and also help to attract direct investment;
- To stimulate the formation of new networks and create an awareness of the fundamental changes required within the public, private and academic sectors for the Biotechnology industry to take root.

Key Research and development areas identified:

- Biofuels (fermentation products);
- Bioinformatics (combining excellence in IT and biology);
- Biomaterials;
- Diagnostic kits and treatment of local prevalent disease including diabetes (centre of excellence for Mediterranean region);
- Environmental biotechnology (waste treatment and purification of contaminated soils of heavy metals and toxic chemicals);
- Genetic medicine and clinical genomics;
- Medicinal plants (maximise turnover from agriland);
- Plant development for arid environment, able to tolerate high salt content in soil and water (including local vine, olive);
- Products of fermentation/cell culture at production scale;
- Specialised foods (including those of marine origin).

It concluded with recommendations as to how to achieve the vision of a thriving Maltese biotech sector by 2015.

The Third Pilot focused on the marine science sector and addressed the following theme "Building the best future for the marine sector in the Maltese Islands" with a sub-title "A vision towards enhancing the marine sector's contribution to the Maltese economy in 2020" (Drago, 2004 and Pace, 2006). This pilot invested strong efforts in bringing together the many authorities whose interests converge in the maritime sector, the maritime authority, the enterprise authority, the tourism authority along with institutions such as the Centre for Fisheries, The Maritime Institute, the Department of Economics of the University of Malta and Malta Dry-Docks. The pilot focused on efforts to quantify and exploit the economic relevance of the marine sector, estimated at generating around one-seventh of total economic activity in Malta but with this however excessively bound to "traditional" activities involving tourism and transport. Its main output was ownership of a shared resource and the challenge of management as well as the identification of guidelines and recommendations for future actions.

Malta's contribution, to the setting up of an integrated system for observing the oceans worldwide was highly emphasised together with the potential of operational oceanography in its many diverse applications in tourism, fisheries, the conservation of marine bio-diversity, coastal zone management, safety at sea, and preventing catastrophic pollution events. The emerging needs and uses of sound marine data (recording the strength and direction of currents, the action of waves near coastlines, wind force, temperature and other factors all have a bearing on our lives) was highlighted. The Pilot also deliberated on the proposal for a common Euro-Mediterranean research space as a "magic formula" for the future of the region. The enlargement of the EU in May 2004 was seen as opening up new opportunities for taking forward the collaborative process with Malta's immediate neighbours in the Southern Mediterranean. Within this context, the marine sector was identified as a key priority area for EU-Mediterranean cooperation. This entailed giving particular consideration to exploring the possibility of introducing common structures for the management of the extensive marine resources in the region. Efforts for regional cooperation were considered as having the potential of dispelling divisive perspectives while focusing instead on the unifying aspects of sharing a common sea.

Key emerging research and development areas identified were:

- Fisheries;
- Aquaculture;
- Transportation;
- Marine biotechnology;
- Marine energy and mineral resources;
- Marine observation, monitoring and forecasting.

CYPRUS

The First Pilot in Cyprus was intended as a tool for agricultural policy reform. Its title and ultimate objective was "Improving the competitiveness of Cypriot agriculture using modern and alternative production methods". The main challenge faced by ARI, the leader of the foresight exercise, was that of bringing together representatives from the main farmers' unions. There are four main farmers' unions in Cyprus. Each is strongly aligned with one of the four main political parties. Far from cooperating, the unions tend to oppose one another to achieve political goals. Nevertheless any major change in sectoral policy would rely on the support of all four unions. ARI realised this from the start and made it one of the main goals of their foresight activity. They were successful in bringing the unions together and eventually developed a constructive dialogue that resulted in a series of recommendations that were adopted into the text of the national development plan. They spent less time experimenting with formal analysis tools such as

SWOT and focused their attention on the drafting of background and position papers that were relevant and meaningful to the various stakeholder groups. As a background activity they consulted regularly with the responsible ministry and with contacts in the Central Planning Bureau. They facilitated open and public discussions with prominent farmers and members of the farmers' unions around issues raised in the background and discussion documents. It is worth noting that academic or research organisations in general have great difficulty finding the right language for communicating with stakeholders. They also have difficulty formulating clear policy-oriented recommendations. This was not the case with ARI. Their foresight exercise was exemplary in the way they handled subtle and complex communication problems in a politically challenging context. From a knowledge management perspective they demonstrated great skill in the "management of conversations".

The Second Pilot conducted in 2003 explicitly aimed to explore agricultural production as a knowledge-intensive industry and the application of knowledge management in solving agricultural problems in Cyprus. They expressly explored a broader agenda than the role of RTD and Innovation and addressed the strategic goal of achieving "A viable agriculture in a developed countryside" over the next decade. The exercise being conducted in Cyprus on the eve of accession put great emphasis on the transformation that would accompany accession, the broader international content of the WTO and the implications of EU membership for such issues, as well as the implications of the "Cyprus Issue" for the future of the agricultural sector. The approach taken was to use the foresight approach to revise and improve a position paper that would lay out a strategic national objective not only for the benefit of the people of Cyprus but for the benefit of the European Union as well. This goal was achieved and the relationship between ARI, the farmers' unions and a broader set of stakeholders was further developed. When asked by the coordinator if he thought that the relationship with and dialogue between the farmers' unions could be considered sustainable, Marinos Markou, the architect of the Cyprus foresight pilots, replied that it is too early to tell. His caution raises an interesting point about the role of continuous foresight. One of the explicit goals of the second foresight exercise was to confirm and strengthen the relationship developed with the farmers' unions. Although this goal was achieved, the leader of the exercise remained unconvinced that these improvements corresponded to a permanent change of culture in organisations that have a history of opposition and fierce partisanship.

It is possible to conclude that the foresight exercise has provided important politico–institutional benefits that require maintenance if they are to be sustained on a short- or medium-term timescale. The benefits established through foresight can only be maintained through continuous foresight. Policy-makers in Cyprus might value the maintenance of good institutional relations between the farmers' unions enough to support continuous foresight in the agricultural sector as a way of maintaining a "public good" that will greatly facilitate their work.

ESTONIA

The First Pilot in Estonia started towards the end of 2002 as anticipated. The later start coincided with the termination of e-Vikings an EU-IST funded analysis of the ICT sector in the region that, as noted above, would provide important inputs for the Estonian exercise. The first pilot dealt with the future of ICT in Estonia. One of the most important issues emerging from this work was the lack of strategic capacity of ICT firms in Estonia, the high level of dependence on subcontracting from firms in neighbouring countries such as Sweden and Finland and the low level of awareness on the side of sub-contractors of the longer-term strategies of their clients. The ICT pilot focused on how a network of centres of excellence could serve the needs of the ICT sector. Its final recommendations included a call for further foresight activities on RTDI in the region.

The Second Pilot dealt with the biotechnology sector (Damvakeraki, 2006). The main input was the results of a Fraunhofer/IST study on the biotech "cluster" at Tartu – a group of about 20 university spin-off companies employing 200 people in the region. This is backed up by an RTD system involving about 300 permanent staff in 15 institutes. Despite the fact that there are no major pharmaceutical RTD performers in Estonia the main emphasis of work within the cluster and the research in the area is on health-medical applications. There is an important misalignment between the RTDI infrastructure and the opportunities presented by local industry in the sense that chemicals, food-processing and wood processing are important local industries in which biotechnology could play an important role. However, players in these industries seem unaware of this in general and the RTDI community is busy being involved in the more trendy medical- and healthcare-related areas. The foresight exercise highlighted this disparity and provided conclusions in the form of recommendations that included a recommendation to increase support for the bio-medical area and to improve biotech-based innovation in the more traditional industry sectors. The results of both foresight exercises were combined in a single report published in Estonian as an input for the 2004–2006 National Development Plan and as a vision document for the subsequent plan for the period 2007–2013.

POLICY EFFECTS AND IMPACTS

The full range of results, insights and learning generated through the eFORESEE project, both targeted and unforeseen, cannot be appreciated until some time after the project ends when the experience gained from the pilots and the information gathered from the eFORESEE international conferences and workshops can be fully digested. Among the tangible benefits of foresight is its facility to make the implementing team and the stakeholders engaged in the exercise aware of the full range of interests and issues involved. At the outset of the project and in particular during the early stage implementation of the foresight pilots, it became clear that the country teams were contending with more than just the set objectives. Indeed efforts had to focus on both a formal and informal set of objectives, the more politically-driven agenda set by the sponsors as against a more socio-cultural agenda driven by societal needs and interests. For example, in the case of Malta, the weak RTDI culture, the lack of open participatory and consensus-building approaches to governance together with the low appeal of and competence for systematic methods for policy-making, reflected a society with diverse interests, divided over the most trivial of issues and with a tendency to debate endlessly rather than move constructively to action (Cassingena Harper and Georghiou, 2005).

These societal attributes were shared by Malta and Cyprus in particular, with Estonia having also to contend with these issues of political and societal maturity in governance and policy development to some extent. There is no indication here that the small country context in policy-making is any more or less complex than its counterparts in the large countries, but size does impose certain constraints, relating to scale, geography and resources, which in turn impact on capacities, mindsets and tangible opportunities for advancement. Yet the microcosm of national governance structures and dynamics made possible from a small country perspective, offers unique opportunities for experimenting and piloting creative new approaches required in adapting foresight to the needs of small countries and regions. The compacting of policy levels and the "many hats" syndrome, whereby stakeholders are playing a range of roles and representing different interests at the same time, provides opportunities – if systematically organised – for deriving new insights and developing dynamics into the foresight process (as evidenced in the first eFORESEE Malta Pilot). Foresight's strength lies in its capacity to facilitate and organise strategic conversations, opening up locked-in mindsets, realign entrenched networks and make transparent barriers to open policy-related discourse.

The eFORESEE project demonstrated this effect in the way the pilots were implemented in the three countries which reflected adaptations and creativity in the methodology and approach used. The three partner countries shared a common challenge of introducing foresight into a transition economy context and into a policy system which was undergoing fast pressures of change in the critical pre-accession phase. The main needs in this context, which have come to the fore during implementation of the project, are the need to activate and support fast policy learning and unlearning processes, and the need to bridge the divide between policy-makers and society by engaging able new actors and moving towards more consensus-oriented dialogue. The main insights generated from the project to date include:

- In the run-up to accession despite the fact that policy-makers in the new member states were very much aware that accession would provide negative as well as positive impacts, they have conducted no organised process of reflection on how to take advantage of opportunities that accession would provide and how negative impacts could best be addressed. In effect the need to deal with the adoption of the Acquis Communautaire, coupled with the need to ensure a positive outcome from general referenda on accession, meant that any such initiative was postponed.
- The negative impacts of accession are often concentrated on specific parts of the population. In many cases they would benefit from a regional rather than a national approach even in small countries. Foresight therefore has a role to play because it puts due emphasis on the role of stakeholders and can lead to a better articulation of needs at regional level.
- Complex regionally situated challenges such as rural unemployment require a variety of policy responses – intervention of different ministries and their agencies. Foresight provides a mechanism for the de-facto coordination of responses because it equips intervening ministries or their agencies with a shared understanding of the issue and a global cross-programme model for its resolution.
- There is a need for a regionally based knowledge infrastructure to complement the national and European knowledge infrastructure. This need is especially acute in certain rural regions across Europe. The approach should be demand rather than supply driven and foresight provides a mechanism to formulate demand on behalf of a group of regionally based stakeholders. To do so effectively requires a better understanding of the role of knowledge in the rural economy than that which exists at present. This raises new research questions that may need to be addressed in future EU Framework Programmes.

METHODOLOGICAL/ORGANISATIONAL HIGHLIGHTS [9]

As indicated above, the small country context within which the eFORESEE project was designed and implemented, was particularly conducive to the introduction of creativity in foresight approaches and methods (see Cassingena Harper and Pace, 2004). A survey of relevant foresight literature highlighted the fact that material on foresight was concerned mainly with the larger more advanced country context; and secondly, with description or comparative analysis of foresight as a phenomenon, rather than foresight as a means to achieve a well-defined goal. In the eFORESEE project, this issue was tackled from a knowledge management perspective (Nonaka and Takeuchi, 1995). CKA developed a definition of "foresight" intended to complement that provided in the FOREN manual.[10] "Foresight is the creation of shared knowledge about the future. Foresight tools are tools for the management of purposeful conversations about the future" (FOREN, 2001).

The project contributed to progress in the domain by developing a knowledge management framework for understanding, planning and implementing foresight activities. The knowledge management framework was systematically developed on the basis of modern concepts of social network analysis, social capital and thinking about the organisation as a network-of-conversations. A new way of thinking about how foresight creates value was introduced by CKA. This relies on concepts for new kinds of knowledge. In particular foresight can be seen primarily as a process to create "political knowledge" and "systemic knowledge".

The eFORESEE project's knowledge management approach to the engineering of foresight activities emerged in first recognising that:

- Foresight is a form of organisational knowledge creation, where the "organisation" in question lies at the level of the nation, the region, the industrial sector, the industrial cluster, or at the level of national and regional "systems of innovation";
- The most important foresight skills were the highly tacit skills of communication, collaboration and conciliation in extended open networks; and that
- Many of the key challenges facing foresight practitioners lay in the handling of day-to-day problems of working with independent, autonomous actors zealously pursuing their own agendas.

The need was identified to define a set of organising principles or fundamental concepts that can help practitioners deal with their unique circumstances by consciously adopting a "problem-solving" approach, rather than a "follow-the-rule" approach, and "engineering" foresight activities to achieve well-defined goals. These principles could provide a basis for:

- Ordering and clarifying foresight methods;
- Choosing appropriate methods and tools for a given foresight initiative;
- Improving the execution of established methods;
- Developing new methods;
- Evaluating performance in the execution of foresight-related tasks.

To better respond to the needs of the partners, eFORESEE therefore developed the concept of the "Foresight Learning Circle". This was a facilitated consultation or problem-solving session, involving only foresight practitioners in the process of mutual learning and mentoring. Such meetings took place in the margin of foresight-related events, so as to take advantage of the presence of experienced foresight practitioners interested in contributing to the knowledge development within the foresight community. In these meetings foresight practitioners from one or more independent initiatives produce their "mission statements" and examine them in a structured way. They may ask questions as to whether or not they will meet their goals, if the tools they have chosen are adequate for the task for which they were intended, if the right people participated, if the intended conversations took place, and if the knowledge generated provides a basis for better decision-making by participants. The focus is on qualitative aspects of the various foresight processes. The reason for doing this in a group of peers is to provide opportunities for learning from the experience of others. It allows for the emergence of conversations about issues that are hard to articulate, and the identification of problems that have not yet "been given a name" but which merit examination if process improvements are to be achieved.

The idea of the FLC or "Foresight Learning Circle" is motivated by a desire to facilitate mutual learning and continuous improvement relating to the design and management of foresight activities. It is loosely based on the concept of "quality circles" that lie at the heart of approaches to excellence in some industries. The pedagogical basis for the FLC is simple: people learn most easily:

- At the moment when they need to learn, when they are pressed by necessity;
- From people who really understand the issues they face, who can empathise with hard-to-explain aspects of a problem that are only known to people who have had similar experiences, or who have faced up to similar problems.

Kolb's learning circle represents the process of learning about complex, practical issues. It represents a process that professionals often unconsciously apply in their quest for better performance. A meeting of an FLC corresponds to the "observation and reflection" part of the circle. In an FLC a group of professionals come together for "observation and reflection" and mutually assisted problem-solving on complex issues related to the design and implementation of foresight tasks.

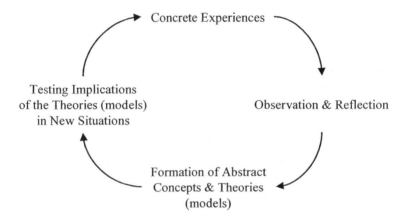

Figure 10.1   Kolb's learning circle

The eFORESEE project also contributed to the conceptual side of foresight in terms of the development of tools to help in the formal organisation of a foresight exercise. Tools such as mission statements, statements of objectives, lists of formal success criteria and the formal modelling of foresight processes were all tried out in the implementation of the pilots. The foresight pilots in each partner country were executed based on this same approach and methodology:

- A preparatory phase in which the core team was assembled, important background research carried out, constituencies assembled and implementation planned;
- The implementation is the part that focuses on stakeholder engagement, the collective ownerships of issues, joint exploration of options and the development of a shared vision;
- A follow-up phase in which the results are further circulated and validated by a broader group of stakeholders but where the main emphasis is on interaction with politicians and administrators with a view to influencing policy.

Work also focused on experimenting with the development of process maps but only Malta was able to make much progress on this. The particularities of the Maltese context necessitated some adjustments to the foresight methodology and resulted in the adoption of a more flexible approach tailored to local circumstances. The first Malta pilot was used as a test case for the other two pilots in terms of identifying appropriate foresight approaches for the Maltese context as well as providing the training for embedding the required foresight skills.

As a systematic approach and input into future strategic policy-making, the pilot was implemented through a number of preparatory phases including stakeholder-mapping and interviews, consultation and networking with strategic players. The phased approach adopted was organised around eight steps, however in reality the process was not so linear, as many of the steps were carried out in parallel and informed each other:

- Scoping the appropriate focus for the first foresight pilot;
- Stakeholder-mapping and enrolling key players in the vision-building process;
- Consultations: embedding the foresight concept among strategic policy-makers and stakeholders;
- Mapping alternative futures: filtering the possible, feasible and preferred;
- Exploring scenarios and training in foresight basics;
- Structuring the consultations: extending and focusing the process through feasibility analyses and SWOTs;
- Steering and driving the process through the Panel and developing recommendations and action lines;
- Dissemination, consultation, and implementation.

Indeed, these phases had no specific end date and continued to evolve throughout the project on an iterative basis, as new contacts were identified and new inputs on focus and priorities were obtained. After an intense phase of stakeholder-mapping to include strategic players as well as a range of experts and other stakeholders, the latter were encouraged to interact through a number of local awareness-raising and training events. The more proactive and enthusiastic stakeholders were invited to participate in the expert Panel which was responsible for refining the inputs from the consultations and enter into a more in-depth phase of the project: the work on the drivers and the recommended actions lines. The more dynamic members started their own foresight initiatives, *MedVision* and *Futurechild*.

CONCLUSION

In conclusion, the analysis and findings from the eFORESEE project outlined above highlight the fact that small countries do generate particular sets of policy needs and priorities particularly during periods of fast transition and policy change, that need to be met through foresight activity specifically designed to fit their context. Smart, creative new approaches are called for to meet the socio-economic challenges faced within the political and societal constraints of the small country context. Such approaches are best developed by local practitioners within a mutual learning framework with other practitioners and with the support of external expertise to facilitate the learning process. Indeed, there are tangible benefits in clustering small countries together through joint transnational foresight activity, where this is well-designed and coordinated through appropriate knowledge management approaches and frameworks such as foresight learning circles, so that mutual learning through benchmarking can be optimised. The learning circles are particularly effective when they focus and address practical needs and provide hands-on support in the critical implementation phases of the foresight exercise. Clustering of small countries in transnational foresight activity can help to overcome the constraints of resources, competencies and lack of critical mass and provide appropriate scale of efforts and synergies in approaches and experiences. Transnational activity does however need to take account of national priorities and interests and an appropriate balance needs to be maintained and designed into the foresight activity. Small country foresight, with its focus on a microcosm of national governance structures and dynamics, provides a unique opportunity for experimenting with novel creative approaches to policy.

Foresight and innovation policy in small countries and regions, particularly for those engaged in the fast-track process of catch-up and transition to the knowledge economy, is set to become a more permanent feature of policy-making, particularly within Europe, where the utility of such activity is becoming increasingly recognised as a key tool for more coherent and sustainable development and supported through structural funds. This should open up an array of appropriate tools and approaches and the identification over time of good practices that can contribute to the knowledge pool for other small countries and regions within and outside Europe.

NOTES

1   This chapter draws on the eFORESEE Project final reports prepared by Patrick Crehan (CKA Consulting Ltd) available at http://results.eforesee.info/.

2   Relevance of EU Regional Foresight Experiences for Small Candidate Countries, (Cassingena Harper, 2002), Malta Council for Science and Technology, Background papers by the STRATA – ETAN Expert Group Action on "Mobilising the regional foresight potential for an enlarged EU": http://www.regional-foresight.de/download/RelevanceofEU.pdf.

3   The objective of the European Research Area initiative combines three related and complementary concepts: the creation of an "internal market" in research, an area of free movement of knowledge, researchers and technology, with the aim of increasing cooperation, stimulating competition and achieving a better allocation of resources; a restructuring of the European research fabric, in particular by improved coordination of national research activities and policies, which account for most of the research carried out and financed in Europe; the development of a European research policy which not only addresses the funding of research activities, but also takes account of all relevant aspects of other EU and national policies.

4   "The foresight process involves intense iterative periods of open reflection, networking, consultation and discussion, leading to the joint refining of future visions and the common ownership of strategies, with the aim of exploiting long term opportunities opened up through the impact of science, technology and innovation on society ... It is the discovery of a common space for open thinking on the future and the incubation of strategic approaches ..." (Cassingena Harper, 2003).

5   The FP5 STRATA programme promoted dialogue between researchers, policy-makers and other societal actors on general science, technology and innovation (STI) policy issues of European relevance. It supported the establishment of networks and expert groups to improve the European STI policy development process at regional, national and international level, as well as interactions with other policy fields. An important part of this work was support for a series of foresight-related actions including – FOREN, FOMOFO, FORETECH, eFORESEE and a project entitled "Integrating Technology and Social Aspects of Foresight in Europe". Details on STRATA projects can be found at http://www.cordis.lu/improving/strata/selected.htm.

6   CAP's aims are to ensure reasonable prices for Europe's consumers and fair incomes for farmers, in particular by establishing common organisations for agricultural markets and by applying the principles of single prices, financial solidarity and Community preference.

7   The results of all the pilots are available at http://results.eforesee.info/Final Deliverables/.

8   Malta carried out three pilots instead of two.

9   This section drarws on the eFORESEE report on Foresight Planning and Learning by Patrick Crehan and a paper 'A Knowledge Management Approach to Foresight and Foresight-Related Learning Processes' presented by Patrick Crehan at the Third International Symposium on Management of Technology and Innovation, Hangzhou, China, October 2002.

10 " ... the systematic, participative gathering of anticipatory intelligence for Vision Building to inform present day decision making, and to mobilise relevant actors".

REFERENCES

Cassingena Harper, J. (2002), 'The Relevance of EU Regional Foresight Experiences for Small Candidate Countries', Paper prepared for EU HLG on Regional Foresight, April.

Cassingena Harper, J. (2003) (ed.), *Vision Document*, Report prepared for the eFORESEE Malta ICT and Knowledge Futures Pilot project, Malta.

Cassingena Harper, J. and Pace, G. (2004), 'The Creative Processes in Policy Making: A Case for Context in Foresight', *Proceedings of the Fifth International Conference on Creative Thinking*, 21–22 June, Malta.

Cassingena Harper, J. and Georghiou, L. (2005), 'The Targeted and Unforeseen Impacts of Foresight on Innovation Policy: The eFORESEE Malta Case Study', *International Journal of Foresight and Innovation Policy (IJFIP)*, 2(1).

Cassingena Harper, J. and Pace, G. (2005), 'Foresight Embedding in Malta', *European Foresight Monitoring Network Brief*, **8**, available at http://www.efmn.eu.

Crehan, P. (2002), 'A Knowledge Management Approach to Foresight and Foresight-Related Learning Processes', *Proceedings of ISMOT02 – Third International Symposium on Management of Technology and Innovation*, 25–27 October, Hangzhou, China: Zhejiang University Press.

Damvakeraki, T. (2006), 'Cyprus 2013', *European Foresight Monitoring Network Brief*, **8**, available at http://www.efmn.eu.

Damvakeraki, T. (2006), 'Biotech Estonia 2020', *European Foresight Monitoring Network Brief*, **70**, available at http://www.efmn.eu.

Drago A. (2004), 'Building the Best Future for the Marine Sector in the Maltese Islands – A Vision towards Enhancing the Marine Sector's Contribution to the Maltese Economy in 2020', Marine Foresight Pilot Final Document, Malta.

Nonaka, I. and Takeuchi, H. (1995), *The Knowledge Creating Company – How Japanese Companies Create the Dynamics of Innovation*, Oxford University Press.

Pace, L. (2006), 'Malta's Marine Sector 2020', *European Foresight Monitoring Network Brief*, **61**, available at http://www.efmn.eu.

Pace L. and Cassingena Harper, J. (2005), *Benchmarking Malta's Foresight Experience*, A Report submitted to the Forsociety ERA-NET Network, Malta.

11. Foresight in Industrialising Asia

Ron Johnston and Chatri Sripaipan

INTRODUCTION

An explanation of the variable rate of economic progress of different countries is the central challenge of economic development theory. The initial focus was almost exclusively on economic growth and industrialisation. In this view, the poorer Latin American, Asian and African countries were "underdeveloped", i.e. primitive versions of developed nations that could, with time, develop the institutions and standards of living of Europe and North America. This thinking was captured by stage theory which argued that all countries passed through the same historical stages of economic development; current underdeveloped countries were merely at an earlier stage in this linear historical progress. The policy conclusion was to patiently imitate the West, and in good time economic benefits would start to flow, though there may be some opportunity for short cuts by which underdeveloped countries might leap over a few stages and catch up more rapidly. The focus was on capital formation, although this term was successively broadened to include human as well as financial capital. This led to an emphasis on education and training as prerequisites of growth and the identification of the problem of the brain drain from what were now called Third World countries to the First.

An alternative structural theory called attention to the distinct structural problems of these countries, arguing they were not merely primitive versions of developed countries, but rather had distinctive features of their own. One of these distinctive features was that, unlike European industrialisation, Third World industrialisation was forced to occur in an economic world dominated by the already-industrialised Western countries.

With the emergence of strong competitive economies in Asia, the growth of technology-based trade and the flowering of the global knowledge economy, the issues of access to and production of technological knowledge have raised new challenges for economic development. Among other consequences, this has lead to strong interest in the tools of foresight that may assist in determining likely directions of technology development, and in developing a capacity to effectively shape national investment in knowledge capacity. This emphasis has an interesting resonance with the arguments made in the 1960s when the contemporary form and practice of foresight emerged. Thus, Bell (1967, p. xxv) attributed this emergence to the effects of economic recovery and growth:

> It arises from the simple fact that every society today is consciously committed to economic growth, to raising the standard of living of its people, and therefore to the planning, direction and control of social change. What makes the present studies, therefore, so completely different from those of the past is that they are oriented to specific social-policy purposes: and along with this new dimension, they are fashioned, self-consciously, by a new methodology that gives the promise of providing a more reliable foundation for realistic alternatives and choices, if not for exact prediction.

This chapter provides an overview of the development and adoption of foresight in the Asian nations. The variety of approaches and contexts is illustrated by more detailed analyses of two distinct situations – the emerging economic superpower of China, and the "almost Asian tiger" of Thailand.

FORESIGHT IN THE INDUSTRIALISING COUNTRIES

Foresight, as a general approach and mode of thinking, was never entirely novel to the industrialising countries, given their largely common emphasis on short- and long-term planning, as exemplified in the typical five-year plans. However, the initial focus of technology foresight on technology, which had to a large extent previously been regarded as "unplannable", and hence omitted from such plans, together with the systematic and structured nature of the foresight approach, made it attractive to those with interests in science policy in these countries. As with most management and planning processes, techniques and fashions that have emerged in the industrialised countries, the adoption of foresight in developing countries has followed a characteristic three-stage process. Even a cursory examination of the history of the development, diffusion and adoption of, to choose a few examples at random – zero-based budgeting, management by milestones, total quality management, organisational re-engineering, knowledge management, KPIs (key performance indicators), Six Sigma Quality, 360 degree evaluation (the list goes on and on) – shows the same pattern:

1. Stage 1 – uncritical imitation – this stage is typically initiated by an expatriate or graduating student returning to their own country, eager to introduce and demonstrate their skill at the latest techniques from the "West". The adoption is driven by the need to emulate what is regarded as the latest "best practice", commonly regarded as the key secret to industrialised countries' economic success. The techniques, and commonly even the focus, are imported uncritically, even though that obviously means accepting assumptions and conditions that apply in the originating country.

2. Stage 2 – learning – in this stage there is a steady development of a more substantial national capability, in both scale and breadth, in the new tool. There is a significant learning from the initial exercises, which are commonly marked by very great difficulty and limited success. As a result there is a growing recognition of the strengths and limitations of the tool, of its design, logistic and infrastructure requirements.

3. Stage 3 – adaptation and innovation – by this stage a well-developed comprehension of the tools has been established, and there is a strong emphasis on adaptation for local issues and considerations. Evidence starts to emerge of effectiveness and positive impacts, and there is a burgeoning of recognition and support. Innovation starts to emerge in methodology, practice and application.

In the case of foresight, Stage 1 of adoption commonly took the form of a direct imitation of a foresight exercise in an industrialised nation. In Asia, the most common form was a copy of the Japanese Delphi exercise, which has been conducted by the National Institute of Science and Technology Policy (NISTEP) at five-yearly intervals since 1970.[1] The imitation frequently took the extreme form of simply distributing a series of questions extracted from the Japanese survey instrument to local "experts". All too frequently, the number of experts in the developing country was so small that statistically significant findings could not be drawn. Furthermore, there was little consideration of whether questions about the estimated date of realisation of a particular technology or technology-based capability had any major relevance for the economy of a small developing country.

Stage 2 was characterised by training of nationals in foresight skills, mostly through recognised international courses in Bangkok (APEC Centre for Technology Foresight),[2] Manchester (PREST),[3] or Eastern Europe and South America (UNIDO).[4] It also typically saw a shift from attempts to engage in an all-encompassing national study to more focused studies, of a smaller scale, on industry sectors, specific technologies of interest, or issues of national economic or social importance. At this stage, with the initial promise of foresight apparently receding, some initiatives lost their impetus in some developing countries. But in others, their value began to be appreciated outside the circle of aficionados, most commonly by institutions with a responsibility for science, engineering and technology (SET). Foresight found a fairly ready application in this case, perhaps not least because of the international orientation which is a feature of the SET community. But in addition, issues of setting priorities for research and gaining insights into the possible trajectories and applications of emerging technologies were themes of common interest across the researcher and SET policy communities, regardless of the size of the national economy.

Economic experiences were also influential (Malee, 2001):

> The recent economic crisis in Asia, especially in Thailand, has confirmed the belief that the regional and global economic systems, especially in relation to finance and credit, and the transnational trading and economic system have as much, if not more, impact on technology development as the national economic system ... The emerging trade and technology development issues, especially related to genetically modified mechanisms, helped to strengthen the case for more effective technology foresight studies and wide public consultations.

Stage 3 was (and is) characterised by significant investment in the adaptation and design of foresight projects to fit local needs and local resources. Large-scale national exercises based on Delphi polling, or other techniques, are much less common. As one of the authors has argued previously (Johnston, 2004), there is a progressive move towards embedding foresight within existing planning, investment and decision-making mechanisms and structures. Hence foresight studies proliferate, but they may be less visible because of confidentiality requirements or simple in-house conduct. This stage also saw an increase in foresight exercises employing multiple tools. Thus, scenarios have been used to generate the questions for a Delphi polling exercise, the latter providing a means of getting a wider judgement on the perspectives developed by the necessarily limited numbers of stakeholders able to engage in a scenario development exercise. Scenarios have also been used to provide starting points, and alternative finishing points, for technology roadmapping exercises. A more detailed categorisation of national foresight studies has been made against the objectives pursued, identified as: (a) national competitiveness; (b) vision building; (c) identification of key or emerging technologies; (d) creation of networks; (e) information dissemination and education and (f) development of a forward-looking culture. The authors concluded that in countries where successive projects have been carried out, one can observe how the evolution in methods employed aims to increase the impact and effectiveness of foresight (Gavigan and Scapolo, 1999, pp. 491–513).

A typology of foresight exercises is applied to describe activities in Asia (Table 11.1).[5] The typology uses four distinct dimensions (Johnston, 2002):

1. The first dimension addresses the **state of evolution**, or experience of foresight in the country. In this analysis, we draw on the categories outlined above – imitation, learning, and adaptation/innovation.
2. The second dimension, **level**, is concerned with whether the focus of the foresight study is national, regional, or sectoral.
3. The third dimension is **focus**. It addresses the objective of the study with the extremes of the dimension being the conduct of the exercise, with its associated benefits, versus policy and planning purposes.
4. Finally the fourth dimension, **objectives**, was developed to address the different outcomes of foresight exercises. The major elements are anticipation, which is concerned with looking to the future and establishing a vision; networking, which incorporates participation; and action – policy, strategy, planning and decision-making.

Table 11.1 National foresight activities in Asia

Country	State of evolution [a]	Level [b]	Focus [c]	Objectives [d]
Brunei Darussalam	*Im*	*Se, O*	*F/s*	*A, Ne*
Cambodia	*F*	*Se*	*F/s*	*A*
China	*Le*	*R, Se*	*F/s, P*	*A, Act*
Indonesia	*Le*	*Se*	*F/s,*	*A, Act-P*
Japan	*A/I*	*N, Se, O*	*F/s, P*	*A, Ne, Act*
Korea	*A/I*	*N, Se, O*	*F/s, P*	*A, Ne, Act*
Lao PDR	*Im*	*Se*	*F/s*	*A, Ne*
Malaysia	*Le*	*N, Se, O*	*F/s, P*	*A, Ne, Ac*
Myanmar	*Im*	*Se*	*F/s*	*A*
Philippines	*Le*	*Se*	*F/s*	*A, Act-P*
Singapore	*A/I*	*Se*	*P*	*Act*
Taiwan	*A/I*	*Se, O*	*P*	*Ne, Act*
Thailand	*Le*	*N, R, Se, O*	*F/s*	*A, Act-P*
Vietnam	*Im*	*Se*	*F/s, P*	*A, Ne, Act-P*

Notes:

a   **State of evolution**: position of foresight/future activities in the country along a spectrum from imitation [Im], via learning [Le] to adaptation/innovation [A/I]
b   **Level**: national [N], regional [R], sectoral [Se], organizational [O]
c   **Focus**: foresight [F/s], policy action [P]
d   **Objectives**: anticipation [A], networking [Ne], action achieved [Act], action proposed [Act-P]

These categories inevitably conceal a great deal of variable detail. Nevertheless, even at this broad level of categorisation, some useful insights are generated. First, the extent of the adoption of foresight is apparent. All countries have some experience of formal foresight projects. The industrialised countries have reached the stage where foresight is embedded in planning and policy processes at a variety of levels. The majority of industrialising countries are still at the early foundational stage, some conducting only their first or second project. However, some have begun to embed foresight processes in regular planning and policy formulation. This stage of development is partly reflected in the focus of the foresight. Those countries at the foundational stage are commonly concerned primarily with the challenge of conducting the foresight exercise, even if some broader national objectives have been identified. With regard to level, even when the study is national, a frequent device to reduce the scope and concentrate the effort is to identify a number of key sectors. These sectors are selected usually on the basis of their present and future value to the national economy. Regional foresight studies are less common, but growing.

FORESIGHT EXPERIENCE IN CHINA[6]

One rapidly industrialising country that has progressively been building its capability in foresight, and applying it to a range of planning activities, is the People's Republic of China. Some insights can be gained for industrialising and industrialised nations, and for the practice of foresight, from an examination of the Chinese experience.

Early History

Long-term planning has played a central role in the development of the People's Republic of China's socialist market economy since its establishment in 1949. While the emphasis has been on more traditional economic development planning, attention has also been paid to long-term planning and macro-strategy formulation for science and technology. Thus, in 1955 under the then Premier Zhou Enlai, a 12-year "National Long-Range Science and Technology Prospective Programme 1956–1967" was prepared. It is reported (Yang Qiquan, 2003) that: "thousands of specialists worked together with the Government economic department chiefs to draw up a long-term development strategy for science and technology". Some 57 research targets, in thirteen focal areas, were identified. Top priority was accorded to 12 "research missions".[7]

This approach, based on involving large numbers of specialists in establishing targets, major missions, key focal areas and technologies, leading to development measures, was continued with the "Science and Technology Plan 1963–1972", the "National Science and Technology Development Plan 1978–1985", the "National Science and Technology Development Plan 1986–2000", the "National five-Year Plan for Science and Technology 2001–2005" and the "National Science and Technology Development Sketch 1990–2000–2020". Other developments included the establishment of a "forecast research association" in 1979, and a major government sponsored future-oriented report "China in 2000" published in 1985 addressing the themes of population, work, consumption, the economy, energy resources, communications, transportation, agriculture, education, the environment, natural resources and science and technology.

Recent Foresight Exercises

A project entitled "Selection of National Critical Technologies", initiated by the State Science and Technology Commission was completed in 1995. The "critical technologies" approach was largely developed within the US, but applied also in France and Germany, and other industrialised countries.[8] The Chinese project was designed both to identify what were the national critical technologies and to establish the research approach and criteria for selecting such priority technologies (Jiayu and Yongchun, 2001). The method was based on the canvassing of candidate technology topics from experts in four areas – information, biology, manufacturing and materials. The resulting 220 topics were evaluated by members of the Chinese Academy and the Chinese Engineering Academy, and 24 critical technologies were selected. These were fed into the Ninth Five-Year National Science and Technology Plan 1996–2000.

In 1999, the National Research Centre for Science and Technology for Development (NRCSTD), an arm of the Ministry of Science and Technology (MOST), led a project entitled "Technology Foresight of Priority Industries in China", focused on agriculture, information and advanced manufacturing.[9] The project was based on three components:

1. The identification of experts, capable of a strategic perspective, and with ten years experience in the relevant research area. More than 1,200 such experts were selected, 80 per cent from research institutes and universities, who were required to complete questionnaires and attend regional workshops;
2. The preparation of a list of candidate technology topics via analyses of current and future markets, and trends and progress within science

and technology. Selection criteria applied were that the technology should be advanced, directly relevant to socio-economic problems, provide the basis of a new industry, and available before 2010;

3.   A systematic basis for evaluation of responses and ranking the prospective technologies against six priority indicators – importance, efficiency, constraints, comparison with international level, route to realisation and time to realisation.

The project produced a list of critical technologies in each of the three fields contained within a broader report on each field, and a technological database in each field, which has been incorporated into a national database.

On the basis of this project, a list of 11 criteria was established for selecting and ranking national critical technologies.[10] Experience with this project served also to establish the role of technology foresight in contributing to national economic growth, science and technology development, social progress and national defence. It did this by:

- Providing important support for decision-making in national S&T and economic development planning, particularly by central Government but also by industry and local government;
- Setting directions for R&D in the institutes and universities;
- Providing a framework for enterprises to select technologies;
- Stimulating new and continuing dialogue between the science base and industry, and forging ties between the institutes, universities and enterprises.

During 2002–2003, NCRSTD has led another Delphi-based foresight project focused on information and communication, life science and biotechnology, and new materials technology.[11] The project involved three stages over 18 months. The first stage addressed project design, analysis of socio-economic needs in China and S&T trends and topic selection. The latter was developed through identification of more than 1,000 experts and 40 consultative seminars, leading to 218 technology topics. The second stage involved a two-round Delphi survey of 1,300 respondents, with a 47 per cent response in the first round, and a 42 per cent response in the second. There were 16 survey items, covering the usual items of expertise, time of realisation, and socio-economic and environmental consequences, but also issues of particular national relevance, such as the gap between China and leading countries, the R&D base in China, IP rights in the next five years, and impact on rebuilding traditional industries. The third stage involved critical technology selection and dissemination of results. The same approach was used in a new Delphi-based foresight project, this time focused on energy, environment and resources, and advanced manufacturing initiated by the same organisations in 2004 and completed in 2005.[12] A total of 28 sub-fields in the three areas, and 261 technology topics were selected (from a total of 500 generated) by the 15 experts in foresight, and three area research groups, each composed of about 20 experts. The two-round Delphi was sent to 1,372 respondents, with a 35 per cent response rate.

In addition to these Government-led foresight activities, the Centre for the Future of China, a not-for-profit futures research institute co-located in Beijing and San Francisco has produced "2002: China Five-Year Forecast".[13] The report, available only to members of the China Foresight Programme, addresses likely paths of economic development, the impact of China's accession to the World Trade Organisation (WTO), the challenges in developing a networked economy, business opportunities along the Old Silk Road, diverging lifestyles, closing the skills gap, cultural innovation, television networks, new consumers, emergence of knowledge workers, and the emergence of consumerism. Evidently, foresight is starting to find a market, and an application, in the private sector, and it is being recognised as having commercial value.

Hence, we can see the steady growth in the experience with, and application of foresight, in China. The initial foundational approaches have quickly moved to direct application in target-setting for research and technology development, involving a wide range of experts and decision-makers.

THE FORESIGHT EXPERIENCE IN THAILAND

A much smaller Asian country which has made a significant investment in foresight is Thailand. It was among the first of the south-east Asian countries to become engaged in foresight. In 1998, foresight was formally introduced in Thailand when the Thai Cabinet agreed to host the APEC Centre for Technology Foresight (APEC CTF) in the National Science and Technology Development Agency (NSTDA) as a project under the APEC Industrial Science and Technology Working Group.[14] After the launch of the APEC CTF, an informal grouping called the Thai Foresight Unit was formed with the aims of involving other organisations and extending beyond technology into the more social aspects of foresight. The initial enthusiasm led to the first sectoral project on "Foresighting Thai Agriculture" in the same year. By 2005, the APEC CTF and Science, Technology and Innovation Policy Research Division of NSTDA had conducted more than ten foresight projects at the national, sectoral and organisational levels.

Since that time other countries with a stronger economic and industrial base, such as Singapore and Korea, have probably made greater strides in the application of foresight to issues of economic development. However, Thailand's achievement is that foresight techniques have been diffused increasingly widely in Thai society, as the examples below will show. They, in turn, describe attempts to apply foresight to policy at a national level, a sectoral level, and within an organisation.

A Vision of Thai Science and Technology in 2020

This project was an initiative of the Ministry of Science and Technology on the twentieth anniversary of its establishment. The timing coincided with the drafting of the 9th National Economic and Social Development Plan (2002–2006). At the time that the project started in 1999, Thailand was in the process of recovering from the economic crisis of 1997, and coming to terms with some major competitive disadvantages. The cost of Thai labour had become too high to compete with labour-intensive products manufactured in China and Vietnam. At the same time, the weak national technology capability precluded competing with newly industrialising countries like Korea and Taiwan in devising new products for the world market.[15]

The project set out to address these challenges, by enhancing the awareness of the importance of S&T in Thai society and linking it more closely to the economic and social development of the country. Since the project was a ministerial initiative, the Minister chaired the Steering Committee composed of eminent persons from both the science and social science communities and a number of high-ranking officials. The Executive Committee was chaired by the Chairman of the National Economic and Social Development Board which is the national planning body. The Science, Technology and Innovation Policy Research Division of NSTDA served as the Secretariat to both Committees.

The project was launched by the First National Congress of Science and Technology for Development in late 1999 to an audience of 1,200 people from the S&T community, the public sector, business and manufacturing, as well as the general public. After introductory lectures, participants were divided into groups of 20 to consider the preferred direction of development of S&T over the next 20 years, in ten key areas that the Steering Committee thought to be of high importance in the future: agriculture and agricultural industry, manufacturing, service industry, management, education and culture, environment, health and welfare, commerce, energy and ICT.

The top 10 desired outcomes for science and technology development were:

1. Useful and correct information any time and anywhere;
2. Self-sufficiency in technology and resources;
3. Enhancement of Thai local wisdom;
4. Reaching international standards;
5. Building networks of cooperation;
6. Value-added to the entire supply chain;
7. Excellence in areas in which Thais have expertise;
8. A general capability to think and act scientifically;
9. Equity and universality of services;
10. Preventative measures for health and the environment.

The second phase involved studies by groups of six to eight people, led by an expert. Their task was to consider the ten desired outcomes, reformulated by the Steering Committee into factors of demand and supply. The demand side was represented by seven economic and social sectors: agriculture; manufacturing; commerce and services; education, culture, health, and welfare; the environment; energy; and communications/telecommunications. The supply side, which would need to realise the objectives of the demand-producing sectors, was divided into six aspects of the S&T system: R&D, technology transfer, human resource development, infrastructure, management and information. The challenge for each of the 13 study groups was not only to address their sectoral focus, but to do so in the context of a policy matrix linking demand and supply, over the next 20 years. The relationship between the demand for science and technology and the strategies for science and technology development is shown in the policy matrix below (see Table 11.2). The 13 reports were submitted to the Second National Congress of Science and Technology for Development held in 2000 with an audience of 1,500 people. After hearing summarised reports from each study, participants were divided into 85 groups to brainstorm on the 13 studies. A third workshop was held in late 2001 for the researchers and the staff of the National Economic and Social Development Board to exchange ideas to ensure that the final synthesis report would fit the framework of the 9th National Economic and Social Development Plan. The output of the project (NSTDA, 2001) formed the major input to the S&T component of the 9th National Economic and Social Development Plan. It also highlighted the need for the coordination of science and technology nationwide, and paved the way for the establishment of the National Science and Technology Policy Committee in early 2001, chaired by the Prime Minister, with NSTDA as Secretariat. Subsequently, this Committee was assigned to draft the ten-year National Science and Technology Strategic Plan (2004–2013) which was approved by the Cabinet in 2004 and was the subject of the Third National Congress of Science and Technology for Development shortly afterwards.

Table 11.2 Demand for S&T and strategies for S&T development

	Telecom	Environment	Education, Culture, Health, Welfare	Services	Manufacturing	Agriculture
Research and Development	-Telecom equipment -Software	-Envir. monitoring -Clean technology	-Multimedia -Post-genomic medicine	-Generic computer programmes -Natural resource remediation -Low emission vehicles	-New materials -Manufacturing process -Machinery improvement	-Plants, animals, microbes strains -Food innovation
Technology Transfer	-Learning from global telecom organisations -Learning from suppliers	-Learning from abroad -Transfer to the people	-Lessons from educational management abroad	-Learning from suppliers, customers, colleagues	-Metals -Polymers -Quality of products	-Local R&D, innovation centres
Human Resource Development	-Computer and electronics engineers Telecom networks	-Researchers with Masters and PhDs	-Network of higher education institutions -System friendly to children of every levels of potential	-Personnel in computer, telecommunications, logistics	-Engineering education -Personnel management	-Technology transfer personnel
Infrastructure	-Telephone and Internet network throughout the country	-National envir. alarm station -National Centre for Environmental Technology	-IT for every educational institutions -Science schools	-60 telephones per 100 population -Electronic commerce	-Testing -Standards -Laws	-Standards -Intellectual property regime
S&T Management	-Fair liberalisation policy	-Institutions in Ministry of Science and Technology	-Educational quality assurance -Career path	-R&D managers -Transportation Network	-Management information system	-National S&T plan
S&T Information	-Movement of other countries -Marketing and commerce	Environmental situation in 76 provinces	-Edutainment -Web-based lessons	-Customer database in tourism and transportation	-Information on materials, machinery -Transaction system	-Commodity prices -Knowledge in agriculture

IT for Community Education

A Technology Foresight project on IT for Education started in 2001 at the time that the Office of the National Education Commission (ONEC) was drafting the Master Plan for ICT in Education. An attempt to link the two initiatives directly was not successful but some key personnel of ONEC were successfully incorporated into the foresight exercise. This project was a local application of the APEC-wide foresight project on Technology for Learning and Culture (NSTDA, 2000), which recommended policies and strategies for adoption by member economies in developing their approach to IT for education.

A range of stakeholders, including local wise men, school owners, community leaders, professors, curriculum developers, NGOs, IT businesses, high-school and university students and youth leaders were brought together in a scenario workshop to consider the key question of "how can the community best develop learning under the influence of IT in the next ten years?":

- The first scenario depicted the failure of educational reform despite the wide spread use of the Internet. Eventually, the communities learned to be self-reliant and how to use IT sensibly.
- In a second scenario, the energy crisis and the economic downturn led to a change of government. Decentralisation of education management and IT businesses encouraged development in software and digital content at the community level.
- The third scenario described an ideal school in the future where students use personal mobile computers as a tool for learning, supported by the infrastructure of software, content and automatic translation. Students are encouraged to research with a scientific mind, learn to be wise in spending by using simulation, and appreciate Thai culture through hands-on experience.

The working group found that the scenarios raised three common issues. First, the development of IT increased community's access to the information and reduced the disparity in opportunity to learn and develop. Second, a process of self-reflection occurred in rural communities, aiming to find the balance between local wisdom and world knowledge. Third, the need for managing the process of change to the desirable future was highlighted.

A subsequent workshop examined the three issues of equity, quality and efficiency. The main recommendations were: to strengthen the community as the base for the development of IT for learning from the grass roots, to develop the content and knowledge relevant to the needs of the community, to understand the changing role of school in education and the learning process, and to have good governance at the local level for the management of IT for learning and development.

There were a number of direct outcomes. First, the "Samka Village" (in the province of Lampang in northern Thailand) model of introducing IT into a community with only one telephone line was widely publicised. The key characteristics were strong collective leaders, a good democratic management system, a database of local information, and a community learning centre. Youth has been helping adults in working on the village accounts, sending e-mails and searching information on the Internet. The village has high potential to further exploit IT for e-commerce, learning and entertainment, and integrating the local curriculum with the main curriculum.

Another outcome arose from the involvement of ONEC staff. The experience has increased confidence in foresight and has led to a request for a scenario planning workshop for headmasters and IT teachers from 30 schools who were competing for IT funding from ONEC. The theme of the workshop was "how can we use IT to support learning reform in schools?": In writing scenarios and forming strategies, the participants developed better ideas on which to base their proposals (Centre for Policy Innovation, King Mongkut's University of Technology, 2002).

Technology Promotion Association (Thailand–Japan)

A vision-setting exercise was carried out by the Technology Promotion Association (Thailand–Japan) or TPA in 2003. The TPA was established in 1973 by a group of Thai alumni from Japan with partial support from the Ministry of Economics and International Trade of Japan. It has the main aim of transferring technology to Thai industry through training courses in technology, business management and languages. Over the 30 years of its operation, it has trained over 350,000 persons in industrial technology, over 100,000 persons in languages, and over 1,000 persons in enterprise diagnosis. It has published more than 830 titles with more than 4.5 million copies, provided calibration services to more than 2,000 companies a year and diagnosed more than 630 companies.

As it approached its 30th anniversary in 2003, the association deemed it appropriate to rethink its vision and strategy seriously, concluding that it would like to develop a five-year strategic plan to guide the yearly budgeting process. After initial consultation with stakeholders and brainstorming sessions of the Governing Board, a more formal foresight process was adopted and the APEC Center for Technology Foresight at NSTDA was contracted to lead the project.

The methodologies used in this project were scenario planning, SWOT analysis and "Balanced Scorecard" (Kaplan and Norton, 1996). After a series of standard pre-foresight activities, workshops were organised with 40 participants including the Governing Board, top executives, core staff and clients.

The first workshop was on scenario planning. After reviewing changing environments affecting the organisation and looking at trends and uncertainties, participants developed three scenarios on the development of TPA over the next ten years. The three scenarios were:

1. **TPA Corporation** – TPA is privatised and run like a corporation. It concentrates on profit-making activities, and emphasises result-based management. The power shifts from the Governing Board to the CEO and there is less synergy among various departments of TPA.
2. **Value of NPO** – TPA realises the value of being a non-profit organisation and has changed its status from an association to a foundation. New dimension for learning, synergy and team work are developed. It receives the Prime Minister's Award in International Cooperation for NPO.
3. **New Synergy** – Competition from universities is increasing, as public universities are being corporatised. TPA develops synergies among its various departments spearheaded by consultancy and practical solution provision. It expands services to the provinces and develops a closer relationship with China.

Two issues that had been under debate in TPA for some time were: "Can TPA survive if Japan cuts off its financial support?" and "Should TPA become a private company since some of its operations are making a profit?" Whilst the scenarios were being written, the threat of dwindling financial support from Japan disappeared when it was resolved that TPA must be financially self-sufficient (while maintaining the close relationship with Japan). It also became clear that TPA would stand to lose more than it would gain by becoming a profit-seeking company.

In the second workshop, the SWOT analysis and the vision of TPA were revisited. The new vision is for TPA to be a leader in creating, storing and disseminating technology in order to provide solutions to Thai industry – a considerably more ambitious role than previously. The Balanced Scorecard was used to develop the five-year strategic plan by moving the process from vision to strategic positioning, core competence, strategic theme, and finally a strategy map. The strategy map was fully debated in a third workshop. The map ensured the coherence and the balance of the objectives, measurements, targets and initiatives. Seven initiatives were chosen to be explored further:

1.   Innovation Programme;
2.   Talent Bank Programme;
3.   Marketing Programme;
4.   IT/Knowledge Management Programme;
5.   Organisation Restructure Programme;
6.   Human Resource Management Programme;
7.   Corporate Identity Programme.

The outcomes were considerable. The present Chairman has stated that strategic planning has changed the way that TPA conducts business, and the change is so ingrained that even with a change of management TPA would not make a U-turn in policy. In the past two years, TPA has been working on all the seven programmes mentioned above. The most significant change is the increased use of IT. Books are now sold on-line and trainees can register through the website. A human resource management package with e-learning has been developed. The databases of members and customers have been integrated.

The first of these case studies shows the progressive introduction of foresight into more familiar planning techniques, as its value for anticipation, networking and policy become apparent. The second case study demonstrates the value of foresight in enabling a very disparate group of stakeholders (some of whom did not even see themselves in that role) to engage in collective consideration of possible futures for a remote community, supported by IT. It produced not only network effects, but a new shared vision of what might be possible, and some mechanisms to pursue some aspects. The third case study shows the use of foresight as a tool for substantially rethinking an organisation's vision and strategy, and, linked with the application of the Balanced Scorecard tool, for implementing significant and long-lasting change.

In each case foresight had a substantial direct impact. The cases also served as models of the effective application of foresight and of the possible outcomes achievable in Thailand.

CONCLUSIONS

This study of the adoption of foresight by industrialising economies reveals a common, three-stage process of imitation, assimilation through learning, and adaptation and innovation. At first the techniques are copied, relatively blindly, with few outcomes beyond the development of a greater awareness of the tool. In the second stage, foresight is progressively assimilated with other planning tools, and into standard planning practice. At this stage, some adaptation is necessary to fit the particular context and conditions of the country. In the third stage, which a number of industrialising countries are now entering into, innovative approaches are developed to better assess and seize new opportunities emerging in a world dominated by rapid change.

A close examination of the tools and procedures used in the industrialising countries, when compared with those more economically advanced, reveals a significantly lower level of specificity, particularly with regard to objective measures of technology and market achievements. Such a capacity, which greatly strengthens the precision and reliability of the findings of foresight studies, would appear to be a product of stronger industrial structures, and greater experience with foresight, to be developed with experience. This capacity could be considered an important component of the infrastructure of a technology-supported knowledge economy, and an outcome from investment in foresight.

A number of key factors for success in foresight projects in industrialising countries have been identified, particularly from Thai experience. These include the level of foresight experience, capability of the participants, support from lead organisations and individuals, and alignment of foresight projects with political or bureaucratic priorities. The analysis of Chinese experience with foresight reveals that the necessary capacity can be built in a relatively short time. The development of this capacity is inherently a learning process, with many aspects of foresight still largely tacit. Under these circumstances, there would appear to be considerable advantage in more active sharing of foresight experience and capability among industrialising countries, and on the part of the industrialised nations.

NOTES

1. Reports available at http://www.nistep.go.jp.
2. http://www.apecforesight.org.
3. http://www.mbs.ac.uk/prest.
4. http://www.unido.org.
5. The Table is based on the broad judgment of the authors, and the specific experience of an ASEAN Technology Foresight and Scan project conducted from 2003–2005 (Sripaipan, 2005).
6. The authors are grateful for the kind assistance of Dr Zhong-Ming Gong, Assistant Research Fellow, Department of Foresight and Development Research, National Research Center for Science and Technology for Development, Ministry of Science and Technology, in researching this section.
7. Basic research; Synthetic fuels; Jet technology; Transport machinery; Nuclear energy technique; Paroxysm prevention and cure; New metallurgical technologies; Modern agricultural technologies; Automation and precision instruments; Petroleum and mineral resources survey and exploration; Yellow

River and Yangtze River exploration and exploitation; and Electronics – semi-conductor technology, computer technology, long-distance control technology.

8. The critical technology approach has most recently been described and analysed by the RAND Corporation; for details see http://istf.ucf.edu/Tools/NCTs/.
9. Based on information contained in Jiayu and Yongchun (2001).
10. These national critical technologies are:
    1. Market need or opportunity;
    2. Contribution to quality of life;
    3. Contribution to environmental protection;
    4. Contribution to elevating technology skills in industry;
    5. Economic efficiency, in terms of increased product quality, reduced cost, new capability, or enhancement of existing capability;
    6. Efficient use of natural resources and reduce energy consumption;
    7. Promotion of other industry or technology development;
    8. Time to realisation – within the next 10–15 years;
    9. Cost to support R&D through to production;
    10. Applicability to a range of industries.
11. This section is based on material drawn from Qi-Qin Yang *et al.*, (2004).
12. Based on information provided as a personal communication by Zhong-Ming Gong, May 2005; the report of the project, available only in Chinese, is "China's Report of Technology Foresight 2004".
13. Accessible at http://www.china-future.org.
14. "Foresight Activities and Strategic Policies of Thailand", 2nd International Conference on Technology Foresight – Third Generation Foresight and Prioritization in S&T Policy, National Institute of Science and Technology Policy, MEXT, Japan, 27–28 February 2003.
15. In 1999, the overall competitiveness of Thailand according to the ranking of the International Institute for Management Development (IMD) in its World Competitiveness Report was 36 among 47 countries while its science and technology infrastructure ranked almost last at 46. In 2005, the science infrastructure and the technology infrastructure were ranked 56 and 45 respectively among 60 countries while the overall competitiveness improved to 27 (World Competitiveness Yearbook, 2005).

REFERENCES

Bell, D. (1967), 'Introduction', in Kahn, H. and Wiener, A., *The Year 2000: A Framework for Speculation on the Next Thirty-Three Years*, New York: Macmillan.

Centre for Policy Innovation, King Mongkut's University of Technology (2002), *Technology Foresight on IT for Education*, final report submitted to the APEC Center for Technology Foresight, National Science and Technology Development Agency, Thailand (in Thai).

Cheng, J. and Zhou, Y. (2001), 'A Brief Introduction of National Technology Foresight in China', in NISTEP, *Proceedings of International Conference on Technology Foresight*, Tokyo, accessible at http://www.nistep.go.jp.

Gavigan, J. and Scapolo, F. (1999), 'Matching Methods to the Mission; a Comparison of National Foresight Exercises', *Foresight*, **1** (6), pp. 491–513.

Johnston, R. (2002), 'The State and Contribution of International Foresight: New Challenges', in IPTS, *The Role of Foresight in the Selection of Research Policy Priorities*, Seville, 13–14 May.

Johnston, R. and Tegart, G. (2004), 'Some Advances in the Practice of Foresight', in IPTS, *EU-US Scientific Seminar – New Technology Foresight, Forecasting and Assessment Methods*, Seville, 13–14 May 2004.

Kaplan, N. and Norton, D. (1996), *The Balanced Scorecard: Translating Strategy into Action*, Harvard: Harvard Business School Press.

Malee, S. (2001) 'Lessons Learned in Technology Foresight from Developing Country Perspectives', in NISTEP, *The Proceeding of International Conference on Technology Foresight*, Tokyo, available at http://www.nistep.go.jp/ achiev/ftx/eng/mat0771e.html.

NSTDA (2000), *Technology for Learning and Culture in the APEC Region to 2010*, APEC Industrial Science and Technology Working Group and the APEC Center for Technology Foresight, Thailand, ISBN 974-7360-12-8.

NSTDA (2001), *Vision of Thai Science and Technology: Status and Strategy: Final Report of the S&T 2020 Project*, ISBN 974-7360-44-6, National Science and Technology Development Agency, Thailand (in Thai).

Qiquan,Y. (2003), 'Technology Foresight and Crucial Technology Focus in China', in NISTEP, *Proceedings of the Second International Conference on Technology Foresight*, Tokyo, accessible at http://www.nistep.go.jp.

Sripaipan, C. and Limsamarnphun, K. (2005), *Foresight in ASEAN – Experience from the ASEAN Technology Foresight and Scan Project*, NSTDA, Thailand.

Yang, Q., Gong, Z., Cheng, J. and Wang, G. (2004), 'Technology Foresight and Critical Technology Selection in China', *International Journal of Foresight and Innovation Policy*, 1(1/2), pp. 168–180 [report of the project: 'China's Report of Technology Foresight 2003', available only in Chinese].

12. Foresight in Latin America

Rafael Popper and Javier Medina

INTRODUCTION

The evolution of science and technology in Latin American countries has often been debated using neoclassical modernisation and dependency theories (Cardoso and Faletto, 1969). This thinking led to a big wave of "structural reforms" and privatisations recommended mainly by the International Monetary Fund (IMF) and the World Bank in the last two decades of the 20th century. However, since the 2000s, emerging economies in the region have introduced more promising "bottom-up" and "contextualised" policy-making processes to (a) identify and promote productive and strategic sectors, (b) support "endogenous development" and capacity-building programmes, and (c) promote industrial/social transformation through technology/policy transfer, technological/organisational innovation and mutual learning (see also Chapter 14). At the same time, regional organisations – such as MERCOSUR, the Economic Commission for Latin America and the Caribbean (ECLAC), the Andean Development Corporation (CAF), the Andrés Bello Agreement (CAB), the Organisation of American States (OAS) and the up-and-coming Union of South American Nations (UNASUR) – have acknowledged that a key challenge of the region is to achieve socio-economic integration with high levels of sustained growth and participation of the civil society, while respecting cultural diversity, reducing poverty and protecting the environment. For this reason, a strong emphasis is thus currently placed on forward-looking and "strategic" initiatives which strengthen the capacity of societies to improve their socio-economic and institutional frameworks.

Against this background, this chapter shows how foresight has become an influential instrument in the region by describing its contributions to science, technology and innovation policy-making. In so doing, the chapter presents an overview of foresight evolution in 13 countries (see Table 12.1)[1] and provides a more detailed description of selected experiences in Brazil, Colombia and Venezuela.

TOWARDS A FORESIGHT CULTURE IN LATIN AMERICA

Foresight entered the Latin American policy environment as a tool for anticipating possible futures but has been characterised by varying interpretations and uses. Some countries adopted the term to refer to traditional "futures studies", "forecasting" or "technology assessment" activities, while some focus on the French "strategic prospective" approach (see Chapter 2 and 3). More recently, in part due to international initiatives – such as UNIDO's Technology Foresight Programme for Latin America and the Caribbean (TFLAC) and a number of European networks and projects – some countries began to practice foresight as a combination of prospective thinking with wide participation to inform policy-making. International foresight experiences, mainly European, had a very strong influence in the last decade, but many Latin American countries have had to introduce new practices (especially cost-effective ways to gather experts' knowledge) to cope with challenges of having limited amount of resources allocated to foresight activities at both regional and national levels.

The degree of international influence and foresight knowledge transfer normally varies depending on the rationales, objectives, time horizon, methodology, territorial scope, research areas and funding scheme chosen by the sponsors and organisers. However, since the 1970s, three major drivers have allowed for the internationalisation of Latin American foresight:

- The creation of major centres, e.g. Bariloche Group[2] in Argentina, Javier Barros Sierra Foundation in Mexico, Centre for Development Studies (CENDES) in Venezuela, the S&T Observatory (OCCT) in Cuba, the Centre for Management and Strategic Studies (CGEE) in Brazil, and the recently created National Institute of Foresight, Innovation and Knowledge Management, in Colombia.
- The support of international organisations, largely with expertise, financial contributions, political support, dissemination and capacity building activities, in particular UNIDO, ECLAC, CAB, CAF and the European Union (EU).
- The emergence of research projects and mapping exercises based on collaborative work between international organisations, government agencies, and research and mobility networks, including the CYTED network, Quo Vadis and the Euro-Latin SELF-RULE[3] network of higher education institutions, among others.

All this experience has produced an extensive literature (mostly in Spanish and Portuguese), interesting methodological innovations and a wide range of organisational forms.[4]

Organisational Forms of Foresight

As in Europe, Latin American foresight could be described using various organisational forms which normally depend on the experience and internal conditions of national environments. These forms vary in terms of their *scope* (more or less inclusive) and *organisational structure* (more or less formal).

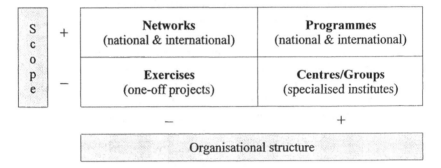

Figure 12.1   Organisational forms of foresight

The prevailing organisational forms are:

- *National and international foresight programmes* with Brazil, Chile, Colombia and Venezuela establishing the first national foresight programmes in the region. UNIDO, CAB[5] and ECLAC have also launched major cross-national initiatives, while other regional institutions have included sectoral exercises in their working agendas; e.g. CAF's Andean Competitive Programme (PAC);
- *One-off projects and exercises* with limited duration. These are often nationally funded and sector-oriented (i.e. Chilean Software Industry 2010,[6] and Colombian Biotechnology 2015,[7] for example). However, international organisations like UNIDO and the European Union have also organised sectoral and cooperation-oriented exercises, such as the Brazilian Civil Construction 2013,[8] and SCOPE-2015,[9] respectively;
- *Specialised centres and groups* organising and supporting foresight research; e.g. CGEE and EMBRAPA[10] in Brazil, the Cali Foresight Observatory in Colombia and 4-Sight-Group[11] in Venezuela. Some countries have also built capabilities in university departments and research institutes;[12]
- *National and International networks* bringing expertise from various countries and building foresight capabilities, such as the CYTED network (2003–2007) and SELF-RULE network.

While Brazil, Chile, Colombia and Venezuela created foresight programmes within national bodies (in particular, ministries of industry, economy and S&T), countries like Peru and Ecuador have constituted "Prospective Consortia" linking universities to the industrial sector. In addition, the Colombian and Venezuelan regional science, technology and innovation (STI) agendas have played a particularly special role in generating STI priorities and achieving consensus and engagement of key stakeholders, including the civil society.

In terms of capacity-building, the region has benefited from the various regional UNIDO courses.[13] These activities inspired certain countries to organise national and sub-national courses, e.g. Colciencias and Univalle University in Colombia, between 2001 and 2006, have organised numerous courses and seminars in collaboration with Futuribles/CNAM (France) and PREST (UK). In Peru, Concytec and the Consortium Prospective Peru (CPP) have promoted international congresses, and CGEE in Brazil is planning to have a role here too.

UNIDO Support to Latin American Foresight

As already mentioned, UNIDO has made significant efforts to promote forward-looking projects to analyse major technologies affecting key productive and industrial sectors in Latin America. Furthermore, in some countries UNIDO has helped the creation of national technology foresight programmes. This has been done through the TFLAC initiative launched in 1999 during its first international TF seminar in Trieste, with the participation of around 150 industrialists, high-level governmental officials and several academicians from 20 countries. The objectives of the seminar were (a) to offer an overview of international TF practices, (b) to discuss the status of TF in Latin America, and (c) to prepare recommendations on how the TFLAC should operate. This initiative led Argentina, Brazil, Colombia, Mexico, Uruguay and Venezuela to launch preparatory activities for setting up national programmes, but only some of these countries managed to institutionalise a TF Programme.

This initiative is still active in some countries where TFLAC has provided funds, technical expertise and logistical support for capacity-building activities (e.g. training courses and seminars) and regional exercises, such as the 2005 study on the "Future of the Fishery Industry in the South American Pacific Coast" (UNIDO, 2005) and the study on the "Future of Andean Products in the High Plateau and Central Valleys of the Andes".[14]

As a result, it is possible to conclude that UNIDO has played a pivoting role in the development of a foresight culture in the region and, at the same time, the promotion of a wide range of cross-national experiences.

NATIONAL FORESIGHT EXPERIENCES

This section presents a short description of national foresight experiences in Argentina, Bolivia, Chile, Cuba, Ecuador, Mexico, Peru and Uruguay.

Foresight in Argentina[15]

In Argentina, the TF programme began with a strong political commitment from the Secretary for Science and Technology (SECYT) to the setting up of a national TF Observatory in 2000 (following an agreement with UNIDO). That year, a small team reviewed several international experiences and prepared various diagnostic studies which included socio-economic and demographic scenarios for the country. But a combination of internal political and economic crises and other external factors affecting the TFLAC project contributed to the interruption of the Observatory and other ongoing studies, such as the Technology, Employment and Technical Training exercise, and three other initiatives concerning the biotechnology, chemical and textile industries (Mari, 2005). In 2003 the National Observatory of Science, Technology and Productive Innovation (ONCTIP) launched the Strategic Mid-Term Plan of STI for 2015, which involved over 4,000 people with a methodology based on expert panels, surveys, workshops and prioritisation of key S&T areas. The process finished in 2005 and its main goals were: (a) to strengthen and enlarge the National System of STI; (b) to improve the quality, efficiency and pertinence of S&T activities; (c) to increase S&T expenditure; and (d) to increase the contribution of the private sector's spend on R&D.

Foresight in Bolivia

In December 1996, the National Academy of Sciences of Bolivia, together with UNIDO and the International Centre for Science and High Technology (ICS), organised an Expert Meeting in Santa Cruz to discuss important developments in technology foresight and main lessons from European and Japanese experiences. In 1999, the Organisation of American States (OAS) supported the attendance of regional experts to the annual PREST Foresight Course in Manchester, and later on a Bolivian delegation participated in the launching of the TFLAC programme in Trieste, Italy (Aguirre-Bastos, 2004). Despite these efforts, various factors, such as lack of political commitment, did not contribute to the creation of a national foresight programme. However, Bolivia has remained active in cross-national studies through UNIDO-sponsored projects, such as the "Future of Andean Products in the High Plateau and Central Valleys of the Andes", which also involves Ecuador and Peru.[16]

Foresight in Chile

The Chilean TF Programme was launched by the President of the Republic together with top business and academic representatives, and was institutionalised within the Ministry of Economy in 2001. The government defined foresight as "a process that allows the discovery of pathways toward a desired future and the identification of strategies or action plan for its achievement".[17] In so doing, national experts have been regularly invited to participate in the formulation of investment strategies for both the public and private sectors. The Chilean programme began with the identification of strategic S&T areas and then focused on specific sectors and themes. Between 2001 and 2005 the programme completed various national exercises and one international study. All of the studies applied Delphi surveys in addition to workshops with targeted experts. Given the national character of most studies, experts have been selected from different regions of the country.

The first study, the largest in scope, covered the whole spectrum of economic sectors, and was aimed at the identification of those "key economic activities" with the largest potential in the next decade.

Subsequent exercises had more sectoral scope, some including:

- "E-ducation industry" (ICT applied to education);
- Biotechnology applied to fruits and horticulture;
- Biotechnology applied to forest industry;
- Chilean software industry 2010;
- Wine production and exports;
- Aquaculture industry.

Each exercise normally requires the creation of an expert database with on average 180 specialists. The TF Programme has also made sure that results get published and reports are downloadable from the web. The methodology is rather conservative, with nearly all exercises following a similar design (i.e. desk research for the diagnosis phase; brainstorming sessions to identify topics and key issues; workshops with targeted experts to discuss the Delphi design; and an on-line Delphi to gather views and opinions about the future).

Three studies were launched in 2006, two related to the agriculture industry (on "Post-graduate knowledge and skills required by the food industry productive chain of crop products" and "Post-graduate knowledge and skills required by the food industry productive chain of livestock products"). The third study was the debut of the technology foresight programme into the regional (sub-national) foresight arena, with an exercise on "Economic activities contributing to the development of the Maule Region in the next 10 years".

Foresight in Cuba

Cuban foresight has been mainly carried out by the Cuban Observatory of Science and Technology (OCCyT) which has given the Cuban exercise a strong emphasis on technology watch and competitive intelligence activities. Monitoring emerging technologies is perceived as crucial for being ready and prepared for technological changes. Technology monitoring is used as a valuable input for developing foresight exercises in key strategic sectors. The three most relevant areas of work are: health, biotechnology and information technologies. Cuban foresight practices have been interesting for many countries in the region mainly because of the effective integration of S&T issues with economic, political, social and environmental factors, thus making Cuban foresight highly interdisciplinary. The Cuban experience is also characterised by the creative use of limited resources and its focus on the development of human capital with proactive attitudes as opposed to reactive. Such proactive thinking has helped Cuban foresight to focus on the identification of technological and social disruptions and to establish research programmes which may facilitate the country's transition towards a knowledge-based economy.

Foresight in Ecuador

In Ecuador there have been foresight-type initiatives in various universities in Quito and Guayaquil. Some training courses have also been supported by international organisations, such as CYTED and CAB, for example. However, applications of foresight to technology management and territorial development have mainly been done by a reduced number of individual consultants. In particular, the Polytechnic School of the Ecuadorian Army (ESPOL) has been keen in providing training courses on foresight and technology watch to army officials. Some thematic and regional experiences include work on professional skills, technological opportunities, and the future of the Pichincha Province. In 2005, an international project on the Fisheries Sector was carried out with the support of UNIDO and OPTI (Spain) and the participation of Colombia and Peru.[18]

Foresight in Mexico

Whilst Mexico has a long tradition in forecasting and futures research, it has yet to organise a fully-fledged national foresight programme. As a pioneer producing some of the first books and exercises in the 1970s, Mexico has focused on capacity-building and teaching programmes mainly located in the Javier Barros Sierra Foundation, the College of Mexico, the Autonomous

University of Mexico, and the Technological Institute of Monterrey. In the 1980s Mexico led the first Latin initiative to promote the integration of foresight efforts in many countries, the so-called Technological Prospective for Latin America (TEPLA), which helped to translate selected European foresight experiences into Spanish. During the 1990s the country showed a decline in regional leadership but since 2000 foresight work has begun to recover. Current activities are mainly related to entrepreneurial foresight and efforts of public institutions focused on education, science and technology. There is also a Mexican node of the Millennium Project, a chapter of the World Futures Society (WFS) and an online journal for prospective studies which has provided a space for disseminating experiences in Spanish.

Foresight in Peru

The Peruvian experience began in March 2001 with the negotiations for the creation of a national programme involving the Ministry of Industry, the National Science and Technology Office (Concytec) and other actors. Among the initial activities that were agreed upon are: the preparation of an Inventory of foresight-type experiences;[19] the promotion of training courses and seminars; and the beginning of a pilot exercise in a strategic industrial sector. The Inventory produced a database of over 1,000 experts and 425 institutions from selected sectors: agriculture, industry, mining, fishing, tourism, energy, construction, health and defence. The exercise was considered useful but commentators suggested that more sectoral representation was needed, given that the agricultural sector alone had 35 per cent of the experts and private sector institutions represented 74 per cent of the sample, leaving the public and academic sectors poorly represented.

In June 2001, TFLAC ran a workshop-type course which produced seven reports on the following sectors: textile and clothing; biotechnology and agriculture; sea products and water; new materials; energy; housing and construction; and tourism. The reports included results on brainstorming and cross-impact exercises and identified a series of drivers and stakeholders shaping the development of the sectors. Later in November a group of academicians,[20] who took part in the TFLAC course, set up the Consortium Prospective Peru (CPP).[21] The main objective of CPP was to improve the interactions between the academic, private and public sectors in foresight-related projects. CPP then organised in 2002 a visioning workshop facilitated by an international team. The main result was a research agenda based on 22 critical factors for the socio-economic development of the country by 2020. In 2004, all four universities in the Consortium joined SELF-RULE network, thus allowing Peruvian researchers to take part in advanced training and mobility programmes in Europe and Latin America.

Equally important have been the activities promoted by Concytec since 2003. Three large-scale international foresight seminars were organised in cooperation with members of CPP and other international institutions.

In terms of foresight experiences, four recent exercises include:

1. A multi-country initiative on the productive chain of the fishery industry of the South American Pacific coast, with a ten-year time horizon. Here, Colombia, Ecuador and Peru followed a particular methodological design proposed by TFLAC;
2. A sectoral study of the Alpaca by 2014, sponsored by the National Commission for South American Camels (CONACS) and supported by Concytec;
3. A pilot on the future of Peruvian agriculture and biotechnology;[22]and
4. A project on the "Future of Andean Products" in Bolivia, Ecuador and Peru, launched in 2006 with the support of UNIDO and OPTI. The main objectives of this study were: to identify strategic technology areas for selected products and industrial sectors; to raise the competitive advantage of national products; and to provide national products with better access to regional and global markets.[23]

Foresight in Uruguay

In the early 2000s, the Uruguayan experience focused on the design of new tools for competitive industrialisation through the integration of knowledge in the value chain. The initiative began with remarkable political support which included the Presidency and various ministries. After a meeting organised by UNIDO in Montevideo in late 2000, the country began three studies in sectors considered highly strategic: energy; biotechnology and agriculture; and transport and logistics. Despite significant strengths in those sectors, the government considered that the effects of globalisation and technological change would bring challenges in the future. Therefore, foresight was seen as a promising approach to improve the country's competitive position and access to international markets.

The programme operated through a network of institutions supporting the exercises. A coordination unit was built whose main task was to manage the interaction with a committee of 260 experts. Each exercise required the creation of a panel of specialists in the sector who worked for two years, supporting Delphi questionnaires and professionally-structured surveys to experts. The programme produced scenarios and recommendations for applying new technologies, training of human resources, new business management strategies, areas for investment in science and technology and modification of regulatory frameworks.

FORESIGHT EXPERIENCES IN BRAZIL [24]

The first national foresight programme (in recent Brazilian history) was launched in the late 1990s as "Brazil 2020". It was conducted by the Secretary of Strategic Issues (SAE) and involved large-scale national dialogue and reflection on alternative development paths Brazil could take in the 21st century. Four scenarios were produced for the year 2020 including the following desirable scenario: "A developed nation with social equity, high quality of life and high educational level. In the international context Brazil would have an important position in the world with a solid and dynamic economy capable of keeping its sovereignty".[25] "Brazil 2020" identified several obstacles preventing achievement of the desired scenario together with a list of recommended actions in 17 thematic areas. Political and State reforms; education and basic social services; redistribution and poverty-fighting policies; economic development; internationalisation; agricultural and rural development; employment policies; culture; citizen's role; planning and public policies; S&T; regional development; human rights protection; society organisation; environmental quality; justice; and ethics.

Two other foresight initiatives were launched in 2000:

1. *Prospectar Programme*: developed within the Ministry of Science and Technology (MCT), examined macro-issues related to STI with a focus on technological trends. The main objective was to inform both government and industry on possible impacts of S&T trends on Brazil's future;

2. *Brazilian Technology Foresight Programme (BTFP)*: launched by the Ministry of Development, Industry and Commerce (MDIC) at the Secretary for Industrial Technology with UNIDO technical and financial support.

In this section we provide a detailed description of these two programmes and then introduce some of the relevant ongoing foresight activities in Brazil.

The Prospectar Programme

Prospectar was a nation-wide S&T foresight programme inspired by international foresight programmes of the mid-1990s. The programme was coordinated by the Ministry of Science and Technology (MCT) and the National Council of S&T (CCT) with the support of Anchor Institutions,[26] the National Research Council (CNPq), the Fund for National Studies and Research (FINEP) and a team of specialists from the Centre of Advanced Studies and Systems of Recife (CESAR).

The overall structure resembled the UK TF Programme, in covering key priorities of the S&T system: agriculture; health; energy; telecommunications and IT; materials; hydro resources; aeronautics; and space. In terms of research issues, while the UK Programme focused on an average of 80 issues per sector, Prospectar had nearly 200 topics per sector. Many topics proved too specific and a more systemic view would have been preferred (i.e. fewer topics clustered around themes and related to inter-sectoral and socio-economic issues). While the UK Office of Science and Technology launched the Programme as a response to the White Paper *Realising our Potential* (see Chapter 4), the Prospectar programme was designed to contribute to the elaboration of the White and Green Papers for Science and Technology (Barros, 2002). The Anchor Institutions, selected on the basis of specialised S&T expertise, were responsible for activities in each sector and tasked with defining the final Delphi topics. They used internal groups and sometimes consultants to write "preliminary" topics (Barros, 2002). The exercise achieved a remarkable response rate (over 10,000 experts) and consultation was largely through an online Delphi developed by the team. Overall, the programme had positive results given that it boosted an interest in long-term planning in the country. The massive mobilisation of scientists was seen as the foremost achievement, mainly because it helped to raise awareness of the challenges facing Brazil as a federation and to enhance foresight's profile in the scientific community.

The Brazilian Technology Foresight Programme (TF Brazil) [27]

TF Brazil was managed by the Ministry of Industry, Development and Commerce (MIDC), and supported by UNIDO.[28] The work focused on four industrial sectors, each of which was assigned to an Anchor Institution[29] supported by EMBRAPA. The innovative part of this programme was the combination of foresight and productive chains methodology.[30] The sectors were selected on the basis of priorities highlighted by the Brazilian Competitive Forum which identified productive chains with strong potential to contribute to development of the country.[31] The sectors chosen were: civil construction; textiles and clothing; plastics; and wood and furniture.

The process involved: (a) the modelling and segmentation of the productive chains and identification of its relations with the market; (b) the identification of performance indicators, e.g. energy use efficiency; (c) the identification of critical factors, e.g. operational costs; (d) the identification of futures events in terms of technology-based factors, internal factors and external factors; (e) the designing and executing the Delphi survey; and (f) other optional activities, such as scenarios and cross-impact and key technologies analysis, were used in some sectors.

Overall, the "productive chains" methodology proved an interesting framework for understanding the complexity of long-term planning on issues concerning a large variety of market segments and stakeholders related to a productive chain (from raw material producers to final consumers). However, the methodology by no means guarantees that the outcomes are easier to implement than those produced by traditional methods (see Chapter 3). This is a challenge faced in equal conditions by any activity aimed at the modification of the *status quo*. Moreover, a great challenge of the methodology is to translate findings into policy recommendations with a long-term vision.

Foresight Outlook in Brazil

In recent years, Brazil's S&T system has undergone a series of reforms which modernised and improved the mechanisms for funding S&T research in the country. Sixteen new instruments were created in 1999 with the aim of strengthening networks and communication channels between the public and private sectors. Within this institutional framework the Centre for Management and Strategic Studies (CGEE) was created in 2001 as a space for: formulating strategies and providing support to activities and projects related to the Sectoral Funds; undertaking evaluation studies in science, technology and innovation; and promoting and executing foresight exercises. Here we briefly describe foresight activities at the CGEE and we introduce a project called Brazil 3 Moments which began in 2004 and is conducted by the Nucleus of Strategic Issues (NAE) of the Presidency of the Republic.[32]

Foresight at CGEE
Most foresight exercises carried out by CGEE are sponsored or contracted by national or regional governmental agencies and ministries. CGEE's theoretical model combines methodological and organisational structures by mainly those proposed in European foresight reports (Dos Santos, 2005, mentions Miles and Keenan, 2002, and Miles *et al.*, 2003; see also Popper, 2006a). The exercises, which have ranged from Energy, Biofuel, Biotechnology, Nanotechnology, Climate Change, and Water Resources, are normally carried out through a platform built in-house, the so-called "project-companion" system – a practical tool which helps: to monitor daily tasks, such as communication with experts; to implement online Delphi surveys; and to track overall progress of a project. CGEE's work, where appropriate, builds on results of foresight initiatives undertaken by the Ministry of Science and Technology. For example, in the Energy and Water Resources exercises CGEE analysed findings of the Prospectar Delphi in those sectors and included relevant topics in its new exercises. In addition, CGEE supported the

Ministry of Industrial Development and Commerce (MDIC) in the final stage of the BTFP by providing additional resources (through Green-Yellow Funds) to complete the exercises.

Brazil 3 Moments Project: 2007, 2015 and 2022
Brazil 3 Moments aimed to define national objectives based on a large-scale dialogue between different stakeholders of society over three different time horizons: 2007, 2015 and 2022 (Oliva, 2007).

The methodology required:

1. *Identification of Trends and Issues.* More than 600 experts engaged in a large-scale brainstorming process on Megatrends within seven dimensions: Institutional; Economic; Socio-cultural; Territorial; Knowledge; Environment; and Global. The dimensions, each coordinated by Anchor Institutions, generated over 1,300 trends and issues;
2. *Identification of Strategic Themes.* Major trends and issues were clustered into 360 issues and further prioritisation activities lead to a list of 50;
3. *Identification of Strategic Goals and Objectives.* Strategic Themes were translated into 50 goals and objectives used in a Delphi Survey (2,080 respondents), which assessed the objectives' probable occurrence by 2015 and 2022, as well as their importance and desirability;
4. *Identification of Strategic Focus.* The project team organised eight sessions with 35 national experts (280 specialists in total) taking part in a cross-impact analysis of 50 objectives. A software tool facilitated the completion of the matrix, and cross-impact and Delphi results were jointly analysed and used to identify strategic foci and alternative futures.

Preliminary results of Brazil 3 Moments Project indicated that "Improving the quality of basic and primary education" would have the highest positive impact in the society. NAE thus launched a specific exercise and a Delphi survey on the topic. Results are still being processed but the response rate has been very satisfactory (around 38,000 participants). One particular feature about Brazil 3 Moments is that the project team holds regular meetings with President Lula da Silva to report on progress and intermediate results.

FORESIGHT EXPERIENCES IN COLOMBIA

Since the early 1970s Colombia has carried out important future-oriented studies including Antioquia 21st Century; Valle 2000; Destiny Colombia; Workshops of the Millennium – Reconsidering the Country; Where is Colombia going?; The Cali We Want; and many others. While these earlier experiences have been important for the development of national foresight skills, at the international level Colombian foresight is mostly recognised for its work on the Colombian Technology Foresight Programme (CTFP). For this reason, we dedicate this section to the evolution of the Programme.

The Genesis of the CTFP (2001–2002)

The CTFP was officially created in 2002 by Colciencias (Colombian Office of Science and Technology), [33] the Ministry of Commerce, Industry and Tourism and the Andean Development Corporation (CAF). During 2001 and 2002 a kind of incubation process began whereby organisational and conceptual schemes were prepared. Initial activities helped raise awareness of national and international foresight practices and key stakeholders via conferences supported by UNIDO and the Ministry for the Economic Development. The management team included a group from El Valle University (Univalle) lead by one of this chapter's authors, and the National Centre of Productivity (CNP), both based in Cali.

The First Cycle of the CTFP (2003–2004)

The first cycle of the CTFP focused on building foresight capabilities both nationally and regionally. Several seminars, congresses and training courses were organised with recognised national and international foresight experts. The overall objective of this phase was to initiate a process whereby key stakeholders of Colombian society become aware of the importance of building a foresight culture in the country. CTFP invested resources in the acquisition of foresight tools and other software packages and provided advanced training courses to a small group of people who later became the technical support team for the Programme. Two calls for proposals were launched for foresight on economic sectors (first call) and regional Clusters and Productive Chains (second call). The first call (2003) supported three sectoral projects:

1. Electrical sector;
2. Food packaging sector;
3. Dairy sector.

The second call (2004) attracted several proposals on Secto-territorial Foresight with 24 of them being submitted to national and international evaluation committees using pre-defined evaluation criteria. The selected projects on Secto-territorial Foresight were:

- Export potential of the health sector cluster in the Cauca Valley;
- Agro-industrial productive chain of vegetable fibre in Santander;
- Horticulture productive chain of the Bogotá plains;
- Making Cartagena a tourist destiny.

The Second Cycle of the CTFP (2005–2006)

In the second cycle, Colciencias led the new management structure and provided technical and methodological assistance provided to exercises approved under the second call. Another key task for CTFP was the identification of strategic economic sectors and important emerging challenges of the country (e.g. Free Trade agreements, productive transformation, etc.). CTFP thus emerged as a decision-support platform for a large variety of national and regional socio-economic and political processes promoting the transformation of the country into a knowledge-based economy. The Programme also fostered national Centres of Excellence by supporting four Technology Watch and Thematic Foresight exercises:

1. New materials;[34]
2. Tuberculosis;[35]
3. Genetic resources and biodiversity;[36]
4. Natural products and essential oils.[37]

Equally important is CTFP's support for Colciencias' new thematic areas related to the current structure of the National System of Science, Technology and Innovation: Fundamental research, Energy and materials, Biological processes, Agricultural products and biodiversity, Human beings and the environment, Education, Institutions and culture, and Technological convergence. CTFP supported other foresight exercises, including the National Biotechnology Programme with the assistance of PREST and the International Fishing Sector exercise sponsored by UNIDO and the Ministry of Industry, Commerce and Tourism. More recently, the CTFP provided methodological assistance to a study on productive transformation towards a knowledge society in Colombia, to help the National Planning Department execute its Internal Agenda and develop a Vision for 2019. This involved a Delphi survey on "Strategic sectors and the Colombian knowledge economy".

The CTFP team has lead various foresight initiatives in Latin America, related to the Programme for the Technological Innovation of the Andean Community (PAITEC); CAB's Prospective Programme; and the foresight training component of CYTED. CTFP support from Colciencias, CAF, UNIDO and the National Training Service (SENA), has contributed to its rapid growth, thus creating the need for a systematic learning and monitoring process of existing experiences. This has been achieved through a cooperation with the SELF-RULE network and 4-Sight-Group which developed an online mapping platform[38] to comprehensively monitor Colombian and Latin American foresight experiences with the support of Univalle and the Cali Foresight Observatory.

Foresight Outlook in Colombia

Colombia has a strong tradition in foresight and is considered a reference point for the Latin America region. The CTFP interacts with large numbers of national and international stakeholders and promotes the "foresight culture" within universities, research institutes and governmental agencies across the country. In 2007 the second cycle of the CTFP began a process of evaluation[39] looking at lessons drawn from the first cycle (Popper and Miles, 2004) and main outcomes from the rich pool of national experiences (Medina, 2006a). These have helped to identify various critical factors affecting current practices in Colombia:

- *Innovation and learning capabilities of practitioners and users.* Foresight needs to focus more on adding value to existing knowledge and on developing complementary techniques;
- *Social participation and interaction.* More effective communication channels and tools to mobilise key social, business and political leaders;
- *Productivity of foresight processes.* Sponsors and organisers need to understand that exercises require flexibility in the way the process is managed, as well as the way in which internationally available methods are chosen and used;
- *Pertinence of foresight exercises.* More focus on key social needs and efficient use of resources, whereby regions share resources and information in the pre-foresight phase;
- *Quality of foresight reflections.* The need for more reliable indicators and deeper conceptual understanding of the "foresight process", together with the strong interregional differences in terms of human capabilities.

FORESIGHT EXPERIENCES IN VENEZUELA [40]

Since the 1970s a large number of "strategic prospective"[41] projects have been launched nationally and regionally with both sectoral and thematic focuses (some relevant examples are: Industrial Prospective 1970–2000; Venezuela: Prospective Study 1975–1990; Tourism 2020; Possible Venezuela in the 21st Century, and Zulia Third Millennium, among others). Several stakeholders and institutions have been responsible for these experiences, including the Centre for Development Studies (CENDES), the Institute for Advanced Administration Studies (IESA), Central University of Venezuela (UCV), Zulia University (LUZ), Los Andes University (ULA) and the state oil corporation (PDVSA), among others.

However, Venezuela has been mainly recognised by its forward-looking culture in the energy sector (including the oil, petrochemical and natural gas industries). Proven oil reserves (approximately 77 billion barrels) together with an estimated 235 billion barrels of oil in the Orinoco Belt mean the country has virtually the largest petroleum reserves in the world. Moreover, at the time of writing, Venezuela ranks eighth among the countries with the largest natural gas reserves (OPEC, 2006). This energy potential has been the main driver of the "2012 Strategic Roadmap" (prepared by PDVSA and the Ministry of Energy), which led to a wide range of long-term initiatives, such as PetroCaribe, PetroSUR, the Energy Security Treaties (TSE) signed by several Latin American countries and the ambitious project of building a South American Gas Pipeline, for example.

Another important accelerator of foresight in Venezuela has been the creation of the Ministry of Science and Technology (MCT) in 1999 and, within it, the Directorate General for Foresight and Planning, with the overall mission of supporting the development of the National System for Science, Technology and Innovation. For this reason, the remainder of this section will focus mainly upon the two most important foresight experiences organised and conducted by MCT in the beginning of the 2000s:

1. The Venezuelan Technology Foresight Programme;
2. The National Science, Technology & Innovation Plan 2005–2030.

The Venezuelan Technology Foresight Programme (TFP Venezuela)

With the setting up of the Foresight Sub-Direction in 2001, the MCT signed an agreement with the TFLAC Programme for launching the Venezuelan Technology Foresight Programme (TFP Venezuela) in which UNIDO agreed to contribute with around 200,000USD, foresight expertise and institutional support.

The first cycle of TFP Venezuela (2001–2002) had four initial objectives:

1. To prepare a National Inventory of foresight-type experiences;
2. To undertake a pilot Delphi exercise on the Oil Industry;
3. To build foresight capabilities through seminars and workshops;
4. To support research programmes in key priority areas.

The National Inventory, which initially covered 43 exercises, identified a large number of lessons learnt from past and ongoing studies and contributed towards the generation of a network of foresight competences and capacities.

In parallel, a pilot study – on existing local potentials to competitively produce the chemical inputs that are required in the drilling and production activities of the petroleum industry (Aguirre-Bastos, 2004) – was conducted by Intevep-PDVSA[42] and MCT. The methodology involved expert consultations via Delphi, preparation of scenarios and identification of potential areas for investment. As a result, several business plans where elaborated together with major chemical input producers.

The other two objectives were led by mathematician Yuli Villarroel who organised various foresight capacity-building workshops for the regional science and technology offices (Fundacites) and other institutions linked to MCT. International foresight practitioners from Brazil, Colombia and Mexico were invited as speakers while national experts brainstormed on key sectors and opportunities for the country. As a result of these and other regional working sessions, TFP Venezuela launched a multi-region foresight exercise on the Yucca (cassava or manioc) sector. The main objectives were: (a) to generate possible scenarios for the next 10–15 years; (b) to design a "success scenario" (attractive for the stakeholders involved); (c) to forecast the future development trends of the Yucca productive chain; (d) to build networks to influence on key drivers of the agricultural system; (e) to identify influential stakeholders and secure their commitment; and (f) to build a "shared vision" with an action plan.

The study involved more than 300 experts and individuals from civil society, and the public and private sectors from five States,[43] and produced the "Yucca in Gondola" report with a success scenario for 2016 (Villarroel *et al.*, 2001). However, as many would expect, after the failed 2002 coup against President Hugo Chávez, the government temporarily suspended its foresight activities at MCT – including TFP Venezuela – and devoted its attention to emerging short-term priorities (i.e. regaining control of key institutions, such as PDVSA). It was only after this was achieved that TFP Venezuela managed to resume its activities. This phase, also known as the second cycle of the programme (2004–2006), provided support to the National STI Plan 2030, as well as other sectoral and thematic projects.

Some of these include:

- Shared vision of the future of the gas industry;
- Scenarios for agro-biotechnological development;
- Foresight on biotechnology for agro-food security by 2011;
- The National Science, Technology & Innovation Plan 2005–2030.

The National STI Plan 2030

In the year 2000, the government launched the National Plan for Economic and Social Development (PDESN) which conceived the development of the country on the basis of five equilibriums: economic, social, political, territorial and international. However, in 2004 a longer-term vision for the country's STI policies was still required and MCT embarked on a process aimed at answering two key questions: (a) What STI was needed?, and, (b) for what type of development? With this in mind, an expert panel and a strategic team were assembled to analyse the type of desired development embodied in the Constitution (CRBV, 1999). As a result, MCT decided to design and formulate a National STI Plan (STI-Plan) using foresight to create shared visions on endogenous, sustainable and human development (MCT, 2005).

STI-Plan time horizon was 2030 and its key objectives were:

- To develop science and technology for social inclusion where social actors actively participate in the formulation of public policies;
- To promote scientific and technological independence and achieve higher levels of technological sovereignty;
- To generate higher STI capacities.

STI-Plan involved two phases supported by virtual discussion forums and feedback platforms. *Phase One* aimed at defining strategic directions for 2030 with experts and public consultations. Over 2,000 people participated in various activities lead by MCT, including:

- A public perception survey;
- Seven visits to different States;
- A methodology validation workshop;
- A survey on the role and SWOT of regional foresight agendas;
- A capacity-building workshop for regional technical teams;
- Interactive focus groups with community stakeholders;
- Several meetings with the central administration.

Phase Two focused on the definition of the STI-Plan's strategies and goals based on contributions from nearly 1,350 people and further analysis of *Phase One* consultations as well as interviews with key stakeholders of the STI system.[44] A total of 1,921 strategies and goals were generated and later clustered into six categories, namely: (a) strengthening the STI system; (b) promotion of research; (c) sustainable development; (d) endogenous development; (e) Ibero-American integration; and (f) science and technology visibility and culture. Some resulting strategies included the creation of STI networks on priority areas; the matching of STI activities to the needs of excluded people; and the creation of R&D centres, among others. In addition, the Plan proposed 17 challenges for the future.[45]

Foresight Outlook in Venezuela

In recent years the number of strategic long-term initiatives promoted by Venezuela has grown considerably – e.g. PetroCaribe, PetroSUR, TeleSUR and the Bank of the South – all with a strong regional integration component. Furthermore, the institutional framework of the 1999 Constitution has increased democracy and participatory policy-making at many levels of the public sector. As a result, forward-looking activities are moving from traditional top-down approaches to more inclusive bottom-up practices, where both theoretical and methodological designs are adapted to the local contexts, thus making it possible to undertake more socially and regionally pertinent studies (with the help of community workshops and citizen panels, for example).

Against this background, a look across the two cycles of TFP Venezuela would suggest that – after the development of the STI Plan – there are some important actions that need to be taken in the near future, for example:

- To create mechanisms capable of shaping citizens' visions and institutions' objectives in accordance to emerging social, economic, scientific and technological needs;
- To promote coordinated activities to achieve major national (and shared cross-national) STI objectives;
- To promote the evaluation of foresight and its policy impacts;
- To ensure that researchers have access to state-of-the-art information systems (e.g. databases, web-libraries, knowledge banks, etc.); and
- To create and strengthen foresight groups and networks capable of (a) supporting national, sub-national and international initiatives, and (b) building a "foresight culture" at universities, research institutions and government agencies (including embassies and consulates around the world).

THE LATIN AMERICAN FORESIGHT PANORAMA

The re-emergence of foresight (or '*la prospectiva*') in the Latin region represents, as in Europe, a challenge for newcomers as well as for experienced practitioners and scholars. For this reason, in this section we expand the typology (state of evolution, level, focus, objectives) proposed by Johnston and Sripaipan (in Chapter 11) to include academic developments and describe the status of foresight activities in 13 Latin American countries.

As observed in Table 12.1, nearly half of the Latin countries considered have well-recognised adaptation and innovation capabilities in foresight. This is reflected in the originality of methodological designs and the use of online tools (e.g. software applications), especially evident in Brazil, Colombia and Venezuela, followed by Argentina, Cuba and Mexico. In terms of levels, these countries have national, regional, sectoral, organisational and academic foresight experiences. Objectives relate mainly to anticipation, networking, action achieved, and action proposed.

Table 12.1   National foresight activities in Latin America

Country	State of evolution [a]	Level [b]	Focus [c]	Objectives [d]
Argentina	*A/I*	*R, Se, O, Ac*	*F/s, P*	*A, Ne, Act-P*
Bolivia	*Im*	*Se*	*F/s*	*A*
Brazil	*A/I*	*N, R, Se, O, Ac*	*F/s, P*	*A, Ne, Act, Act-P*
Chile	*Le*	*N, R, Se*	*F/s, P*	*A, Act, Act-P*
Colombia	*A/I*	*N, R, Se, O, Ac*	*F/s, P*	*A, Ne, Act, Act-P*
Cuba	*A/I*	*R, Se, Ac*	*F/s, P*	*A, Ne, Act, Act-P*
Ecuador	*Le*	*Se, Ac*	*F/s*	*A*
Panama	*Im*	*Se*	*F/s*	*A*
Paraguay	*Im*	*Se*	*F/s*	*A*
Peru	*Le*	*N, R, Se, O, Ac*	*F/s, P*	*A, Ne*
Mexico	*A/I*	*N, Se, O, Ac*	*F/s, P*	*A, Ne*
Uruguay	*Le*	*N,R, Se*	*F/s*	*A*
Venezuela	*A/I*	*N, R, Se, O, Ac*	*F/s, P*	*A, Ne, Act, Act-P*

Notes:

a   **State of evolution:** position of foresight/future activities in the country along a spectrum from imitation [Im], via learning [Le} to adaptation/innovation [A/I]

b   **Level:** national [N], regional [R], sectoral [Se], organisational [O], academic programmes [Ac]

c   **Focus:** foresight [F/s], policy action [P]

d   **Objectives:** anticipation [A], networking [Ne], action achieved [Act], action proposed [Act-P]

Another group of emerging practitioners is lead by Chile with a large variety of objectives mainly focused at the national level (with some sectoral and regional projects too). In contrast, Peru shows activities at different levels but mainly oriented towards anticipation and networking objectives. The growing number of foresight capacity-building activities in countries like Peru and Chile may soon result in better adaptations and innovations in practices. Finally, activities in Bolivia, Panama and Paraguay have mainly focused on sectoral projects. To sum up, Latin American foresight is very rich and varied.[46]

CONCLUSIONS

The evolution of foresight activities in Latin America has been slow but progressive given that many countries have launched national programmes and projects. Latin America still lacks a coherent and ambitious capacity-building strategy, but this can also be said for other regions, including Europe. Countries like Colombia have systematically organised theoretical and methodological training courses for practitioners and users of foresight (governmental agencies, research laboratories, and the scientific community). These courses have contributed to raise awareness of the important role that different stakeholders play in a foresight process.

Latin America incorporates concepts and techniques from a wide range of international foresight exercises, mainly from Europe. However, the region has also managed to achieve its own "Latin Flavour" given that the creative use of limited resources has sometimes resulted in effective innovations in practices and tools (from new management systems and support tools to new ways to reach stakeholders' commitment to original marketing strategies for presenting and disseminating results).

Some experiences, such as the Cuban, are especially interesting because of the effective integration of S&T issues with economic, political, social and environmental factors, thus making foresight highly interdisciplinary. The Colombian and Chilean experiences have focused on emerging technological developments in order to inform governmental and industrial policy-making – especially about potential impacts of S&T trends and drivers. Equally important is the Venezuelan experience which uses a "societal model" of foresight (see Chapter 5) to promote technological sovereignty and social inclusion through endogenous development programmes and participatory processes. Finally, in Brazil one of the major objectives of foresight has been to raise high-level political awareness of the future of key sectors through internationalisation and networking of the major stakeholders of the S&T system.

Unlike Europe, Latin America does not yet have regional instruments (e.g. EU Framework Programmes for RTD) supporting foresight research on strategic sectors and thematic areas. Some efforts are underway by regional organisations but these have been, generally, one-off exercises and therefore it is not surprising that experiences are mainly dominated by national programmes. However, the emergence of specialised centres, consortia, think-tanks and networks are tracing an interesting path of development with innovations in practice and support tools. This leads to an increase in the number of countries systematically using foresight as an instrument for (a) the identification of emerging areas, strategic sectors and future opportunities, as well as (b) the development of coherent national and sub-national STI policies.

How "regional" foresight will evolve in Latin America will depend on the willingness of citizens and policy-makers to support regional integration initiatives, such as the promotion of a Latin American Research Area through improved coordination of national STI policies (Popper, 2005).

NOTES

1   Space limits do not allow an extensive discussion of foresight activities in these countries, and thus some country specific features cannot be dealt with either.
2   See Herrera *et al.*, (1976).
3   SELF-RULE is an academic network funded by the European Commission's ALFA Programme and several Latin American universities.
4   For further information on:
   - *Latin experiences*, see also: Montañolas (1987); Yero (1989, 1991); Medina (2000); Dagnino and Thomas (1999); Leone (1999); Suarez (2000); CYTED (2003); Mari (2003); Medina and Ortegón (2006); Popper and Villarroel (2006); see cases at http://www.4-sight-group.org/case_studies/latin_america/.
   - *General vision of the region*, see also: Valenzuela, S. and Valenzuela, A. (1978); Matus (1993, 2000); Gallopín (1995); ECLAC/CDCC (2000); Melo (2001); Chomsky (2004); Castro *et al.*, (2005); Popper (2005); see also http://prest.mbs.ac.uk/prest/SCOPE/documents/LA_Scenario_4_Success.pdf.
   - *Recent international collaborations*, see also: UNESCO-Gregoriana-Univalle (Angulo *et al.*, 2000); Nuevo Paradigma (Castro *et al.*, 2005); and SELF-RULE (http://www.self-rule.org).
   - *Caribbean*, see also: Downes (2000).
   - *Argentina*, see: François (1977); Mari (2005).
   - *Brazil*, see: Moura (1994); Marcial and Grumbach (2002); Cristo (2003); Santos *et al.* (2004); Caruso and Tigre (2004); Aulicino and Kruglianskas (2004); Canongia *et al.*, (2006); Oliva (2007).
   - *Colombia*, see: Mojica (1990, 2005); Medina (1996, 2000); Angulo *et al.*, (2000), Suarez (2000); Popper and Miles (2004); COLCIENCIAS (2006); De Peña *et al.*, (2006); Medina and Rincón (2006).

- *Cuba*, see: Díaz Otero (2005).
- *Peru*, see: Popper (2002); see also http://www.unido.org/doc/56607.
- *Mexico*, see: Hodara (1984); Miklos and Tello (1991).
- *Uruguay*, see Cabrera (2003).
- *Venezuela*, see: Del Olmo (1984); Yero (1989, 1991); Villarroel *et al.*, (2001); La Rosa (2004); Romero (2004); Villarroel and Popper (2006).

5   CAB's Science and Technology Foresight Programme is a joint initiative with the main S&T governmental agencies of twelve (12) countries: Bolivia, Colombia, Cuba, Chile, Dominican Republic, Ecuador, México, Panama, Paraguay, Peru, Spain and Venezuela. Since 2003 the programme, under the leadership of Colciencias (Colombia) and SECAB, has supported a wide range of activities, including publications, seminars and a foresight project on "Higher Education for Productive Transformation and Social Equity by 2020". For further information see http://ciencia.convenioandresbello.org/prospectiva/.

6   See Wilson *et al.*, (2004).

7   See De Peña *et al.*, (2006).

8   See Escola Politécnica da USP (2002).

9   SCOPE 2015 developed 4 scenarios for RTD cooperation between Europe and Latin America, see http://prest.mbs.ac.uk/prest/SCOPE/ and success scenario at http://prest.mbs.ac.uk/prest/SCOPE/documents/LA_Scenario_4_Success.pdf.

10 EMBRAPA is a Brazilian Agricultural Research Corporation with recognised expertise in Productive Chain methodology, see: http://www.embrapa.br/.

11 See http://www.4-sight-group.org/.

12 E.g. UFRJ, USP and Unicampi (in Brazil); Univalle and Externado (in Colombia); the National Autonomous University (in Mexico); and UCV, ULA, UNEFM and UNEFA (in Venezuela).

13 Some of these course include Venezuela (1999); Uruguay (2000); Colombia, Brazil and Peru (2001); Peru (2004) and Ecuador (2005), among others.

14 See http://www.unido.org/doc/56607.

15 Argentina was a pioneer in the production of forward-looking reflections in Latin America. Amilcar Herrera and the Bariloche Group lead the first global model produced in Latin America in the 1970s. This model was considered by many scholars as one of the deepest and most creative reactions to the *Limits of Growth* study produced by Club of Rome. Another pioneering study lead by Herrera was the Technological Outlook for Latin America project (PTAL) (see Herrera *et al.*, 1976; Albornoz, 2006). However, the Argentinean leadership in foresight has been disbanded over the last years.

16 See http://www.unido.org/doc/56607.

17 See http://www.ppt.cl/.

18 See http://www.unido.org/doc/56607.

19 UNIDO promoted the elaboration of National Inventories in various countries. This activity contributed to the creation of a database of experiences, institutions and people involved in projects with a long-term time horizon.

20 Victor Guevara (National Agrarian University La Molina), Edwin Dextre (National University of Engineering), Isaías Quevedo (University of Lima) and Sandro Paz (Pontific Catholic University of Peru).

21 The Consortium included seven institutions: four universities (ULIMA, UNI, UNALM and PUCP), the National Service of Training in Industrial Work

(SENATI), the National Institute of Agricultural Research (INIA), and the National Society of Industries (SNI).

22 This was a demostrative activity which takes up previous work initiated by UNIDO in 2001 with the leadership of the National Agrarian University La Molina, the National University of Engineering (UNI) and the methodological support of the CPP, PREST and SELF-RULE. Univalle University in Colombia and 4-Sight-Group in Venezuela helped to design the Pre-Foresight Phase of the project (see http://www.4-sight-group.org/self-rule/peru/uni/).

23 See http://www.unido.org/doc/56607.

24 We are grateful for the documents – provided by Carlos Cristo (MDIC), Dalci Dos Santos, Lelio Fellows, Marcio de Miranda and Lucia Melo (CGEE), Adelaide Antunes (UFRJ) and the information available in Portuguese on the Brazilian Ministry of S&T website – which helped to research this section.

25 Our translation from Brazil 2020 project reports available at the website of the Brazilian Nucleus of Strategic Issues (Núcleo de Assuntos Estratégicos da Presidência da República – NAE), see "Cenários Brasil 2020" and "Projeto Brasil 2020 Prospectiva" at http://www.nae.gov.br/.

26 Prospectar was supported by Anchor Institutions in key sectors: Agriculture (led by EMBRAPA); Health (INCOR and FIOCRUZ); Energy, (CENPES, CEPEL and CNEN); Telecommunications and IT (CPqD, ITI and SOCINFO); Materials (INT, CETEM, INPE, ITI); Hydro resources (CPRM and ANA); Aeronautics (CTA); and, Space (AEB, INPE, CTA).

27 See Castro *et al.*, (2002); Cristo (2003) and Popper (2005a).

28 UNIDO's support was mainly financial (USD250,000); political (high-level meetings); and logistical (training courses).

29 BTFP was supported by Anchor Institutions in four sectors: Civil construction (led by Escola Politécnica, University of São Paulo); Textiles and clothing (SENAI-CETIQT); Plastics (Sistema de Informações sobre a Indústria Química – SIQUIM/EQ/UFRJ); Wood and furniture (Instituto de Pesquisas Tecnológicas IPT-SP / Divisão de Produtos Florestais).

30 The combination of foresight and productive chains methodology involves: (a) describing the production chain, (b) analysing the institutional and organisational environment, (c) identifying the needs and aspirations of the production chain partners, (d) analysing their performance and identifying critical factors, and (e) forecasting the behaviour of critical factors and visualising the future performance (see Castro *et al.*, 2002).

31 Especially by increasing or enhancing: employment, wealth creation, exports, and technological and regional development.

32 See http://www.nae.gov.br/brasil3tempos.htm.

33 See http://www.colciencias.gov.co/.

34 Lead by UNIVALLE University.

35 Lead by Antioquia University.

36 Lead by Pereira Technical University.

37 Lead by Industrial Santander University.

38 See http://www.4-sight-group.org/mapping/.

39 See Popper *et al.*, 2007.

40 We are grateful for the information provided by Yuli Villarroel, Yadira Córdova, Luis Marcano, Tibisay Hung, Grisel Romero, Rubén Reinoso, Hilda González,

Irama La Rosa, José Cruces, Omar Ovalles, Alirio Martínez, Nina Sánchez and the great amount of documents available in Spanish in the Venezuelan S&T Ministry website which helped to research this section.

41 See Chapter 2 for discussion about the Strategic Prospective.

42 A research institute of the State oil company PDVSA.

43 Anzoátegui, Aragua, Cojedes, Miranda and Zulia.

44 MCT, FONACYT, FUNDACITES, IVIC, INTEVEP, CDCH, and major universities, among others.

45 Some of these include: increasing S&T expenditure to 2 per cent of GDP by 2010; increasing by 50 per cent students in S&T related careers; and creating a technological park to produce medicines with local technology, etc.

46 This tradition have been enriched with:
   - *the adaptation and contextualisation of European practices* (see Miles, 1981; Masini; 1993; Godet, 1994; Godet and Gabiña, 1996; Georghiou, 1998; Palop and Vicente, 1999; OPTI, 1999; Rodriguez Cortezo, 1999; Escorsa and Maspons, 2001; Popper, 2003, 2006; Miles and Popper, 2004);
   - *the mapping of European and Latin American foresigh experiences* (see Popper *et al.*, 2005, 2007a);
   - *the dissemination of practical guides* (see Díaz, 1994; Miles *et al.*, 2003; Miles and Keenan (2003); Medina and Ortegón (2006); Keenan and Popper, 2007);
   - *the promotion of mobility programmes for European and Latin American researchers* (see http://www.self-rule.org); and
   - *the contributions of Brazilian experiences* (at the CGEE: see Coelho, 2003; Santos *et al.*, 2004, 2004a; Dos Santos, 2005; at the SENAI: see Caruso and Tigre, 2004; and at the SIQUIM/UFRJ: Antunes and Canongia (2006).

REFERENCES

Aguirre-Bastos, C. (2004), 'Technology Foresight in Latin America: A Review of Ongoing Efforts and Future Developments', Paper presented at the 13th International Conference for the International Association for Management of Technology (IAMOT2004), 3–7 April, Washington, DC, USA.

Albornoz, M. (2006), 'Applying FTA Methods in Less Developed Countries', Paper presented at the International Seville Seminar on Future-Oriented Technology Analysis (FTA): Impacts on Policy and Decision-Making, Seville: IPTS.

Angulo, A., Barbieri Masini, E., Conversi, P. and Medina, J. (eds) (2000), *Verso una società multiculturale, possibili scenari in Italia e Colombia*, Pontificia Università Gregoriana, Rome: UNESCO, CIDS.

Antunes, A. and Canongia, C. (2006), 'Technological foresight and technological scanning for identifying priorities and opportunities: the biotechnology and health sector', *Foresight*, **8**(5), pp. 31–44.

Aulicino, A. and Kruglianskas, I. (2004), 'A Contribuição de Foresight Tecnológico na Formulação de Políticas de CT&I do País – Estudo de Caso: MCT-Estudo Prospectar do Brasil', in *Simposio de Gestão da Inovação Tecnológica*, **23**, pp.

2337–2350, Curitiba: Anais/Núcleo de Políticas e Gestão Tecnológica da Universidade São Paulo.

Cabrera, R. (2003), 'La Experiencia del Uruguay', Presentación en Seminario Internacional sobre Programas Nacionales de Prospectiva Tecnológica e Industrial, Bogotá, Colombia: Colciencias.

Canongia, C., Pereira, MNF. and Antunes, A. (2006), 'Modelo de Estratégia de Prospecção de Setores Intensivos em P&D: Sinergias entre Inteligência Competitiva (IC), Gestão do Conhecimento (GC), e Foresight (F)', *DataGramaZero–Revista de Ciência da Informação*, 7(1), February.

Cardoso, F. and Faletto, E. (1969), *Dependencia y desarrollo en América Latina.* Ensayo de interpretación sociológica, Mexico, Siglo XXI.

Caruso, L.A.C. and Tigre, P.B. (eds) (2004), *Modelo SENAI de Prospecção: Documento Metodológico*, Montevideo: CINTERFOR/OIT.

Castro, A.M.G. de, Lima, S.M.V. and Maestrey, A. (2001), *La Dimensión de Futuro en la Construcción de la Sostenibilidad Institucional*, Serie Innovación para la Sostenibilidad Institucional, Proyecto ISNAR "Nuevo Paradigma", San José.

Castro, A.M.G. de, Lima, S.M.V. and Cristo, C.M.P.N. (2002), 'Cadena Productiva: Marco Conceptual para Apoyar la Prospección Tecnológica', *Espacios*, 23(2), Caracas.

Castro, A.M.G.de, Lima, S.M.V., De Souza, J., Maestrey, A., Ramirez, J., Santamaría, J., Mengo, O. and Ayala, A. (2005), *Proyecto Quo Vadis: El futuro de la Investigación Agrícola y la Innovación Institucional en América Latina y el Caribe*, Red Nuevo Paradigma, Quito.

Chomsky, N. (2004), *Hegemonia o Supervivencia*, Barcelona: Ediciones B.

Coelho, G. (2003), *Prospeccao Tecnológica: Metodologias e Experiéncias Nacionais e Internacionais*, Rio de Janeiro: INT/FINEP/ANP.

Colciencias (2006), *Programa Nacional de Prospectiva Tecnológica e Industrial. Direccionamiento Estratégico Programa Nacional de Biotecnología*, Bogotá, Colombia.

Costa Filho, A. (1988), *Planificación y Construcción de Futuro*, Instituto Latinoamericano y del Caribe de Planificación Económica y Social, Santiago de Chile: IILPES.

CRBV (1999), *Constitución de la República Bolivariana de Venezuela, República de Venezuela*, published in Gaceta Oficial n° 36.860, 30 December 1999.

Cristo, C.M.P.N. (2003), *Programa Brasileiro de Prospectiva Tecnológica Industrial*, Brasilia, Brazil: Secretaría de Tecnología Industrial, Ministério do Desenvolvimento, Indústria e Comércio Exterior.

CYTED (2003), *Discusion Final I Jornada Iberoamericana de Vigilancia y Prospectiva Tecnológica*, Santa Cruz De La Sierra, 31 March/4 April, 2003

Dagnino, R. and Thomas, H. (1999), 'S&T Forecasting in Latin America: The Democratization Scenario and the Research Community Role', *Technology Foresight: A UNIDO-ICS initiative for Latin America and the Caribbean*, International Workshop, Trieste, Italy: UNIDO-ICS, 7–9 December.

De Peña, M., Castellanos, O., Carrizosa, M., Jiménez, C., Clavijo, P. and Del Portillo, P. (2006), *La Biotecnologia, Motor de Desarrollo para la Colombia de 2015*, Bogota, Colombia: Colciencias.

Del Olmo, E. (1984), *Métodos Prospectivos*, CENDES Publicaciones, Caracas, Venezuela: Vadel Hermanos Editores.

Díaz Otero, S. (2005), 'La Prospectiva en Cuba uno de los Soportes hacia la Sociedad del Conocimiento', presentation of the Observatorio Cubano de Ciencia y Tecnología at Prospecta Andina 2005, Lima, Peru.

Díaz, P. (1994), *Métodos de Análisis Prospectivo: Reseña y su Utilidad para Proyectos de Inversión*, ILPES, Dirección de Proyectos y Programación de Inversiones, Santiago de Chile: ILPES.

Dos Santos, D. (2005), *Uma Visão das Atividades Prospectivas no CGEE*, Centro de Gestão e Estudos Estratégicos em Ciência, Tecnologia e Innovação, October, Brasilia: CGEE.

Downes, A. (2000), 'Long-range Thinking, Institutional Downsinzing and Action', Paper presented at the High-level Seminar on Basic Planning Functions, 5–6 October, ECLAC/CDCC, Port of Spain, Trinidad & Tobago.

ECLAC/CDCC (2000), *Report of the High–Level Seminar on Basic Planning Functions*, Economic Commission for Latin America and the Caribbean, Sub-regional Headquarters for the Caribbean, Port of Spain, Trinidad & Tobago.

Escola Politécnica da USP (2002), *Estudo Prospectivo da Cadeia Productiva da Construção Civil. Produção e Comercialização de Unidades Habitacionais Urbanas. Diagnóstico Preliminar*, São Paulo, Brazil: USP.

Escorsa, P. and Maspons R. (2001), *De la Vigilancia Tecnológica a la Inteligencia Competitiva*, Madrid: Prentice Hall.

François, C. (1977), *Introducción a la Prospectiva*, Buenos Aires: Pleamar.

Gallopín, G. (1995), *El Futuro Ecológico de un Continente. Una Visión Prospectiva de la América Latina*, Buenos Aires: FCE.

Georghiou, L. (1998), 'Issues in the Evaluation of Innovation and Technology Policy', *Evaluation*, **4** (1), pp. 37–51.

Georghiou, L. (2003), 'Evaluating Foresight and Lessons for its Future Impact', *The Second International Conference on Technology Foresight*, Tokyo, 27–28 February.

Godet, M. (1994), *De la Anticipación a la Acción. Manual de Prospectiva Estrategica*, Barcelona: Editorial Marcombo.

Godet, M and Gabiña, J. (1996), *Caja de Herramientas de la Prospectiva Estratégica*, Paris: Lips-Prospektiker.

Herrera, A., Scolnik, H., Chichilnisky, G., Gallopin, G., Har-doy, J., Mosovich, D., Oteiza, E., De Romero Brest, G., Suarez, C. and Talavera, L. (1976), *Catastrophe or New Society? A Latin American World Model*, Ottawa: International Development Research Centre.

Hodara, J. (1984), *Los Estudios del Futuro*, México: Instituto de Banca y Finanzas.

Keenan, M. and Popper, R. (eds) (2007), *Practical Guide for Integrating Foresight in Research Infrastructures Policy Formulation*, Brussels: European Commission.

La Rosa, I. (2004), *Modelo metodológico para la formulación del Plan Nacional de Ciencia y Tecnología e Innovación*, Caracas, Venezuela: MCT.

Leone, A. (1999), 'Experiencias sobre Prospectiva Tecnológica en América Latina', in *Technology Foresight: A UNIDO-ICS initiative for Latin America and the Caribbean*, International Workshop, Trieste, Italy; 7–9 December.

Marcial, E.C. and Grumbach, R.J.S. (2002), *Cenários Prospectivos – Como Construir um Futuro Melhor*, Rio de Janeiro: Editora FGV.

Mari, M. (2003), *Materiales de curso de Prospectiva*, I Jornada Iberoamericana de Vigilancia y Prospectiva Tecnológica, Santa Cruz de La Sierra, 31 March–4 April.

Mari, M. (2005), *Escenarios y Visiones sobre Estrategias de Desarrollo desde la Perspectiva del Sistema Científico y Tecnológica Argentino*, Plan Argentino de Ciencia y Tecnología, Buenos Aires: SECYT.

Martin, P. (1997), *Prospectiva Tecnológica: Una Introducción a su Metodología y a su Aplicación en Distintos Países*, Madrid: Fundación COTEC para la Innovación Tecnológica.

Masini, E. (1993), *La Previsión Humana y Social*, Ciudad de México: Fondo de Cultura Económica.

Matus, C. (1993), *Política, Planificación y Gobierno*, Caracas: ILPES–Organización Panamericana de la Salud.

Matus, C. (2000), *Los Cuatro Cinturones de Gobierno*, Caracas: Fundación Altadir.

MCT (2005), *Plan Nacional de Ciencia, Tecnología e Innovación: Construyendo un futuro sustentable – Venezuela 2005–2030*, Caracas: Ministerio de Ciencia y Tecnología, 158 pp.

Medina Vásquez, J. (1996), *Los Estudios del Futuro y la Prospectiva: Claves para la Construcción Social de las Regiones*, Documentos ILPES.

Medina Vásquez, J. (2000), 'Función de Pensamiento de Largo Plazo: Acción y Redimensionamiento Institucional del ILPES', *Cuadernos del ILPES*, **46**.

Medina Vásquez, J. and Suárez, F. (2005), 'Proyecto "CONCILIO": Imágenes y Visiones de Futuro en Situaciones de Conflicto. Software de Cooperación', *Revista Electrónica Prospectiva*, **6**, México.

Medina Vásquez, J. and Ortegón, E. (eds) (2006), *Manual de Prospectiva y Decisión Estratégica: Bases Teóricas e Instrumentos para América Latina y el Caribe*, ILPES–CEPAL, Serie Manuales, **51**, Santiago: ILPES–CEPAL.

Medina Vásquez, J. and Rincón, G. (eds) (2006), *La Prospectiva Tecnológica e Industrial: Contexto, Fundamentos y Aplicaciones*, Bogotá: Conciencias–CAF.

Melo, A. (2001), 'The Innovation Systems of Latin America and the Caribbean', Inter-American Development Bank, Research Department, Working Paper, **459**, Washington, DC: IDB

Miklos, T. and Tello, M. (1991), *Planeación Prospectiva. Una Estrategia para el Diseño del Futuro*, Noriega Editores, México DF: Ed. Limusa.

Miles, I. (1981), 'Scenario Analysis: Identifying Issues and Ideologies', in Acero, L., Cole, S. and Rush, H. (eds), *Methods for Development Planning: Scenarios, Models and Microstudies*, Paris: UNESCO.

Miles, I. and Keenan M. (eds) (2003), *Practical Guide to Regional Foresight in the United Kingdom*; European Commission, available at: http://www.cordis.lu/foresight/CGRF.pdf.

Miles, I. and Popper, R. (2004), 'La Meta del Foresight es la Integración Regional', Boletín del Programa Andino de Competitividad, Corporación Andina de Fomento, Diciembre, **13**.

Miles, I. and Popper, R. (2005), 'Foresight: Un camino de Largo Aliento', *Encuentro CAF por la Competitividad*, Caracas, Venezuela: CAF, Mayo, pp. 33–37.

Miles, I., Keenan, M. and Kaivo-Oja, J. (2003), *Handbook of Knowledge Society Foresight*, European Foundation for the Improvement of Living and Working Conditions, Dublin, Ireland: EFL.

Mojica, F. (1990), *Prospectiva. Un Telescopio para Visualizar el Futuro*, Bogotá: Legis.

Mojica, F. (2005), *La Construcción del Futuro*, Bogotá: Convenio Andrés Bello–Universidad Externado de Colombia.

Montañolas, J. (1987), 'Prospectiva Económica y Social', in *La Construcción del Futuro en América Latina*, UNESCO, Caracas.

Moura, P. (1994), *Construindo o Futuro. O Impacto Global do Novo Paradigma*, Rio de Janeiro: MAUAD Editorial.

Oliva, O. (2007), *Agenda para o Futuro do Brasil*, Cadernos NAE/Núcleo de Assuntos Estratégicos da Presidência da República, Edição especial (May), Brasília: Núcleo de Assuntos Estratégicos da Presidência da República.

OPEC (2006), 'Venezuela: Petroleum for the People', *OPEC Bulletin*, **37**(3), May/June, ISSN 0474–6279, Vienna: OPEC, pp. 6–7, see http://www.opec.org.

OPTI (1999), *Futuro Tecnológico en el Horizonte del 2015, Primer Informe de Prospectiva Tecnológica Industrial*, Observatorio de Prospectiva Tecnológica e Industrial-Ministerio de Industria y Energía, Madrid: OPTI.

Palop, F. and Vicente, J. (1999), 'Vigilancia Tecnológica e Inteligencia Competitiva. Su Potencial para la Empresa Española', *Serie Estudios Cotec*, **15**, Madrid: Fundación COTEC.

Popper, R. (2002), *Cross-impact Method for Detecting Key Drivers in Peru*, Report of the foresight workshop organised by the Consortium Prospective Peru, 17–18 October, Lima, Peru: CPP.

Popper, R. (2003), *The Knowledge Society EUFORIA Delphi*, European Foundation for the Improvement of Living and Working Conditions, Dublin, Ireland: EFL.

Popper, R. (2005), *Cooperating with Latin America: The Potential of Latin American RTD*, SCOPE 2015 report to the European Commission DG Research, Manchester, UK: PREST, The University of Manchester.

Popper, R. (2005a), 'Production Chains 2016 – The Brazilian Technology Foresight Programme', *European Foresight Monitoring Network Brief*, **15**, available at: http://www.efmn.eu.

Popper, R. (2006), 'Diseño y Preparación de un Ejercicio Delphi: El Caso EUFORIA', in Medina Vásquez, J. and Rincón, G. (2006) *La Prospectiva Tecnológica e Industrial: Contexto, Fundamentos y Aplicaciones*, Bogotá: Colciencias–CAF.

Popper, R. (2006a), *Evaluation of Foresight Activities at the Brazilian Centre for Management and Strategic Studies (CGEE)*, PREST/SELF-RULE Technical Note, CGEE, Brasilia: CGEE.

Popper, R. (2006b), *Evaluación de actividades de Vigilancia y Prospectiva Tecnológica relacionadas con las Áreas Temáticas Estratégicas de los Centros de Excelencia en Colombia*, PREST–SELF-RULE report, Bogota: Colciencias.

Popper, R. and Miles, I. (2004), *Recomendaciones al Programa Nacional de Prospectiva Colombiano*, PREST report to Colciencias, Manchester.

Popper, R. and Miles, I. (2005), *The FISTERA Delphi: Future Challenges, Applications and Priorities for Socially Beneficial Information Society Technologies*, Manchester, UK: PREST, The University of Manchester,

Popper, R., Keenan, M. and Butter, M. (2005), *Mapping Foresight in Europe and other Regions of the World*, EFMN Annual Mapping Report 2004–2005, prepared by PREST–TNO for the European Commission DG Research, Manchester.

Popper, R. and Villarroel, Y. (2006), 'Euro-Latin Foresight Network: SELF-RULE', *European Foresight Monitoring Network Brief*, **66**, http://www.efmn.eu.

Popper, R., Keenan, M. and Medina, J. (2007), 'Evaluating Foresight – The Colombian Case', *European Foresight Monitoring Network Brief*, **119**, available at: http://www.efmn.eu.

Popper, R., Keenan, M., Miles, I., Butter, M. and Sainz, G. (2007a), *Global Foresight Outlook 2007: Mapping Foresight in Europe and the rest of the World*, The EFMN Annual Mapping Report 2007, report to the European Commission, Manchester: The University of Manchester–TNO.

Rodríguez Cortezo, J. (1999), 'El observatorio de prospectiva tecnológica industrial, una herramienta al servicio de la política tecnológica', *Technology foresight: A UNIDO-ICS initiative for Latin America and the Caribbean*, Internacional Workshop, Trieste, Italy; 7–9 December.

Romero, G. (2004), *Visión de la ciencia y tecnología en el marco del plan nacional*, Ministry of Science and Technology of the Bolivarian Republic of Venezuela, Caracas: MCT, available at http://www.octi.gov.ve.

Sánchez, J.M. and Palop, F. (2002), *Herramientas de Software para la Práctica de la Inteligencia Competitiva en la Empresa*, Madrid: Triz XXI.

Santos De Miranda, M, Coelho Massari, G., Dos Santos, D. and Fellows Filho, L. (2004), 'Prospeccao de Tecnologias de Futuro: Métodos, Técnicas e Abordagens', *Parcerias Estratégicas*, **19**, December, Brasilia: CGEE.

Santos De Miranda, M., Dos Santos, D., Coelho Massari, G., Zackiewicz, M., Fellows Filho, L., Morelli Tucci, C., Cordeiro Neto, O., De Martino Jannuzzi, G. and de Carvalho Macedo, I. (2004a), 'Prospeccao em Ciencia, Tecnologia e Innovacao: A Abordagem Conceitual e Metodológica do Centro de Gestao e Estudos Estratégicos e sua Aplicacao para os Setores de Recursos Hídricos e Energia', *Parcerias Estratégicas*, **18**, August.

SELF-RULE (2006), *The Euro-Latin Foresight Network: Mobility Scheme*, available online at http://www.self-rule.org/mobilities/.

Suarez, J. (2000), *Aplicación de la metodología prospectiva en un sistema territorial. Construcción de escenarios futuros para tres comunas*, Documento de Trabajo ILPES, Santiago de Chile: ILPES.

UNIDO (2005), *Estudio de Prospectiva para la Cadena Productiva de la Industria Pesquera en la Región de la Costa del Pacífico en América del Sur*, available online at http://www.unido.org/es/doc/12937.

Valenzuela, S. and Valenzuela, A. (1978), 'Modernization and Dependency: Alternative Perspectives in the Study of Latin American Underdevelopment', *Comparative Politics*, **4**, pp. 535–557.

Villarroel, Y. and Popper, R. (2006), *SELF-RULE: Strategic European and Latin American Foresight for Research and University Learning Exchange*, Intermediate Report 2006, Coro, Venezuela: Universidad Nacional Experimental Francisco de Miranda (UNEFM).

Villarroel, Y. Mojica, F. and Popper, R. (2001), *Visión Prospectiva de la Agenda Raíces y Tubérculos: Modelo de Prospectiva en Agroalimentación*, Caracas: MCT-UCV-ONUDI, at http://www.4-sight-group.org/documents/yuca_en_gondola.pdf.

Wilson, F., Briones, A., Troncoso, M. and González, L. (2004), *Prospectiva Chile 2010. La Industria Chilena del Software*, Programa de Prospectiva Tecnológica, Chile.

Yero, L. (1989), *Estudios Prospectivos en Países Desarrollados*, Caracas: CENDES Publicaciones.

Yero, L. (1991), 'Los Estudios del Futuro en América Latina', in *Estudios del Desarrollo*, **1**, Caracas: CENDES Publicaciones.

13. Foresight in CEE Countries

Attila Havas and Michael Keenan

INTRODUCTION

The main challenge for the countries of Central and Eastern Europe (CEE) has been the transition to market economies and, for several of them, it has been coupled with accession to the EU and cohesion with the more advanced EU Member States.[1] Given the planned economy heritage, not only has the "usual" macroeconomic stabilisation been required, but a much more challenging and more complex modernisation programme, involving fundamental structural and institutional changes, has also been requisite. Systemic changes have been necessary in order to transform CEE countries into viable economies, capable of economically, socially and environmentally sustainable development. Some countries have progressed further than others, so that across the region, the demanding and socially rather costly process of political and economic transition has been achieved to a different degree.

An inherent contradiction of the transition and catching-up process has lay in the tension between short-term and long-term issues, which have had to be tackled simultaneously, while intellectual and financial resources have been insufficient to deal with all these issues at the same time. Besides establishing the fundamental institutions of market economies and political democracies – which undoubtedly have had long-term impacts – up until the mid- or late 1990s, most efforts had been directed towards solving short-term problems. Thus, it had neither been possible to pay sufficient attention to emerging global trends, nor to devise appropriate strategies to improve long-term competitiveness in these new settings. However, in recent years, several CEE countries have started to consider the longer term more systematically. Among these still rather limited efforts are a number of national technology foresight exercises, launched with a view to setting S&T priorities that take account of longer-term developments, and, in some instances, to improving dialogue between scientists, business and policy-makers. These national level exercises are the subject of this chapter.

The chapter begins with a brief description of developments in science, technology and innovation in the transition era, highlighting the many challenges faced and the main problems that remain. It is suggested that the CEE countries suffer from a serious innovation "deficit" to varying degrees and that this acts as a drag on future sustainable development. The prospects of technology foresight addressing some of these problems and challenges are then briefly examined, followed by descriptions of national technology foresight efforts in six countries: Hungary, Czech Republic, Russia, Ukraine, Romania and Poland. The chapter finishes with a discussion of these experiences, highlighting their strengths and weaknesses, and speculating on how technology foresight might develop in the region over the next few years.

SCIENCE AND TECHNOLOGY IN THE TRANSITION ERA

The legacy of central planning and the transition process together have caused a number of problems in the RTDI systems of CEE countries.[2] Both public and private R&D funds have been cut severely since the early 1990s, due to austerity measures, worsened by the weak position of the S&T funding bodies in the contest for budgetary resources. Public R&D expenditures are 20–30 per cent below the EU15 average in most CEE countries. An especially worrisome feature is that the level of business R&D expenditures (BERD) measured as a percentage of GDP is a mere 13–30 per cent of the EU15 average in six CEE countries; even the best performers are 20–30 per cent below the EU15 mean.[3]

Given these financial constraints, the number of researchers has decreased in most CEE countries up to the mid- or late-1990s, albeit with non-negligible country differences, and then started picking up. A lack of funding also made equipment increasingly obsolete – again, with a considerable variation across countries, and sometimes even within the same research organisation – while the 1990s witnessed a strong need for ever more expensive equipment to keep up with other countries. These unfavourable trends are reflected in the unpopularity of science and engineering (S&E) studies in several CEE countries: young people are simply reluctant to specialise in S&E when job prospects are gloomy in this field.[4] Since 2003–2004 an increasing share of young people (aged 20–29 years) obtain S&E degrees in five CEE countries, but even in these cases there is a 20–25 per cent gap with the EU25 average. Furthermore, the region has suffered extensive brain drain, both internally (in the form of skilled and experienced researchers leaving the R&D sector for other types of jobs) and externally (i.e. trained people leaving their country altogether, either for R&D or other types of jobs abroad).

The educational attainment of the workforce in CEE countries is quite often praised as an important asset, yet the region cannot be complacent here either. From the point of view of catching up, i.e. the cohesion of the enlarged EU, it is a major issue that research and higher education are still somewhat isolated in most cases. This is in spite of the well-documented fact that the most important contribution of academic research conducted at universities to socio-economic development is training skilled labour, who can then work in various sectors of the economy, exploiting not only their scientific knowledge, but their problem-solving skills as well (Pavitt, 1991, 1998; Salter and Martin, 2001). Another severe problem is the lack of relevant managerial skills in academia, in particular the ones required for project development, managing international projects and IPR issues, as well as the exploitation of results. Furthermore, academy–industry links are still weak in all CEE countries, albeit to a somewhat different degree (Inzelt, 2004). Finally, the intensity of patenting activities are way below the EU25 average in the CEE countries, while their innovation performance – measured as the share of new-to-firm and new-to-market products/services in total sales – is above the EU25 average with the exception of Hungary, lagging considerably behind (EIS, 2006).

Many of these problems were perhaps to be expected as part of the process of transition. However, mindsets and perceptions remain that are less than conducive to the exploitation of R&D for socio-economic benefit.[5] Whereas in the technologically advanced countries RTDI is understood as one of the major tools for enhancing international competitiveness and improving quality of life, in CEE countries it is largely regarded as a luxury item. More precisely, it is worth distinguishing two groups of policy-makers, with different roles and responsibilities.

First, there are those who are in charge of fiscal policies, who focus on short- (or, at best, mid-) term financial target indicators, and who tend to pay less attention to long-term issues and impacts. Moreover, they have not seen domestic innovation efforts as a contributor to economic growth either in the planned economy period or in the recent past of market economy. The mindsets of such decision-makers are partly influenced by the legacy of the planned economy period, when return on R&D expenditures was a non-issue: R&D activities were primarily conducted for military purposes and the remaining, much smaller part was financed to boost national prestige. A second factor is that foreign investors have introduced innovations – new products, services, processes, as well as new organisational and managerial techniques – "in bulk" and fast, and thus somewhat eclipsed the role of (domestic) R&D in the innovation process. Under these circumstances, RTDI is perceived by such decision-makers as a "burden" on the budget. They, therefore, are inclined to cut RTDI spending whenever there is pressure to reduce budget imbalances, without taking into account the wider, longer-term socio-economic consequences.

Second, and as a "mirror" of this ignorance, their colleagues running various S&T organisations only think in terms of isolated science priorities, advocating more spending on R&D. The latter think that R&D results would automatically lead to socio-economic benefits, i.e. they make a connection between RTD and catching-up in their special way. The main deficiency of this view is that it disregards the fact that (a) innovation is a much more complex process; (b) innovation systems are hampered by systemic failures; and thus (c) there is a strong need for a number of policy tools to remedy systemic failures – besides the ones aimed at increasing R&D spending. It is not uncommon for high-ranking officials of the Academies of Sciences to claim that the only source of knowledge is basic research. This way of thinking neglects the fact that there are different types, forms and sources of knowledge – not only R&D conducted in certain, glamorous fields of science. It fails to realise the significance of exploitation of knowledge for socio-economic purposes, by emphasising solely the importance of knowledge creation. From a different angle, it neither takes into account the "other facet" of R&D, namely learning (Cohen and Levinthal, 1989, 1990), nor other ways and channels of learning, which are all crucial to strengthen innovation competences, among them so-called dynamic technological capabilities (Bell and Pavitt, 1993), and make informed decisions as to what technologies would be appropriate for a given firm or country, how to operate and adapt those technologies to new settings, and most importantly, how to improve upon them. In sum, this way of thinking clearly cuts innovation from R&D, considering the latter to be a luxury, or a privilege, for a narrow elite, by ignoring the abundant evidence on the nature and economic relevance of the innovation process, and the concomitant policy implications. From a practical point of view it also means that scarce resources would be spent on prestige projects, as opposed to focusing R&D and innovation efforts to address relevant socio-economic challenges, e.g. health-related and environmental problems, international competitiveness of domestic firms, etc.

INNOVATION: THE MISSING LINK

As already mentioned, the practical repercussions of these misconceptions have been rather severe: whenever austerity measures have had to be introduced in CEE countries in the last 15–20 years to balance the central budget, RTDI expenditures were always among the first targets. In other words, it is a counter-productive strategy to put innovation into the shade and talk only about "science": instead of securing more funding, the likely outcome is that RTDI activities would always be financed from the residue of the central budget, once all the "important" objectives are funded. More importantly, the real issue, i.e. exploiting RTDI results to enhance competitiveness and improve quality of life, is eclipsed by this way of thinking. This inappropriate perception of RTDI, therefore, needs to be changed.

Catching up can only be achieved if it is supported by technological, organisational and behavioural innovations. Thus, RTDI policies are of crucial importance for CEE policy-makers when they are trying to devise adequate responses to the above challenges. Without devising and implementing sound policies to foster both knowledge creation and exploitation (diffusion) of knowledge, these countries would continue to lag behind the advanced EU members; moreover, the current development gap is likely to widen and deepen. Innovation can and should play an important role in solving some of the major transition challenges. Loss of former markets, and hence the need to find new ones, necessitates the introduction of new products, production processes and services, as well as modern managerial techniques and other types of organisational innovations to raise productivity. Pressures at the macro-level, notably severe budget, trade, balance of payment deficits, also call for a successful, competitive economy, capable of "growing out" from these traps. Poor quality of life (considering its economic, health, environmental aspects) cannot be improved without thousands of incremental and radical innovations in a large number of fields. Finally, brain drain, which is rather harmful both from an economic and social point of view, can only be reversed, or at least slowed down, by offering attractive conditions for researchers and engineers; i.e. interesting projects, appropriate funds, much better equipment and higher income.

One can also rely on a sort of indirect reasoning when analysing innovation activities in CEE countries. In brief, fierce competition, in both export markets and the open, liberalised domestic ones, compels domestic firms to innovate. Indeed, case study evidence strongly suggests that they introduce new products and/or processes; they would have not survived otherwise. In most cases, however, these innovations are not based on domestic R&D projects. Quite often they rely on technologies provided by parent companies or other foreign partners, e.g. under a subcontracting agreement. Foreign firms are also encouraging their local suppliers to introduce new managerial techniques and other organisational innovations.[6] Joining the international production networks, especially in electronics and automotive industries, has also opened up the gates of the global markets for several endogenous firms in a number of CEE countries. Domestic innovative activities outside the domain of formal R&D do play an important role, too, e.g. engineering and redesigning to adjust to local needs and production facilities, as well as upgrading production equipment and tooling up to increase efficiency and/or to introduce new products and processes.

In sum, not only the various elements of CEE national systems of innovations (NSI) are underdeveloped, but their NSI are poorly integrated, too. On top of that, a number of observers have identified a further obstacle to development, namely the persistence of the linear model of innovation in mindsets of policy-makers; that is, the ignorance of, and/or resistance to, up-to-date, relevant policy knowledge. To conclude, drastic restructuring, institution-building, learning and unlearning are required in various sectors and at all levels (policy-making, research organisations, firms, individuals), i.e. a sort of "planned, policy-assisted creative destruction" is needed. Yet, in most CEE countries the innovation policy constituency is small, fragile and somewhat disorganised. Moreover, the RTDI policy framework is bipolar (S&T or Education vs. Economy Ministries), and thus in most cases communication and coordination among the ministries responsible for various elements of RTDI policies are either lacking altogether, or rather weak. Public spending on RTDI can only be inefficient in these settings.

THE POTENTIAL OF FORESIGHT

Foresight can be relevant even in small or medium-sized countries, not being at the forefront of technological development but rather in the semi-periphery. A number of factors seem to contradict this observation at first glance. Foresight can be a costly project in terms of money, and even more so if participants' time required by meetings, workshops and surveys is taken into account. Moreover, advanced countries, whose experts, in turn, know more about leading-edge technologies, regularly conduct foresight programmes, and their "products" – reports, Delphi-survey results, and the like – are readily available. Yet, only a local programme can position a country, region or sector in the global context and spark a discussion on how to react to major trends. Similarly, strengths, weaknesses, opportunities and threats (SWOT) of a given territory/sector would not be analysed by others, let alone broad socio-economic issues. Process benefits cannot be achieved without a local programme either. Without these, a territory would be unable to improve the quality of life of its population and enhance its international competitiveness.

Considering these points, CEE countries – faced with a number of similar challenges when trying to find their new role in the changing international settings, while still characterised by their own distinct level of socio-economic development, organisational structures and set of rules, norms and attitudes – could benefit significantly from conducting foresight programmes, and in a number of ways. For instance:

- Foresight is a useful decision-preparatory tool, as suggested by its widespread use across continents, as well as by theoretical considerations. Foresight can assist decision-makers in tackling a number of complex challenges: it can reduce technological, economic or social uncertainties by identifying various futures and policy options, make better informed decisions by bringing together different communities of practice with their complementary knowledge and experience, obtain public support by improving transparency, and thus improve overall efficiency of public spending.

- Quite a few widely occurring pressures – especially the need to change attitudes and norms, develop new skills, facilitate co-operation, balance budgets – are even stronger than in the case of advanced countries. Moreover, most of these countries also have to cope with additional challenges: they need to find new markets; enhance fragile international competitiveness; improve relatively poor quality of life; stop or at least slow down brain drain; and so on. These all point to the need to devise a sound, appropriate innovation policy, and even more importantly, to strengthen their respective systems of innovation. Foresight can be an effective tool to embark upon these interrelated issues, too, if used deliberately in this broader context.

- Foresight can also contribute to tackling yet another challenge of emerging economies: most of them are struggling with "burning" short-term issues – such as pressures on various public services, e.g. health care, education, pensions and thus severe budget deficit; imbalances in current accounts and foreign trade; unemployment; etc. – while faced with a compelling need for fundamental organisational and institutional changes. In other words, short- and long-term issues compete for various resources: capabilities (intellectual resources for problem-solving); attention of politicians and policy-makers who decide on the allocation of financial funds; and attention of opinion-leaders who can set the agenda (and thus influence discussions and decisions on the allocation of funds). These intellectual and financial resources are always limited, thus choices have to be made. A thorough, well-designed foresight process can help identify priorities, also in terms of striking a balance between short- and long-term issues.

- Further, foresight can offer additional "process benefits" in the CEE countries. By debating the various strengths, weaknesses, threats and opportunities of a country posed by the catching-up process, and the role of universities and research institutes in replying to those challenges, the process itself is likely to contribute to the realignment of the S&T system (including the higher education sector) to the new situation. An intense, high-profile discussion – in other words, a wide consultation process involving the major stakeholders – can also be used as a means of raising the profile of S&T and innovation issues in politics and formulating economic policies.

TF IN THE REGION

As in many West European countries, the adoption of technology foresight has been dependent upon the efforts of academics and enlightened officials who have come to learn about foresight's benefits from international experiences and then sought to transfer the approach at home. However, in addition to these localised efforts, several international agencies have also been active in promoting foresight's diffusion across CEE countries. These include the European Commission (EC), which has funded scoping work for a national technology foresight exercise in Poland (see below), funded pilot foresight studies in Estonia, Bulgaria and Romania,[7] and run foresight training programmes through the ForLearn initiative;[8] and UNIDO, which has run numerous foresight training courses across the region, e.g. in Hungary, the Czech Republic, Russia, Ukraine, Bulgaria and Turkey (these courses are open to participants from all over the CEE region and beyond), funded pilot foresight studies in sectors such as automotive and biotechnology, and, at the time of writing, established a Regional Virtual Centre for Technology Foresight for the CEE region.[9] In addition, national bodies from Western countries have also provided support: for example, the British Council has supported national foresight efforts in the Czech Republic, Hungary, Russia and Ukraine through the organisation of workshops and exchange visits between national and UK experts. The German Federal Ministry of Education and Research (BMBF) has also assisted several foresight initiatives in the region.

In the remainder of this section, national technology foresight experiences in six CEE countries – Hungary, the Czech Republic, Russia, Ukraine, Romania, and Poland – are summarised. The exercises are presented chronologically, starting with Hungary.

Hungary

The first national foresight programme in a former planned economy was launched in Hungary in 1997, called Technology Foresight Programme (TEP). The country was undergoing fundamental economic and social changes, and major organisations and institutions were being shaped when TEP was designed. The national innovation system was weak and fragmented, in general, and academia–industry links were patchy and *ad hoc*, in particular. R&D results were poorly exploited for economic and social purposes, in spite of a relatively successful research system (reflected by publication and citation indices). Against this backdrop, it was a must to think about medium- and long-term issues: how to improve quality of life and enhance long-term international competitiveness, and how RTDI processes could contribute to achieving these overall aims. Moreover, the size of the country and the level of socio-economic development played a decisive role in setting the objectives of the programme: it was driven by broad socio-economic needs and problems, rather than a narrow S&T agenda.

Given the legacy of central planning, it was decided to launch a "bottom-up", expert-driven professional programme rather than a "top-down", centralised, politically laden one. TEP relied on panel activities, and a large-scale Delphi survey, and was conducted in three stages: pre-foresight (July 1997–March 1998), main foresight (April 1998–May 2000), and dissemination and implementation (June 2000 onwards). The Steering Group (SG) and seven thematic panels (covering the areas listed in Table 13.1) assessed the current situation, outlined different visions for the future, and devised policy proposals, emphasising the significance of both knowledge generation and exploitation (for further details, see Havas, 2003a). The SG consisted of business people, researchers and policy-makers. It was therefore not directly influenced by the government agency that initiated and financed the programme (OMFB, National Committee for Technological Development).[10] All the major decisions were taken by the SG (meeting once a month on average), the panels themselves or at joint meetings of the SG and panel chairs and secretaries. Methods were also adjusted to the Hungarian context, e.g. a large number of non-technological Delphi-statements were framed by the panels, as opposed to, for example, the Japanese or British questionnaires.

The foresight process proved to be more challenging than originally envisaged. It was truly a learning process in many respects, for all the interested parties. A number of difficulties arose during the scenario-building process. The most severe one was the unexpected, but sometimes rather strong, resistance to this way of thinking. Two reasons might explain this opposition. First, it was openly stated that "being scientists, we should think about the future in a scientific manner, and apply scientific methods to identify the optimal future"; hence, there would be no need for alternative visions. The other, more context-specific reason relates to the legacy of central planning, which did not promote thinking in terms of alternative futures. Plans only had "optimist" and "pessimist" versions of a single, "socially optimal" future. Influenced by this legacy, most TEP participants could only think of "optimist", "pessimist" and "business-as-usual" futures at the beginning. Developing qualitatively different visions, therefore, took several rounds in the case of some panels.

Table 13.1   Hungarian TEP (1997)

Formal Objectives	Areas Covered
• Devise viable R&D strategies and identify technological priorities; • Strengthen the formal and informal relationships among researchers, business people and civil servants; and • Support the preparation for the accession negotiations with the European Union.	• Human resources • Health and Life Sciences • Information Technology, Telecommunications and the Media • Natural and Built Environments • Manufacturing and Business Processes • Agribusiness and the Food Industry • Transport

For the SG, a major methodological difficulty was to align the macro and panel futures (visions). The uncertainties of the overall transition process called for the development of macro visions as a "reference frame" for the seven thematic panels. Given the different levels of analyses, as well as the unique, inherent logic and structure of the panel issues, there were obvious constraints to harmonise the macro and meso (panel) visions.[11] Some policy-makers, more accustomed to the linear model of innovation, and hence the dominance of technological issues, also found it hard to interpret and utilise the results of TEP. Most business people, however, were rather quick in understanding the significance and benefits of these methods, e.g. establishing new contacts with each other and researchers through the programme.

With the benefits of hindsight, TEP can be seen as a mixed case. It is considered a success from a methodological point of view by experts, confirmed also by an independent, external evaluation conducted by an international panel. Furthermore, practitioners and policy-makers in CEE countries and other emerging economies are still interested in the Hungarian experience, and lessons learnt. However, TEP has not had a substantial, immediate policy impact; and its results and recommendations have been implemented with some delay, in several cases in an indirect way.[12] According to the evaluation by the International Panel (2004, p. 6):

A careful analysis indicates an impact both on the climate of thought in many policy areas and a series of indirect but significant effects on policy in several domains. It seems that TEP created a reservoir of knowledge that entered the policy system in a non-linear fashion, either through personal networks of participants or simply by having cogent text available when policies were being drafted. (…) The reasons for lack of direct implementation lie, we believe, in the implementation environment in which the programme was situated. Its origins within the OMFB may initially have given it a welcome degree of freedom but with the radical change in nature of that organisation and a change of government, there was no natural channel, nor an obvious champion in government able to act upon the results. Even if OMFB had been unchanged, it was itself at a distance from some of the political decisions implied in the recommendations. (…) The added value of TEP came from being able to take a holistic view of sectors which a purely sectoral exercise could not have achieved. While greater engagement by some ministries would have been beneficial, reporting to them directly could have constrained thinking and lost the benefit of multidisciplinarity within panels and learning generated through interaction between them.

It is a sobering fact that although the eight final reports of TEP were published in 2001, and thus a new round of foresight would be clearly needed, it is not on the agenda at the time of writing.

Czech Republic

The first Czech national foresight exercise was initiated by the Czech Ministry of Education (responsible for R&D) in 2001 and coordinated by the Technology Centre of the Czech Academy of Sciences. Its purpose was to populate the first National Research Programme (NRP I) with "oriented" research priorities that would guide public spending on R&D.[13] The foresight was a multi-sectoral, technology-focused exercise, using a key technologies approach.[14] Expert panels were set up in 13 "application areas" plus three cross-cutting themes (see Table 13.2), with approximately 350 experts involved altogether, drawn from universities (37 per cent), the Academy of Sciences (14 per cent), industrial research (17 per cent), the application sector (27 per cent), and government (5 per cent). Their tasks included listing important research directions and then prioritising these through an online voting procedure to identify key research directions for the Czech Republic. The panels were guided in their prioritisation efforts by a complex set of 35 criteria that would add up to an assessment of the importance and feasibility of the research areas being considered. Besides their expertise, panels could draw upon "input information" when making their judgements. This information had been previously prepared by the project team and consisted of the results of almost 300 interviews carried out in the "application sphere", sectoral SWOT analyses, and various statistical data. Through online voting, the panels prioritised an initial 612 important research directions to yield 163 key research directions. These were further prioritised by a cross-panel working group, resulting in a final list of 90 key research directions (for a detailed account, see Klusacek, 2004).

After some delay, the National Research Programme I (NRP I) was eventually launched in January 2004. However, due to (temporary) restrictions of public spending on R&D in 2004, additional foresight work was carried out, again by Technology Centre, to focus the NRP more on practical research results. This was an even quicker process than the first exercise, and involved a multidisciplinary panel of 56 experts defining and clustering areas of the most pressing societal needs that research may contribute to. These areas were then prioritised by a smaller group of experts. The updated NRP was formally renamed the National Research Programme II (NRP II) and was launched at the start of 2006 (Klusacek, 2007).

Table 13.2   Czech Technology Foresight (2001)

Formal Objectives	**Areas Covered**
• Identify the most important technologies (research priorities) likely to be demanded by Czech industry and the service sector over the next ten years; • Accordingly, to identify priorities for the first National Research Programme (NRP I); • Make recommendations on the structure and function of the cross-cutting programmes to be organised as part of the NRP I; and • Identify basic principles for the implementation and management of the NRP I.	• Agriculture and Food • Environment • Healthcare and Pharmaceuticals • Information Society • Building Industry, Urbanism and Housing • Materials and Technology of their Production • Discrete Manufacturing • Instruments and Devices • Machinery and Equipment • Chemical Products and Processes • Transport Systems • Energy and Raw Materials • Social Transformation • Human Resources for R&D • Integrated R&D • Regional and International Cooperation in R&D

Shortly after the completion of the second exercise, the Technology Centre was awarded a framework contract to conduct foresight activities and strategic studies for the Ministry of Education over a five-year period (2004–2008). With a budget of around €450,000 per annum, the main activities have revolved around preparing analytical and future-related information for decision-makers involved in shaping national research and innovation policies.[15] This investment is perhaps testament to the perceived utility of foresight results for decision-makers in the Ministry.

The latest, flagship study has been the third national foresight exercise, which is intended to inform the third National Research Programme (NRP III) starting in 2009. As in the earlier exercises, the main objective has been the identification and formulation of key technologies (research themes) to populate the NRP III. However, the methodology being used for this exercise is more elaborate than the earlier rounds of foresight, and involves a combination of analytical desk research, work of a multidisciplinary panel of 70 experts, exploratory and normative scenario building, and a prioritisation procedure based upon multi-criteria assessment of technologies suggested by experts. Using this approach, 36 thematic areas have been defined and grouped into four thematic priorities for the purpose of administering the NRP III. At the time of writing, the latter has still to be formulated, but if previous experience is anything to go by, it is likely that the foresight results will play a major part in shaping the NRP III.

Russia

The modern era of technology foresight in Russia started in the mid-1990s when the Russian State Committee on Science and Technology initiated a critical technologies exercise. Previously, during the Soviet era, a series of "long-range S&T forecasting" exercises had been carried out, approximately every five years, with a 15–20-year time horizon. These exercises were, in some respects, similar to modern era technology foresight: for example, thousands of experts were typically consulted on future developments in S&T. However, they also differed in many important ways: for example, they largely failed to make sufficient links between R&D and production, while they also tended to consider singular rather than multiple futures.[16] By the late 1980s, interest in these exercises had largely waned: in the crises that accompanied perestroika and the transition period, there was believed to be little scope for effective long-range S&T planning.

In 1996, however, a critical technologies exercise was launched, inspired by S&T prioritisation approaches employed in the United States and parts of Europe. A list of 70 critical technologies was identified, but had little impact on funding, partly due to the large number of areas identified and their generality (Sokolov, 2006a). With the purpose of reducing the number of priority areas, the exercise was updated in 2002, resulting in a shorter list of 52 critical technologies across nine priority areas. The latter were intended to set the general trends of the country's S&T development and included the usual sorts of topics, including information technologies, living systems technologies, new materials, etc. Again, the results were considered too broad to become real targets for the implementation of S&T policies by the Russian government. Accordingly, the Russian Ministry of Education and Science (MES) organised a new critical technologies exercise (in 2004) to revise and "correct" the 2002 results (Sokolov, 2006b). By this time, the overall economic climate had substantially changed for the better, and along with it, a new government commitment to innovation-led growth. Thus, perhaps for the first time, the government "needed" the results to provide background information and analysis for defining budgeting priorities and for informing the Federal S&T Programme. According to Sokolov (2006b), the new list of S&T priority areas and critical technologies has been used by the Federal Agency for Science and Innovation as a background for the Federal Science and Technology Programme. The detailed results have also been used for selection of large-scale innovation projects jointly funded by the government and the private sector. The new National S&T Programme to be implemented in 2007–2012 is also reported to have been designed in line with the results of the exercise (see Table 13.3).

In addition to the holistic critical technologies exercises at the federal level, other exercises have been carried out on particular sectors (e.g. education, power generation) and technologies (e.g. nanotechnology).[17] Moreover, taking into account the sheer scale of Russia, the MES has carried out a pilot foresight study on regional innovation priorities, where the aim has been to develop and test methodologies for building long-term innovation strategies at the regional level. The Republic of Bashkortostan was chosen as the pilot region, and a set of regional critical technologies developed on the basis of interviews, expert surveys and brainstorming seminars. During this time, the Bashkortostan government was involved in developing a mid-term (to 2015) strategy for social and economic development, and the foresight project's findings were used as background for the strategy components related to S&T and innovation. Furthermore, it is reported that an important effect of the study was the building of informal networks between major stakeholders in the region (Sokolov, 2007).

Finally, in 2007, a new national technology foresight programme is to be launched, costing several million dollars over two years. The exercise has four core modules covering (a) macroeconomic scenarios, challenges and opportunities; (b) technology foresight for industrial sectors; (c) S&T foresight, using a Delphi approach; and (d) scenarios for S&T development.

Each component is being carried out by a different organisation, necessitating the creation of an additional fifth component for synthesising the results of the four core modules. The programme represents an attempt to move beyond the limits of the past critical technologies exercises, and promises to be much more extensive in scope and to involve many more people in processes of consultation and deliberation.

Table 13.3   Russian Key Technologies (2004)

Formal Objectives	Areas Covered
• Develop criteria for evaluating technologies; • Analyse components of existing critical technologies, through expert assessment of their input in developing innovation products that would be competitive in domestic and foreign markets; • Identify research areas within critical technologies with the greatest potential for developing such products and therefore capable of making the greatest contribution to accelerating GDP growth and increasing the competitiveness of the national economy; • Create revised lists of priorities and critical technologies together with recommendations on their use; • Evaluate the innovation capacity of critical technologies; and • Develop proposals concerning practical implementation of the selected S&T priorities.	• Information and Telecommunications Systems • Nanosystems Industry and Materials • Living Systems • Rational Nature Utilisation • Power Engineering and Energy Saving • Safety and Terrorism Counteraction • Prospective Armaments, Military and Special Equipment

Ukraine[18]

As with other Soviet Republics, the Ukrainian Soviet Socialist Republic used to run long-term (20-year time horizon) forecasting on S&T issues. These efforts typically engaged more than one thousand researchers, but were severely limited by fixed assumptions on the future socio-economic system. These activities were followed by technological forecasting exercises by various ministries, the National Academy of Sciences of Ukraine (NASU) and even large companies in the military–industrial complex. However, due to fundamental social and economic changes from the late 1980s onwards, almost all these forecasts had no serious impact on developments in Ukraine.

In recent times, national S&T priorities have been set by the Ukrainian Parliament on five-yearly cycles. However, these priorities have been criticised for their broad formulation. Moreover, the funding programmes that are initiated in light of national priorities are considered to be poorly designed, unclearly targeted, and to lack sufficient financial support – essentially symptomatic of an under-resourced and unreformed research system. Partly in response to these problems, the Ukrainian Ministry of Education and Science (MES) agreed to supplement existing priority-setting mechanisms with a new foresight-type programme. Other drivers for the MES to move in this direction included anticipated joint-funding from UNIDO for a foresight programme and the promotion of foresight by the Kyiv office of the British Council.

The "Foresight Programme of S&T and Innovation Development of Ukraine", which eventually got underway in late 2004, had the aim of defining a list of critical technologies important for national economic development and national security. This critical technologies list was to be underpinned by consideration of alternative pathways of national development in S&T and innovation and was to take into account opportunities for exploiting new external markets. The Programme also had to make recommendations for the Ukrainian government on the exploitation of public R&D results, and to propose a permanent system of foresight-type studies for the country.

The Programme soon hit problems, however, due to political turmoil in Ukraine. An early consequence was the failure of UNIDO to provide the anticipated funding for the exercise. This funding shortfall was further exacerbated when the new government slashed the Programme's budget by 80 per cent in 2005. Despite these setbacks, the Programme had already developed its own dynamic and continued largely as planned. In a pre-foresight phase, the Scientific Council for the Programme was formed, and included around 30 prominent Ukrainian scientists and senior state officials. The day-to-day management and operation of the exercise was carried out by the STEPS Centre of the NASU and a team in the MES. After broad consultation, 15 thematic panels of scientists and other specialists were formed, covering several themes as shown in Table 13.4.

Table 13.4 Ukrainian National Foresight Exercise (2004)

Formal Objectives	Areas Covered
• Elaborate the basic and alternative variants of S&T and innovation development of the country; • Form the list of the most prospective technologies and innovations, which will create opportunities for opening new external markets; • Form the list of so-called critical technologies, which will have exceptional importance for the stable development of the national economy and for the national security; and • Prepare recommendations for the Ukrainian government on how to use effectively R&D results, financed by the state, and to create the background for the permanent system of foresight-type studies in the country.	• State Support of Basic Sciences and Research Infrastructure • Biotechnologies • Medical Technologies • Information and Telecommunication Technologies • Energy • Agri-food Technologies • Metallurgy • Lasers and Ionisation Technologies • Nanotechnologies, Functional and Instrumental Materials • Chemical Materials and Technologies • Environmental Protection and Sustainable Development • Macroeconomic and Demographic Trends • Applied Earth Sciences • Innovation in Construction and Architecture • Innovation in Transportation Systems • Space Technologies and "Dual-use" Technologies

Each panel consisted of 25–40 specialists from different research institutes, universities or leading industrial companies, usually from different parts of Ukraine. Special questionnaires were prepared and distributed among these experts in a two-round Delphi exercise. For the second round, about 20 per cent of the experts were replaced by other specialists as a result of the analysis of the initial responses. A third round of consultation got underway in late 2006 to determine policy recommendations. Each cycle of consultation has been rounded off with special conferences and round tables of experts and invited "external" specialists, who have discussed the emerging results of the Programme. Furthermore, results have been published and widely distributed among specialists within the country.

In a technical sense, the Programme can be considered a success, particularly under the difficult circumstances. However, continuing political instability and the slow pace of reform has meant the results of the Programme are unlikely to be extensively taken up by the MES as originally hoped for. Bluntly stated, clientelism and system inertia remain unhappy facts of life – a common problem in many states of the former Soviet Union. On a more positive note, the Programme has contributed to an interest in foresight in Ukraine, where regional and sectoral activities are planned or underway, and possibly a new national exercise might be launched in the near future.

Romania

Like most RTDI systems in the region, Romania faces the common problems of RTDI underinvestment, limited resources spread too thinly, brain drains, and so on. These problems are to be addressed by the second National Plan for Research, Development and Innovation (NPRDI II). The main objective of the Romanian Science and Technology Foresight Project (ROST), launched in 2005, was to underpin NPRDI II by identifying strategic and specific goals for the Romanian RTDI system (EAHERF, 2005).

ROST was a two-year exercise sponsored by the Ministry of Education and Research and the National Authority for Scientific Research, and was managed by the Executive Agency for Higher Education and Research Funding (EAHERF). It involved a large consortium of organisations, representing a wide variety of stakeholders. The project established a rather complex structure, with a steering group, four high-level expert groups (one for each of the four phases of the project), a scientific council, an international advisory board, and seven expert panels. The project was run over four phases: topic-identification, intelligence-gathering, vision-building and policy-making (Curaj, 2006).

In the initial topic-definition phase, consultations were first undertaken with key persons and organisations in science policy-making. This was followed up with an online survey of experts from science, public administration, industry and civil society. The aim of the consultations was to identify (and specify) a limited number of "directions of investigation" that would reframe S&T in terms of national need. Through these consultations, five such directions were identified (see Table 13.5), plus an additional sixth direction dedicated to transformational priorities in science management. In a follow-up negotiation workshop involving around 150 experts, these directions were confirmed, but a seventh direction was also added, on

"Science, frontier science and knowledge development". This last development might be considered as "regressive", since it provided a space for science priorities to be considered in the "traditional" way without much reference to users. Moreover, it has been suggested that a disproportionate time was given to this initial phase of the exercise, necessitated by the need to first negotiate the appropriate governance models of science for Romania (Grosu, 2007).

Table 13.5 Romanian Science and Technology Foresight (2005)

Formal Objectives	Areas Covered
• Contribute to the National RDI Plan and Strategy (2007–2013); • Reconstruct the RDI system around long-term perspectives; • Broaden participation in RDI strategy processes to include those key players affected by strategy implementation; • Improve the information base on which political decisions are grounded, elaborated, and implemented; and • Establish a national core of competences in prospective analysis that can become an efficient instrument of change management.	• Information Society Technologies; • Competitiveness through Innovation; • Quality of Life; • Social and Cultural Dynamics; • Sustainable Development • Institution Building/ Empowering; • Science, Frontier Science, and Knowledge Development.

During the main intelligence-gathering phase, expert panels were set up for the seven "directions of investigation", with each panel having fifteen members. Through a combination of desk research, workshops, and stakeholder interviews, each panel identified 400–800 possible RTDI priorities, which were subsequently clustered into a maximum of 99 areas. These were then discussed at a series of further negotiation workshops, each having 60–100 participants, with the aim of reducing the number further (Curaj, 2006). The methodology used was based upon approaches developed for conflict resolution in international relations. In practice, the process worked less than optimally, since "negotiators" tended to aggregate rather than select priorities (Grosu, 2007), a common problem in prioritisation processes. The result was 15 RTDI priorities per panel, while the seventh panel identified over thirty priorities for "basic science". A wider consultation on the priorities was then carried out by an online survey.[19] Almost five thousand people participated in this survey, with the results analysed by a small group of experts using the Analytic Hierarchy Process (AHP). In parallel, the "basic science" panel conducted its own online survey, with the participation of around 700 researchers. Twenty-six research priorities were identified plus priorities for the RTDI system's transformation (Curaj, 2006).

In the vision-building phase, the expert panels were asked to synthesise debates by defining a visionary objective for each of the 26 research priorities. These visions highlighted the resources and science governance mechanisms required in order to evolve toward stated objectives (Curaj, 2007). In addition, five of the RTDI system transformation priorities were selected and used as the basis for the generation of four normative scenarios. Although this step was supposed to be participative and involve workshops, it was instead carried out by a small group of experts, due to time constraints. A dominant scenario was selected and used as the basis for a one-page vision on the transformation of the Romanian RTDI system. In the final policy-making phase, a "strategic vision" was produced, describing the most desirable path into the future.

Whilst the government has indicated a strong commitment to use ROST results for the NPRDI II, they sometimes had unrealistic expectations as to what could be achieved, perhaps due to a vague understanding of how foresight could contribute to setting national strategic objectives (Curaj, 2006). Moreover, although there was the aim of trying to better link S&T to other policy areas, in reality, dialogue rarely moved beyond S&T policy, despite participation of actors from other policy areas – a common problem of much technology foresight in a national setting. Furthermore, innovation policy considerations were very much secondary to those of S&T, reflecting the dependency of the former on other (non-sponsoring) governmental actors, notably the Ministry of Trade and Industry (ibid.).

As is common with many foresight efforts in the CEE region, the sponsor's expectations were predominantly product-oriented, i.e. supply of analytical reports and lists of priorities, with the purpose of informing the NPRDI II. However, the project team held a wider set of expectations (based upon international learning in foresight), which included facilitation of networking and communication between different stakeholders and enhancement of a science–society dialogue. These process-oriented features are often essential for the necessary enrolment of different groups to agendas of change. According to Grosu (2007), the exercise paid insufficient attention to the variety of participation required (due to the urge for convergence in getting results) for the process to deliver on its promises. On the other hand, it is clear that the exercise has resulted in the roots of a foresight culture now taking hold in Romania from which further activity is expected to grow over the coming years.

Poland

A national technology foresight programme similar to those carried out in other parts of the CEE region was only initiated in Poland during 2006, although the idea of conducting such an exercise had been considered earlier. For instance, British and German experts were engaged by the Ministry of Science and Higher Education (using EC funding) to conduct a scoping study for a national technology foresight exercise already in 1999 (see Nedeva *et al.*, 2000). Yet, little activity followed, essentially for political reasons, including the election cycle and prolonged discussions of the organisational location of a national foresight effort. The idea for such an exercise remained on the agenda, though it was not until 2004 that the government launched a pilot study to test the feasibility of a large-scale national effort, and even this was partially paid for using EC funding dedicated to the restructuring of S&T in EU New Member States.

The pilot study focused upon the area of Health and Living, and had a ten-year time horizon (for details, see Łepeta, 2006). The choice of topic was dictated partly by public interest, and partly by the anticipated convergence of many fields of S&T in the area. Besides piloting foresight as a set of practices, the study had the aim of defining S&T priorities in light of future opportunities, of widening participation in debates on the future, and of better demonstrating the links between scientific achievements and economic growth. Despite its relatively narrow scope, the study had a rather complex organisational structure consisting of a Steering Committee appointed by the Ministry, which in turn appointed a Main Topic Panel to coordinate the study. Eleven thematic panels in the health area were then appointed, with each panel having 10–18 experts drawn from science, industry, and public policy. Several methods were piloted in the study, including SWOT analysis, expert panel meetings, key technologies for prioritisation and "social consultation" using a mix of public focus groups, interviews and a survey. The expert panels used SWOT and key technologies (over 3–4 meetings) to arrive at lists of "critical areas". Panel results were then processed by the Main Topic Panel, which distinguished between "high priority areas" and "priority areas". These priorities were juxtaposed with those emerging from the "social consultation", highlighting some mismatch between the public's concerns around health and the research interests of the scientific community. The pilot study was judged to have been successful, largely in terms of the breadth of participation and the subsequent use of some of the results in decision-making.

Consequently, a full-scale national exercise, called "Poland 2020 National Foresight Programme", was launched in 2006 (see MNSIW, 2006). Its stated objectives are summarised in Table 13.6. Anticipated outcomes include a rationalisation of public expenditure on S&T around areas contributing to medium- and long-term national economic growth; and the development of a language for public debate and a more future-oriented way of thinking, which in turn should lead to coordinated, joint efforts towards economic growth.

Table 13.6   Poland 2020 National Foresight Programme (2006)

Formal Objectives	Areas Covered
• Lay out the development vision of Poland up to the year 2020; • Set out – through a consensus with the main beneficiaries – the priority paths of scientific research and development which will, in the long run, have an impact on the acceleration of social and economic growth; • Put research results into practice and create preferences for them when it comes to allotting funds from the budget; • Present the significance of scientific research to economic growth and how it can be absorbed by the economy; • Adjust Polish scientific policy to the requirements of the EU; and • Shape science and innovation policy towards the development of a Knowledge-Based Economy.	**Sustainable Development of Poland** • Quality of Life • Sources and Use of Power Resources • Key Ecological Problems • Environmental Protection Technologies • Natural Resources • New Materials and Technologies • Transport • Integration of Ecological Policy with Sectoral Policies • Product Policy • Sustainable Development of Regions and Areas **Information and Telecommunications Technologies** • Access to Information • ICT and Society • ICT and Education • e-Business • New Media **Security** • Economic Security • Intellectual Security • Social Security • Technical and Technological Security • Development of Civic Society

Many of the features of the successful pilot study have been carried over to the design of the national Programme, including some of the methods and organisational arrangements. Accordingly, the Programme has a Steering Committee, responsible for advising on the design and performance of the Programme; a Main Panel, responsible for developing a baseline 2020 vision linking S&T to the economy and for providing guidelines for the work of other panels; three Research Area Panels, responsible for analysis and synthesis associated with the formulation of scenarios; and 20 Topic Panels responsible for a variety of tasks, including the formulation of Delphi topic statements (see Table 13.6 for the list of Research Area Panels and Topic Panels).

The project is being carried out by a consortium of two institutes of the Polish Academy of Sciences plus an organisation specialising in opinion and market research. Programme participants include domain experts in the topic areas covered, experts in the social and economic sciences, and representatives of potential beneficiaries and users of foresight including industry, politics and the media. Besides expert panels, the Programme will see the use of several other methods, including SWOT and PEST analysis, cross-impact analysis, Delphi and scenarios. With much emphasis placed upon achieving broad-based participation, the Delphi is intended to broaden the Programme's reach to a wide group of experts, whilst open conferences, an on-line survey and a web forum will be used to engage the wider public. As the Programme has just started at the time of writing, it is too early to make any assessments of its conduct or impacts.

CONCLUSION

There have been no formal evaluations carried out of the foresight exercises described above, besides the international panel review of the Hungarian TEP. This makes it rather difficult to assess whether foresight has delivered on its potential. Moreover, as the Hungarian (and other West European) examples show, it can take some time for foresight to have impacts. Since many of the exercises have been completed only recently, it is perhaps too early to pass judgement on their worthiness in addressing the fundamental challenges facing the region. Nevertheless, it is clear that the national foresight exercises described here have attempted to address, to a greater or lesser extent, many of the aforementioned weaknesses present in national research and innovation systems. For instance, they have sought to focus research efforts by setting spending priorities; they have generated fora bringing industrialists and policy-makers (the "application sphere") into a dialogue with scientists; and they have sought to emphasise the importance of innovation, thereby strengthening the link between RTDI and socio-economic development. Indeed, to varying degrees, exercises have attempted to address broad socio-economic needs and problems, rather than a narrow S&T agenda. Topics being considered – for example, by expert panels – are often framed in terms of industrial sectors or some sort of socio-economic challenge, rather than by areas of science or technology, which has resulted in an innovation agenda gaining importance in many of the exercises.

It is clear that foresight in the region has borrowed heavily from a variety of Western experiences, but exercises have also tended to adapt approaches to suit local needs. Certain approaches seem to be favoured, particularly those that generate "solid", expert-based, codified outputs, such as critical technologies and Delphi. We therefore see critical technology exercises performed in Russia, Ukraine and the Czech Republic (inspired by experiences in the United States, France and the Netherlands), and the Delphi method used in Hungary and Russia (inspired by Japan, Germany and the UK).[20] Scenario approaches have also been used, but have tended to play a less prominent role than in West European exercises. As has already been suggested with reference to the Hungarian case, this might be attributable to the influence of a central planning legacy, where the mindsets of policy-makers and foresight participants have been more accustomed to dealing with single rather than qualitatively different futures.[21]

Without exception, all exercises have tried to open up deliberative processes to a wider variety of actors than would normally be the case. Some have gone further than others in this regard, while few, if any, exercises have tried to engage citizens (a possible exception here is Poland). Furthermore, some exercises have emphasised foresight process benefits more than others, with Hungary, Romania and Poland particularly strong in this respect. Yet, it seems sufficient attention is seldom paid to processes in the rush to generate codified results. In some instances, even the value of the process itself remains largely unacknowledged, with foresight viewed merely as a source of intelligence for governmental policy-making. This is especially evident in Russia and Ukraine, where perhaps the least amount of attention has been paid to harnessing the benefits of foresight processes – possibly a reflection of a still lingering legacy of central planning.

As already mentioned, many of the foresight exercises discussed in this chapter have attempted to give greater prominence to innovation. Yet, in many, if not all, countries in the region, a wide appreciation of the importance of innovation remains elusive. RTDI policies continue to be isolated (often "innovation" does not even exist as an explicit policy field), and major socio-economic policies are uncoordinated with RTDI policies. This gives rise to an apparent paradox: foresight is needed in the region more than ever to challenge prevailing attitudes and institutional arrangements that prevent the mainstreaming of RTDI policies with socio-economic policies. But exactly because of these inhibiting factors, research and innovation systems – if they exist at all – are poorly equipped to accommodate foresight. Subsequently, it is often difficult for foresight and its results to find a "natural" home. This maybe explains, at least in part, why exercises are only "scattered" across the region when compared to the much richer landscape of activity observed in Western Europe.

Perhaps unsurprisingly, exercises have had most visible impact when they have been initiated with the purpose of informing some sort of newly created national policy or funding programme. This is most obvious in the cases of Czech Republic, Romania, and the 2004 update of the Russian critical technologies exercise. But even where this has not been the case, exercises have often generated externalities that have seen, for example, spin-off foresight studies initiated, as in Ukraine and Poland, where regional and sectoral foresight is becoming increasingly popular.

If one wanted to be pessimistic, then the region's commitment to "profound" foresight exercises could be called into question, i.e. its commitment to exercises that encapsulate serious consideration and determined implementation of policy recommendations, and that see an acceptance and introduction of a new decision-making culture, along with a new way of thinking, with more emphasis on communication, cooperation, consensus among the major stakeholders, and joint commitments to take action. But then this may be too much to ask at this relatively early stage. On a more positive note, national TF exercises have already been repeated in Russia, Czech Republic and Poland, suggesting that the approach is perhaps becoming increasingly embedded. Moreover, in several countries, there is a flourishing of regional and sectoral foresight activities. As foresight is, above all else, a learning process, its diffusion to other levels and sites may be cause for optimism.

NOTES

1 This chapter draws on Havas (2006b) and Keenan (2005), as well as on their "tacit" knowledge obtained from being involved in foresight activities in the CEE region in various capacities: managing and/or advising foresight programmes, participating in EU and UNIDO methodological and mapping projects on foresight, and delivering training seminars.
2 Space limits do not allow an extensive discussion of these issues here, and thus crucial country specific features cannot be dealt with either, although a considerable diversity can be observed even among the Central European countries, let alone Russia and Ukraine, given their size and different development path over many centuries. More details can be found, e.g. in Acha

and Balázs (1999); Bucar and Stare (2002); Chataway (1999); Havas (2002, 2003b, 2006a); Inzelt *et al.* (2005); Keenan (2005); Kubielas (2003); McGowan *et al.* (2004); Meske *et al.* (1998); Müller (2002); Nauwelaers and Reid (2002); Reid *et al.* (2002); Piech and Radosevic (2006); Radosevic (1994, 1998, 1999, 2002, 2003a, 2003b); Radosevic and Sadowski (2004); and Stephan (2005). In addition, further information can be found in the recent ERAWATCH country profiles and TrendChart reports at http://cordis.europa.eu/erawatch, and http://trendchart.cordis.lu, respectively.

3   All figures used in this section are from OECD Main S&T Indicators 2006/2 and the European Innovation Scoreboard 2006, or results of our own calculations based on these sources.

4   Moreover, the scarcity of business and law graduates, especially up until the late 1990s, and hence the much brighter opportunities and higher salaries in those types of jobs, has made those professions – and thus university courses – much more attractive.

5   The following discussion draws on Havas (2006a).

6   For a more detailed analysis of the major automotive cases in Central European countries, see e.g. Havas (2000).

7   Estonian pilot studies were funded through the eForesee project – see Chapter 9 for details. Bulgarian and Romanian pilot studies were funded through the ForeTech project – for details, see http://foretech.online.bg/index.php.

8   For details, see http://forlearn.jrc.es.

9   For further details on this initiative, see Keenan (2006) and Havas (2006b).

10   This agency has been substantially reorganised and renamed several times since 1997. During the lifetime of TEP, in 1997–2001, it was headed by four different people, with changing official titles.

11   As macro scenarios had not been developed in any other country engaged in foresight activities at that time, there was no way to learn from international experience while resolving this problem.

12   The futures outlined in the first National Development Plan (2004–2006), heavily rely on the so-called macro visions published in the Steering Group report of TEP, as well as the overall vision of the new STI policy strategy, approved by the government in March 2007.

13   About 25 per cent of public expenditure on R&D in the Czech Republic is allocated through the National Research Programme (NRP). The main objective of the NRP is to support key research directions (research in key technology areas) having a strong potential to contribute to a favourable economic development and to the satisfaction of societal needs. The first National Research Programme (NRP I) was launched in January 2004, the second (NRP II) in 2006, and the third (NRP III) will be launched in 2009 (Klusacek, 2007).

14   The idea for the exercise – and its philosophy – were especially influenced by national technology foresight exercises in the UK and Germany. However, with very limited time available to conduct the exercise – around one year – the approach taken was heavily influenced by "quicker" key technologies approaches, such as the Dutch Technology Radar exercise (see Rand Europe and Cooper & Lybrand, 1998) and the French Ministry of Industry's Key Technologies exercises (see, for example, Ministère de l'Industrie, 1995).

15 Additional activities have also included the development of methodology, training, and international cooperation/coordination.
16 For an account (in English) of Soviet S&T long-range planning, see Fortescue (1985) and Ksenofontov (2001).
17 For further information on the National Nanotech Foresight Programme, see Gaponenko (2006).
18 The following account is based largely upon Yegorov (2006) and personal correspondence between one of the authors and participants in the Ukrainian Foresight Programme.
19 The survey was referred to as a Delphi. Whilst it was not a Delphi, it had the appearance of a Delphi, since it was one of the expectations of the project's sponsor that Romania should conduct a Delphi similar to those carried out in other national technology foresight exercises (Grosu, 2007).
20 At the same time, there is little evidence of learning from previous communist-era long-term planning and forecasting work. This can be attributed to the simple fact that the current generation of foresight practitioners in the CEE region, for the most part, come from a science, technology and innovation studies background – as opposed to a forecasting or long-term planning tradition – and have found it more natural to learn from colleagues abroad than from communist-era forecasting work. Moreover, in many countries, the latter tends to be tainted by association with a central planning past and is widely considered to be unsuitable for the new demands and conditions of a modern market economy (this is especially the case in the countries of Central Europe).
21 Yet, current structural changes in the world economy and the emergence of new, global concerns related to environmental, health and demographic issues, imply that the scenario method may be relevant – not only in transition countries, per se, but also in countries with long-established, crystallised institutional systems. A growing body of literature suggests that technological and socio-economic changes are intertwined. Scenario workshops, therefore, can contribute to a better understanding of these complex relations, leading to policy proposals, which help in making appropriate choices in an increasingly complex environment. In many cases, technical experts are aware of the importance of non-technological issues (human resources, institution-building, legislation, regulation, organisational innovation). Also, taken alone, the Delphi-method can facilitate the foresight process only to a limited extent, and thus the process benefits are bound to be limited, too.

REFERENCES

Acha, V. and Balázs, K. (1999), 'Transitions in Thinking: Changing the Mindsets of Policy Makers about Innovation', *Technovation*, **19**(6–7), pp. 355–364.
Bucar, M. and Stare, M. (2002), 'Slovenian Innovation Policy: Underexploited Potential for Growth', *Journal of International Relations and Development*, **5**(4), pp. 427–448.
Chataway, J. (1999), 'Technology Transfer and the Restructuring of Science and Technology in Central and Eastern Europe', *Technovation*, **19**(6–7), pp. 345–353.

Curaj, A. (2006), 'Foresight in Policy Cycles: Case of Romania', presentation to the UNIDO Technology Foresight Training Module, Gebze, Turkey, November 2006.

Curaj, A. (2007), 'Foresight for National RTDI Strategy: Technology Foresight in Romania', paper presented at the UNIDO Technology Foresight Summit, Budapest, September 2007.

EAHERF (Romanian Executive Agency for Higher Education and Research Funding) (2005), *Elaboration of RDI Strategy within the Framework of the National Foresight Exercise*, EAHERF, Bucharest.

Fortescue, S. (1985), 'Project Planning in Soviet R&D', *Research Policy*, **14**, pp. 267–282.

Gaponenko, N. (2006), 'Russian Nanotechnology 2020', *European Foresight Monitoring Network Brief*, **75**, available at http://www.efmn.eu.

Grosu, D. (2007), 'Romanian Foresight Programme: Beyond Predictability', Futura 2/2007.

Havas, A. (2000), 'Changing Patterns of Inter- and Intra-Regional Division of Labour: Central Europe's Long and Winding Road', in J. Humphrey, Y. Lecler and M.S. Salerno (eds) *Global Strategies and Local Realities: the Auto Industry in Emerging Markets*, Houndsmill, London and New York: Macmillan and St Martin's Press, pp. 234–262.

Havas, A. (2003a), 'Evolving Foresight in a Small Transition Economy: The Design, Use and Relevance of Foresight Methods in Hungary', *Journal of Forecasting*, **22**, pp. 179–201.

Havas, A. (2003b), 'The Relevance of Foresight for Accession Countries and Possibilities for Co-operation', *IPTS Report*, **73**, pp. 4–11.

Havas, A. (2006a), 'Knowledge-intensive Activities vs. High-tech Sectors: Traps and learning options for Central European policy-makers', in: K. Piech, S. and Radosevic, S. (eds), The *Knowledge-Based Economy in Central and East European Countries*, Basingstoke: Palgrave, pp. 259–279.

Havas, A. (2006b), 'Developing UNIDO TF Programme for the CEE/NIS Region', available at http://www.unido.org/file-storage/download/?file_id=57728.

International Panel (2004), *Evaluation of the Hungarian Technology Foresight Programme (TEP)*, available at: http://www.nkth.gov.hu/main.php?folderID=159.

Inzelt, A. (2004), 'The Evolution of University–Industry–Government Relationships during Transition', *Research Policy*, **33**, pp. 975–995.

Inzelt, A., Zaman, G. and Sandu, S. (eds) (2005), *Science and Technology Policy Lessons for CEE Countries*, Romanian Academy, Bucharest.

Keenan, M. (2005), *Key Drivers and Outlooks for RTD in the CIS Region*, report of the SCOPE 2015 project, PREST – The University of Manchester, available at: http://prest.mbs.ac.uk/prest/SCOPE/CIS_reports.htm.

Keenan, M. (2006), *Establishment of a Technology Foresight Regional Virtual Centre for the CEE/NIS Countries: International Aspects*, Vienna: UNIDO.

Klusacek, K. (2004), 'Technology Foresight in the Czech Republic', *International Journal of Foresight and Innovation Policy*, **1**(1–2), pp. 89–105.

Klusacek, K. (2007), 'Key Technologies for the Czech National Research Programme', paper presented at the UNIDO Technology Foresight Summit, September, Budapest.

Ksenofontov, M. (2001), 'Technology Foresight in the Russian Federation: Background, Modern Social, Economic and Political Context, and Agenda for the Future', paper presented at the UNIDO Regional Conference on Technology

Foresight for Central and Eastern Europe and the Newly Independent States, 4–5 April, Vienna, Austria.

Kubielas, S. (2003), 'Polish Macroeconomic and S&T Policies: Interlinkages for Growth and Decline', *Journal of International Relations and Development*, **6**(2), pp. 156–184.

Lepeta, M. (2006), 'The Polish Foresight Pilot: Health and Living 2013', *European Foresight Monitoring Network Brief*, **38**, available at http://www.efmn.eu.

McGowan, F., Radosevic, S. and von Tunzelmann, N. (eds) (2004), *The Emerging Industrial Structure of the Wider Europe*, London: Routledge.

Meske, W., Mosoni-Fried, J., Etzkowitz, H. and Nesvetailov, G.A. (eds) (1998), 'Transforming Science and Technology Systems – The Endless Transition?', *NATO Science Series*, Amsterdam: IOS Press.

Ministère de l'Industrie (1995), *Les Technologies Clés pour l'Industrie Française á l'Horizon 2000*, Paris: MI.

MNISW (Polish Ministry of Science and Higher Education) (2006), *National Foresight Programme Poland 2020 Brochure*, Warsaw: MNISW.

Müller, K. (2002), 'Innovation Policy in the Czech Republic: From Laissez Faire to State Activism', *Journal of International Relations and Development*, **5**(4), pp. 403–426.

Nauwelaers, C. and Reid, A. (2002), 'Learning Innovation Policy in a Market-based Context: Process, Issues and Challenges for EU Candidate-countries', *Journal of International Relations & Development*, **5**(4), pp. 357–379.

Nedeva, M., Loveridge, D., Keenan, M. and Cuhls, K. (2000), *Science and Technology Foresight: Options for Poland*, report to Polish Ministry of Science (KBN).

Pavitt, K. (1991), 'What makes Basic Research Economically Useful?', *Research Policy*, **20**(2), pp. 109–119.

Pavitt, K. (1998), 'The Social Shaping of the National Science Base', *Research Policy*, **27**(8), pp. 793–805.

Piech, K. and Radosevic, S. (eds) (2006), *The Knowledge-Based Economy in Central and East European Countries*, Basingstoke: Palgrave.

Radosevic, S. (1994), 'Strategic Technology Policy for Eastern Europe', *Economic Systems*, **18**(2), 87–116.

Radosevic, S. (1998), 'The Transformation of National Systems of Innovation in Eastern Europe: Between Restructuring and Erosion', *Industrial and Corporate Change*, 7, pp. 77–108.

Radosevic, S. (1999), *Restructuring and Reintegration of S&T Systems in Economies in Transition*, final report, TSER project, Brighton: SPRU.

Radosevic, S. (2002), 'Introduction: Building the Basis for Future Growth – Innovation Policy as a Solution?', *Journal of International Relations & Development*, **5**(4), pp. 352–356.

Radosevic, S. (2003a), 'A Two-Tier or Multi-Tier Europe? Assessing the Innovation Capacities of Central and East European Countries in the Enlarged EU', *Journal of Common Market Studies*, **42**, pp. 641–666.

Radosevic, S. (2003b), 'Patterns of Preservation, Restructuring and Survival: Science and Technology Policy in Russia in Post-Soviet Era', *Research Policy*, **32**, pp. 1105–1124.

Radosevic, S. and B. Sadowski (eds) (2004), *International Industrial Networks and Industrial Restructuring in Central and Eastern Europe*, Dordrecht: Kluwer.

Rand Europe and Coopers and Lybrand (1998), *Technology Radar: Methodology*, Delft, The Netherlands.

Reid, A., Radosevic, S. and Nauwelaers, C. (2002), 'Innovation Policy in Six Applicant Countries: The challenges, European Commission', Innovation Papers No. 16, Luxembourg, also available at http://www.cordis.lu/innovation-smes/src/studies.htm.

Sokolov, A. (2006a), 'Identification of National S&T Priorities Areas with Respect to the Promotion of Innovation and Economic Growth: The Case of Russia', in *Bulgarian Integration into European NATO*, NATO Security through Science Series: Human and Societal Dynamics, IOS Press, pp. 92–109.

Sokolov, A. (2006b), 'Russian Critical Technologies 2015', *European Foresight Monitoring Network Brief*, **79**, available at http://www.efmn.eu.

Sokolov, A. (2007), 'Foresight in Bashkortostan: Identification of Regional Innovation Priorities', paper presented at the UNIDO Technology Foresight Summit, Budapest, September.

Stephan, J. (ed.) (2005), *Foreign Direct Investment and Technology Transfer in Transition Countries: Theory – Method of Research – Empirical Evidence*, New York: Palgrave Macmillan, Basingstoke: St. Martin's Press.

Yegorov, I. (2006), 'Ukrainian RTDI 2025', *European Foresight Monitoring Network Brief*, **74**, available at http://www.efmn.eu.

PART THREE

Policy and Management Issues in Foresight

14. Policy Transfer and Learning

Luke Georghiou and
Jennifer Cassingena Harper

INTRODUCTION

This chapter examines the phenomena and roles of international transfer and learning of policy in the specific case of foresight. In so doing we will discuss the degree to which foresight in a given situation has been influenced by other countries' approaches and experiences and the role of key vectors such as international organisations and other providers of advice in exerting such influence. With reference to foresight we will identify some of the broader issues and difficulties involved in policy transfer and explore the role of mutual learning in a transnational setting.

At a superficial level research and innovation policy measures and instruments in different countries often appear rather similar. Efforts to produce generic categorisations of these measures, for example, the European Commission's Trendchart on Innovation Policy in Europe[1], reinforce this impression of similarity and at the same time seek to identify or distil success and to facilitate its transfer from one country to another. Such transfers have long been a feature in this area. The Organisation for Economic Cooperation and Development (OECD) states that its mission is to provide governments with the analytical basis for policy formulation. In practice this normally means facilitating exchanges of experience or providing information about the environment for research and innovation, such as reliable comparative statistics, that allows the relative effects of policies to be considered.

More recently, the European Commission has also made transfer and learning central elements of its research policy, as part of its activities to promote the growth of the European Research Area through improved coordination of national policies (through what is known as the "open method of coordination"). Specific measures being promoted include developing a "knowledge platform" for systematising and exchanging foresight experiences in Europe, benchmarking of national research policies and the

ERA-Net instrument to bring together those who are responsible for national or regional programmes. As well as these multi-lateral efforts, research and innovation policy also are subject to numerous bilateral linkages through visits, circulation of people and joint activities, either by those who design and implement policies or by the academics who sometimes provide the conceptual basis for their work.

Against this broad background, this chapter first seeks to develop a general framework for the understanding of mutual learning and policy transfer and then considers how this may be applied to foresight. Three short case studies of learning in foresight are presented to illustrate these issues.

MUTUAL LEARNING AND POLICY TRANSFER

Learning forms a central tenet in the "systems of innovation" perspective that now forms one of the main guiding posts for the design of innovation policies. Lundvall's pioneering work on this topic[2] gives a central role to interactive learning in his explanation of how new knowledge is produced and diffused through the relationships between actors in those parts of the economy and society engaged in innovative activities, in this case within the borders of a nation-state and extending to user–producer interactions. Transnational comparison within the systems of innovation framework followed soon after with Nelson's 15-country study which sought to explore the significance of the similarities and differences between the institutions supporting innovation in each case (Nelson, 1993). Implicit in this approach is the idea that a more complete understanding of the complexities and interdependencies of policies and institutions in a national context would make more realistic the prospect of successful learning from one to another.

More recent thinking on the systems of innovation perspective has emphasised that a system could be conceptualised in multiple dimensions rather than simply that of the nation. Systems may relate to sectors of economic activity, regions or technologies. Connections to global industries and knowledge flows also point to the need for transnational conceptualisations of systems. Metcalfe argues that the systems themselves are a transient and evolving feature of "restless capitalism" and exist to solve specific and local innovation problems by connecting a more fundamental unit, the "innovation ecology" (Metcalfe, 2006). This he defines as the set of individuals and organisations which generate, store and build connections around knowledge and information. In Chapter 1, we showed how the systems perspective provides an important component for the rationale for foresight. For the time being the key aspects we may take forward are the central role of learning and the inherent complexity of the environment for learning when policies are intricately related to multiple and overlapping systems of innovation. The transient and evolving nature of systems and their context underpins the adaptive features of foresight as it interacts dynamically with the context, shaping and being shaped by it.

If we take the two themes of this chapter, learning and transfer, they may be distinguished partly by timing and partly by intention and action. In the normal course of events, learning could be seen as an integral part of any policy process and something that would precede any attempt to transfer the lessons to another context. The most natural environment for policy learning is the one in which the original measure was conceived and implemented (see Table 14.1). The expectation is that here, largely the same actors will be dealing with extensions of policies, or their successors in an environment which has changed only incrementally. Under these conditions learning may be both codified, that is written down by practitioners or external evaluators, or tacit, being the resultant change in the light of experience in the contextual knowledge or routines of practice held by the individuals responsible for the policy.

Table 14.1 Conditions favouring mutual learning

Foresight Setting	Similar structural environment – possibly deriving from EU membership obligations
	Compatibility with political direction
	Sense of dissatisfaction with current solutions
	Similar available resource levels
	Similar available expertise levels in foresight methods and expert domains
Foresight Instrument	Not so over-specified as to be idiosyncratic
	Well-documented and evaluated
	Compatible with but complementary to existing instruments

From this baseline situation we may now begin to elaborate on the challenges involved in transfer to another context on the one hand and the conditions favouring learning on the other. These concern the level of learning and its means of transmission together with the type of instruments and the environment. In terms of the level, in the situation described above policy-makers have no need to make explicit many of the relevant factors. They will all be familiar with the "rules of the game" in their own national and organisational context and will be sensitive only to what is new or deviant from established practice. However, as we shall discuss below, much of this accepted context remains relevant and may be crucial to the transfer of the policy to another context where different individuals may not experience the same background knowledge or conditions. Mutual learning functions best on the other hand where the countries or organisations share a similar institutional setting and resources and expertise levels and where there is low specificity and clarity in the policy instrument selected.

Moving on to the means of transmission, while codified learning may be available, this too may be restricted in its scope. An evaluation carried out in support of an ongoing system can again embody shared assumptions that do not hold elsewhere. A new situation may not involve access to the tacit knowledge held by individuals, and even if those individuals move with the policy, or are present in an advisory capacity, they in turn may lack the equivalent contextual knowledge of the new situation. The assumption that learning precedes transfer may not hold in all cases. In a fast-moving and novel situation, there may be a need to apply lessons from one dynamic context to another. The situation of most interest to us here is that where the flow is in both directions, what we term "mutual learning". With two or more actors in an innovative policy situation the challenge is to elucidate and apply lessons as far as possible in real time. In some ways this situation can be more favourable to learning. The presence, even part-time, of actors unfamiliar with the background assumptions is likely to create pressure to make these more explicit. Direct exposure to another's situation can help in accessing tacit lessons. However, the risks are also high and less satisfactory or superficial characterisation is achieved because of the dominance of short-term issues that turn out not eventually to have been the critical factors.

Rose has defined "lesson drawing" as a detailed cause-and-effect description of a set of actions that government can consider in the light of experience elsewhere, including a prospective evaluation of whether what is done elsewhere could someday become effective in one's own circumstances (Rose, 1993, p. 27). He characterised learning in terms of five different types of lessons that may be drawn: copying, adjusting, hybridisation, synthesis and inspiration.[3] For our purposes these can be amended and extended as follows:

- Imitation – In this case learning involves an attempt to do almost exactly the same as was done in the original situation where relevant and possible;
- Adaptation – Again the original model is followed but in this case efforts are made to adapt it to the new context and/or time and resources;
- Combination – Here the novel element is multiple learning, to combine elements of two or more programmes, almost certainly from different contexts;
- Inspiration – Here the experience of others is taken as a starting point for new creation of design. For example it may be sufficient only to know that a policy of a certain type can be made to work in the circumstances of interest; and
- Elimination – Here the experiences of others create a realisation that the approach is not valid for one's own circumstances. Lessons can be negative as well as positive. Knowing that an approach is inappropriate can save resources on a large scale. It is important always to maintain this option and not to become so engaged in learning that one is blinded to the possibility that action may not be desirable;
- Iteration – Any of above may be moderated in light of one's own experiences and interaction with dynamics of other players.

There have been efforts to characterise learning in terms of the state of evolution in different regions, see for example Tables 11.1 and Table 12.1. As mentioned above, benchmarking is now seen as a key tool for policy improvement and learning (Barré *et al.*, 2002). The process has been defined as a "continuous, systematic process for comparing performance of for example, organisations, functions, processes of economies, policies, or sectors of business against the 'best in the world', aiming … to exceed [these performances]".[4] The origins of benchmarking are as a management tool in the private sector for the comparison of industrial processes and mutual learning of best practice among those taking part. Key elements include a comparison with those perceived to be the best, the use of indicators to characterise and measure performance, and ensuring that the resultant learning process is based upon an understanding of the underlying processes that cause different levels and types of performance. Recent efforts to extend the approach to research and innovation policy have recognised that taking the nation as the unit of analysis implies a more complex environment for learning.[5] As Lundvall and Tomlinson (2001, p. 122) put it:

Countries are characterised by systemic differences and therefore what is best practice in one country or region will not be best practice in another. Therefore the more modest aim to develop "good" and "better" practices through "learning by comparing" is more adequate.

Implicit in this approach is a move away from a simple array of quantitative indicators, and also from an overall concept of best practice. Instead it is accepted that only partial aspects of a system can be examined and that the ensuing learning process is likely to be both continuous and mutual.

To complete this short review of the conceptual background to learning and transfer in foresight we may consider the broader political science literature on the subject of policy transfer. This offers the definition by Dolowitz (2000) initially conceived in the context of transfer if welfare policies but easily generalised:

> The process by which knowledge of policies, administrative arrangements, institutions and ideas in one political system (past or present) is used in the development of policies, administrative arrangements, institutions and ideas in another political system (Dolowitz, 2000, p. 3).

Dolowitz points out that transfer may not always be synonymous with learning. While learning is an essential element of transfer in voluntary situations, policy transfer may also be coercive as when, for example, the International Monetary Fund imposes structural adjustment policies as a condition of assistance (even here it could be argued that learning is implicit). While foresight is most unlikely to be imposed in this sense, it is possible to imagine its adoption as part of a package of other policies which the affected community has little choice but to accept.

Writing with Marsh, the same author has identified a framework for understanding the reasons for failure in policy transfer, all of which feasibly could apply in the context of foresight (Dolowitz and Marsh, 2000, p. 326):

- **Uninformed transfer**: The case in which the borrowing country has insufficient information about the policy that is being transferred with the result the policy is imperfectly implemented;
- **Incomplete transfer**: Where crucial elements of a policy or programme that made the policy or programme a success are not transferred; and
- **Inappropriate transfer**: Where insufficient consideration is given to social, economic, political and ideological differences between the borrowing and the transferring country leading to programme failure.

It should be noted that this framework has been criticised on a number of counts. Most obviously the judgement of success and failure in policy terms is highly prone to subjectivity even if it is reduced to correspondence with the aims of those seeking the transfer. James and Lodge (2003) argue that none of the failure categories described above refer to the process of transfer, being rather judgements about the transferred policy itself. They also argue that lesson drawing and transfer are in effect variants upon a rational model of policy-making that disguise or conflate more complex processes that may be going on.

For the purposes of analysing learning and transfer in foresight the key points to take forward from this debate are to recognise that a distinction needs to be made between the process and outcomes of transfer. It is likely that the process of learning and transfer will itself have an influence upon the recipient programme, or programmes in the case of mutual learning. There is also a need to avoid judgements about the success or failure of transfers. Rather, the aim is to ascertain whether the transfers actually took place, what circumstances favoured them, and insofar as it is possible, what effects the transferred elements of the policy instruments had in their new contexts.

LEARNING IN THE CONTEXT OF FORESIGHT

Tables 1.1 and 1.2 in Chapter 1 sets out a chronology of foresight exercises, recording mainly those carried out at the national level. In a sense this maps the historical opportunities for learning and transfer. As we shall see in the case study below there is a clear family tree in terms of the use of large-scale Delphi surveys which also spills over into the hybrid exercises. Another explicitly-related family tree is that of critical technologies exercises. Broadly speaking the earlier exercises have been the most influential, partly because of their pioneering nature and partly because some of their key participants have become expert in the process of policy transfer itself.

Learning is implicit in any foresight activity, how ever extensive or restricted the exercise is in terms of funding or time available, stakeholder involvement, or other resource. The systematic efforts expended in the design, scoping, implementation, diffusion and follow-up activity are driven by intense processes of interaction and learning – these contribute to qualifying the activity for the label of foresight exercise! What may be more difficult to ascertain is the extent of learning, tangible and intangible, targeted and unforeseen, which merits the qualification since different foresight activity generates alternative modes and levels of learning.

Recent experiences in international policy transfer and learning generated through joint foresight exercises involving more than one region or country complement and contrast with the more traditional national exercises. The modes and levels of learning generated by countries/regions adapting the foresight experiences of other countries/regions to their own context stand in contrast with the more intense mode of collaborative learning evident in the implementation of joint foresight initiatives or studies with other countries/regions.

Table 14.2 below aims to highlight key similarities and differences in the frameworks for learning by adapting and learning by collaborating. In either case, the question of context is a critical factor in the learning process, since joint exercises are simultaneously enhanced and hindered by the challenge of addressing in parallel highly different contexts (e.g. advanced economy vs. transition economy). Where the element of mutual learning is deliberately designed/built into the foresight activity at the start during the scoping and planning phase, the potential for learning is increased and alternative modes and levels can be explored and generated, including:

- A well-defined communications strategy aimed at engaging a range of stakeholders through deliberate and targeted efforts to promote an ongoing process of mutual learning. These efforts may take the form of awareness and foresight training activity in response to the needs and requirements raised in real-time while implementing the exercise; or a diffusion activity to engage those not directly involved in the exercise but potentially affected by its impacts or results;

- A strategy for drawing in knowledge and know-how from outside from more experienced foresight practitioners and experts through the organisation of workshops, conferences and courses; by participating in relevant fora outside the country/region; or through the use of web-based fora to elicit expert opinion at a distance from the diaspora;

- Joint foresight activity affords greater opportunity for learning through the joint undertaking of scoping and planning work and deeper engagement in discussion on appropriate methods and approaches. The level of mutual learning generated ultimately depends on the extent to which the learning element is deliberately designed into the foresight exercise, the open framework put in place to encourage iterative learning processes and the flexibility in reacting to emerging learning needs and opportunities as they arise.

Table 14.2   Learning modes in foresight

	Learning by Adapting	Learning by Collaborating
Who	Foresight managers Sponsors and clients Government (national and regional) Private sector Society Inputs from international players	Foresight managers Sponsors and clients International players – stronger role Government (national and regional) Private sector – multinationals Society – emphasis on multi-cultural inputs
Why	Fast-track learning State-of-the art Good practices and ensuring quality An external perspective	Common concerns Need for coherence in policy approaches Critical mass and value added Fast learning by pooling diverse inputs Assigning tasks on the basis of competencies A European/global perspective
What	Orientation and approaches Tools and methodologies Implementation structure Thematic focus and content	Consensus-based approaches Creative tools and methodologies Shared tasks and responsibilities Themes of common interest
How	Studying final and background reports /results Engaging foreign experts on panels International workshops and training activities Distributed learning using web-based tools	Joint studies Joint panels Joint workshops and training activities Common platform for web-based learning
When	Preparatory and implementation phase Evaluation and follow-up	Through the whole process

CIRCUMSTANCES FOR LEARNING IN FORESIGHT

Mutual learning in foresight can be enabled potentially in a number of contexts including:

- National foresight with international inputs before or during the programme;
- Global/international foresight, e.g. the Millennium Project;
- Inter-governmental foresight studies within frameworks such as COST and the European Science Foundation;
- National-Transnational foresight as found in a variety of European Union-sponsored projects;
- Regional foresight, notably in the FOREN network which translated and adapted guidelines to national contexts across Europe;
- Private sector foresight, e.g. the Corporate Foresight Network; and
- Other configurations such as foresight for cities or NGOs.

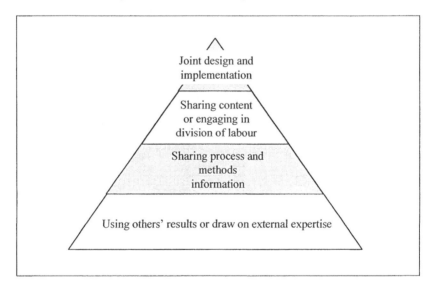

Figure 14.1   A hierarchy of cooperation options

Figure 14.1 presents a simple hierarchy of options for international cooperation ranging from use of published outputs through to joint work. In this chapter we present three short case studies of learning in different circumstances. The first case concerns sequential transfers of methodology, specifically within the national Delphi surveys predominant in the first half of the 1990s.

In the second case, the emphasis switches to mutual learning in the context of the eForesee project described in Chapter 10. As noted there, part of the concept was that the three participating small countries acceding to the European Union could share their experiences in the application of foresight and learn from each other as the project proceeded.

Finally, the third case study describes some recent and ongoing efforts to systematise learning about foresight, with a particular focus on "candidate countries", that is those which were or are seeking accession to Membership of the European Union. All of these cases involve projects sponsored by the European Union. They include the construction of databases of foresight activity, efforts to promote training and benchmarking activities.

National Delphi Surveys

As already noted in the discussion of "family trees" earlier in this chapter, there has been a connected sequence of Delphi surveys.[6] While the technique of Delphi originated in the USA, for our purposes, the sequence begins with the Japanese technology forecasts carried out approximately every five years since 1971 and now in their eighth iteration. In this short case study of transfer we will focus only on the "topics" or "statements" that form the core of Delphi exercises. An extended discussion could also have examined the learning and variation in the questions asked about these topics – for example time of realisation, national benchmarking, and of questions about the expertise of respondents.

As Kuwahara *et al.* (1994) have noted, perhaps the more remarkable feature of the Japanese surveys is that for the first two decades of this large-scale activity very little notice was taken of it outside Japan, despite some of the reports being translated into English. A key vector in terms of raising European awareness of this activity was the international review carried out by Irvine and Martin in 1984. This described the forecasts in some detail and gave them a positive evaluation in terms of:

- The effect they had on making researchers and decision-makers think strategically;
- The fact that their broad scope allowed "holistic visions" to become apparent (for example the emergence of mechatronics);
- Providing a general framework for more systematic sectoral forecasts by Japanese industry; and
- Helping the government establish priorities for the allocation of resources.

With hindsight this would be seen to some extent as an overstated view of the impacts of the surveys (though they have gained in influence in recent years through a much more systematic connection to policy-making through their present home in the National Institute of Science and Technology Policy). Nonetheless, this endorsement, reiterated in the conclusions of the book and associated with the then apparent dominance of Japanese technology, gave a significant profile to these activities in Europe and elsewhere.

The first concrete realisation of that influence came when the German government decided to experiment with the technique by translating the second round questions of the fifth Japanese survey into German and implementing it in Germany (as described by Cuhls in Chapter 6). Several issues involved in the transfer were described in Kuwahara *et al.* (1994). Many were linguistic – how exactly to translate semi-technical terms such as "capsule" which could have meant a chamber or a pill. Some specialist terms only had currency in Japan (e.g. bioholonics) and adapted words which had apparently the same meaning but where drift had occurred. More complex still were cultural references embedded in questions (the example given referred to entertainments provided by employers for their employees' leisure time). Heroic efforts at translation (cultural and linguistic) resulted in only three questions being dropped – one for example referring to cultivation of rice – not a crop in Germany!

A similar translation exercise was undertaken in France in 1993/94, on the same scale though with much less of a public profile than in Germany. For various reasons, including a change of government during the project's life, the project effectively halted before there was a dissemination phase. Some of the criticisms levelled at the study (recounted in an article by one of the team responsible for it)[7] indicate problems of policy transfer. Foremost was a cultural rejection by French experts of a questionnaire related to "culture and lifestyle". While concerned with technological issues, these had a high sociological content and were hence potentially meaningless outside their Japanese context. Unlike the German exercise it was not an explicit goal to confront experts with visions from Japan, and hence the prospect of resistance was higher. Experts not only struggled with the cultural context where it was apparent but also missed its presence in a number of other technological topics. Heraud and Cuhls (1999) suggest that the experts steeped in the French foresight tradition of *prospective* (see Barré, Chapter 5) were looking for linkages and frame conditions not available through the Delphi method. *La Prospective* favours contrasting but internally coherent scenarios, perhaps more in line with the national psyche.

Another transfer at that time was to South Korea which carried out its first large-scale technology Delphi in 1993–1994 (Shin *et al.*, 1994). This again was influenced by its Japanese predecessors which were familiar, having been translated into Korean. However, it was decided that national factors should be taken into account in the design and selection of topics. The result was that three-quarters of the topics were different from the most recent Japanese Delphi (Shin, 1998). For the 327 topics that were the same, comparison was carried out with Japan and Germany. In retrospect the different priorities afforded to topics in each country was also used to justify the decision to develop national topics.

The United Kingdom began its first Technology Foresight Programme in 1993 and close attention was paid to the Japanese and German experiences. However, reuse of the Japanese topics was never contemplated. The view of UK Foresight managers was that these topics reflected the agenda of Japanese industry and scientists and would not necessarily correspond to the aims of the UK Programme (Georghiou, 1996). The strong emphasis upon promoting networking between science and industry favoured the more interactive panel-based approach that was chosen. A Delphi survey on a similar scale to those already mentioned was constructed but its use was as a consultative instrument. Topics and issues raised were generated by the panels and reflected directly their concerns. It was also seen as a tool for communication as the survey would reach an order of magnitude of more people than would normally attend workshops or other events. However, several issues of transfer emerged. Firstly, the control of the exercise by the government body managing foresight meant that methodological considerations were subordinated to political imperatives. In particular the timescale for the exercise (two rounds) was compressed to three months, a fraction of that available for the Japanese and German surveys. Even on this timescale results could only be delivered towards the end of the process. There were also cultural factors relating to postal questionnaires. Many respondents objected to the entire concept of spending time filling in a complex form (made more complex and lengthy at the insistence of panels). In the end, with some exceptions, the main benefit to panels was considered to be the discipline imposed by formulating topics rather than the anonymous views about these from the responding experts. Interestingly one panel managed to miss out from its report an accurate Delphi forecast of one of the most significant scientific events in the UK in subsequent years – the cloning of Dolly the sheep. One UK innovation, the use of the co-nomination technique to identify foresight participants has become a fairly standard feature of subsequent foresight activities across Europe.[8]

While Germany embarked on collaborative exercises with Japan, Delphi continued to roll out through other European foresight exercises, all of which drew explicitly upon those that had gone before. The Austrian foresight programme of 1996 brought a small country perspective to bear which was interpreted as a need to identify market segments and niches in which the country could prosper rather than to search for emerging technologies whose implementation would be largely dictated by bigger countries.[9] There was also a desire to be problem- and demand-oriented and to respond explicitly to societal needs. This latter goal was pursued by running two linked Delphi surveys – one on technological and organisational innovation and the other on society and culture.

Hungary explicitly reviewed all of the exercises mentioned above, though its programme, TEP, most resembled the first UK Foresight Programme (Havas, 2003; see Chapter 13). Delphi statements were formulated by identifying the major trends in Hungary and studying foreign questionnaires. With panels free to choose their statements, a different national context again produced a different outcome, with over half of the topics being non-technological even though the authors were themselves mainly technologists. An evaluation of TEP (PREST, 2004) found that, as in the UK case, much of benefit of the Delphi came from its role as a discipline for the panels.[10] Again, the combination of Delphi and panels had led to an inflation of length and complexity with consequences for response rates.

As a footnote to this case, it seems that the era of the large-scale across-the-board technology Delphi has come to an end in Western Europe, although the approach appears to be popular among the countries of Central and Eastern Europe, many of which have only recently initiated national technology foresight exercises (see Chapter 13). More targeted exercises continue to flourish in the West, but the holistic approach is now continuing mainly in its country of origin, Japan. Some reverse learning has also taken place as the Japanese Delphi is now closely linked to other foresight approaches as a means to create a more interactive use of results.

All in all, the transfer of Delphi exercises has exhibited the full range of lesson drawing discussed above. We can see imitation (deliberately) in the translation of the Japanese survey, adaptation (by Korea and the UK), combination in the Hungarian case which drew upon elements from several of its predecessors, and inspiration in the Austrian case which took earlier efforts largely as a point of departure. Though not documented here, it is also true to say that elimination featured as a form of learning – some of the countries engaged in scenario-type exercises examined the experiences of Delphi surveys before deciding that these would not meet their needs.

Learning in eForesee

In contrast with the previous case which reflects learning through adaptation, the eFORESEE project[11] is an example of knowledge transfer through collaboration and mutual learning. This aspect was factored into the project at the start and this influenced the whole project cycle, from design to implementation. The eFORESEE project was aimed at the exchange of foresight experiences among three small candidate countries, Cyprus, Estonia and Malta, in their drive to improve the effectiveness of their RTDI policies in support of the national pre-accession strategy. The three participating countries shared the common challenge of introducing and catalysing fast processes of policy learning and unlearning in the critical pre-accession phase. The exchange of foresight experiences between the three country teams was considered as constituting the core rationale of the project and therefore efforts were made to extend this exchange and learning process as widely as possible throughout the project.

The main tools for instituting and generating this learning curve were:

- The implementation of two foresight pilots per country (although Malta carried out a third pilot). Each country was however left free to consult on and choose its own pilot themes and whereas Malta and Estonia focused on common themes of ICT and biotechnology, Cyprus opted for addressing different aspects of agriculture. Moreover, the precise thematic focus and approach adopted in Malta and Estonia differed considerably and made subsequent comparison and sharing of results less viable. The impact of context was clearly visible in the thematic choice and the results generated;

- The adoption of common tools, procedures and templates, in particular for the mission statements, statements of objectives, the setting of success criteria and, to a lesser extent, the resulting vision documents. Thus, the mission statements for the Malta and Cyprus pilots addressed a common set of clearly-defined objectives, namely: (a) to elaborate a vision for the country as an advanced-knowledge economy in 2020 whose main resource is its ability to develop human capital in new economy skills all round the world from a Mediterranean base (adapted for the two other pilots); (b) to guide the decision-making and input into the National Development Plan (2003–2006); and (c) to mobilise public-private sector partnerships to take action on business opportunities;

- The development and use of the project website and dedicated pilot mailing lists allowed the sharing of documents, news, experiences and contacts between the three teams. The website helped to make the process underway in the three countries more transparent and facilitated open learning in real-time. It is difficult to map and quantify all the learning generated as often cross-country transfer of ideas, approaches and tools is triggered through a comment or reaction visible to all connected through a common mailing list. On the other hand, more learning could have been generated through the web tools, but constraints of time, distance and lack of face-to-face contact detracted the full potential of mutual learning from being exploited;

- Finally, but most importantly, the regular project events were critical in encouraging mutual learning through dedicated sessions, termed "Foresight Learning Circles", for exchanging experiences between the team members facilitated by the coordinator. The sessions focused on problem-sharing and problem-solving and the exchange of good practices. The main insights and learning were, however, generated through participation in each other's events, in particular the joint training activities.

Table 14.3   Mapping collaborative learning in eFORESEE

	●●	Foresight managers
	●	Sponsors and clients
	●●	International players
	●●●	Government (national)
Who		Government (regional)
		Large firms and multinationals
	●●	Small and medium-sized enterprises
	●●●	Academia and students
	●●●	Non-governmental
	●●●	Individuals and laypersons
	●●●	Societal concerns
	●●	Common concerns
Why	●●●	Coherence in policy approaches
	●●●	Critical mass and value added
	●●	Fast learning by pooling diverse inputs
	●●	Assigning tasks on the basis of competencies
	●●	A European/global perspective
	●●	Consensus-based approaches
What	●●	Creative tools and methodologies
	●	Shared tasks and responsibilities
	●●●	Themes of common interest
	●	Joint studies
How	●	Joint panels
	●●●	Joint workshops and training activities
	●●●	Common platform for web-based learning
	●●●	Design Phase
When	●●●	Running the exercise Phase
	●●	Follow-up Phase

Whilst eFORESEE was specifically designed to generate international policy transfer and learning, and this was deliberately factored into the project design through the introduction of common approaches and tools, the level of learning resulting from the project did not reach its full potential. This mainly related to factors extending beyond the influence of the coordinator and teams involved:

- The impact of context and in particular socio-cultural factors which influenced the choice of themes, but also the implementation style;
- The constraints of distance and time which discourage ongoing exchanges as not all aspects of the process can be recorded;
- Insufficient face-to-face contact at critical points of the project implementation which allows for sharing of more tacit, "confidential" knowledge.

Efforts to Codify, Classify and Benchmark Foresight Experiences

In this case study, an alternative aspect of international policy transfer and learning is presented, namely EU efforts to provide a policy learning platform to help candidate countries cope with the accession challenge. This group of countries consisted mainly of Central and Eastern European countries in transition to market economies from their communist past, plus two small Mediterranean island states, Malta and Cyprus.

Since 1999, the EU Joint Research Centre Institute for Prospective Technology Studies (JRC-IPTS), a research organisation which is part of the European Commission, has undertaken a number of activities to promote policy transfer and mutual learning in response to the emerging European context. This has led to a number of actions in preparation for enlargement of the European Union in 2004 and more recently, in relation to the challenge of foresight in the context of an enlarged Europe.

Foresight and the Enlargement challenge
The Enlargement Project was designed and developed as "an instrument for improving the level of information about the Pre-Accession Countries (PACs) in the European Commission and for strengthening co-operative activities between the EU Member States and the Candidate Countries as well as among themselves". These activities centred on the organisation of a number of prospective seminars on S&T policy and its possible impact on socio-economic development and on the setting up of dialogues on techno-economic issues relevant to Enlargement. The Project was aimed at providing support to European decision-makers with their foresight activities and the integration process. The high-level dialogues generated a number of results,

in particular the development of contacts and networking between foresight practitioners and experts in EU Member States and the PACs. A follow-on initiative, the Futures Project, in 2001 aimed at continued support for national foresight activities in the PACs, given that the latter were facing common issues requiring a common approach.

The "Futures" Project, which focused on the techno-economic and societal impact of Enlargement, had the aim of examining the major contemporary technological, economic, political and social drivers in PACs and their impact on technology/science, competitiveness and employment until the year 2010. The project brought together experts and policy-makers, both from the PACs and EU countries to work together in thematic expert panels, supported by background research. Both the Enlargement and Futures Projects provided policy-makers in the PACs with the opportunity to forge valuable contacts with their counterparts and to tap strategic intelligence in meeting the accession challenge. In general the projects helped to enhance the mutual learning curve through the open dialogue which developed between EU Member States and the PACs.

Development of a foresight knowledge platform
In 2003, in response to the continuing need for foresight training and support in the enlarged Europe, the JRC-IPTS launched a pilot programme of foresight teaching and awareness-raising. Developed in collaboration with a network of European STI policy organisations, the activity was aimed at "designing and testing some training modules in the field of foresight and strategic prospective studies with the final goal of establishing a permanent international programme".[12] A nine-month pilot study[13] was carried out, which involved the organisation of a foresight training course and two awareness workshops, a review of foresight training supply and demand in a number of European Union (EU) Member States and PACs, and development of proposals for a future foresight learning platform.

The foresight platform proposals were subsequently taken up in two new initiatives sponsored by the EC's Directorate-General Research. The first is the European Foresight Knowledge-sharing Platform (the development of an open knowledge facility for diffusing useful results from forward-looking activities), which includes the development of a database of foresight exercises carried out in Europe and other parts of the world. At the time of writing, this database has almost 2,000 entries.[14] The second initiative is FOR-LEARN, which includes the development of an on-line foresight guide and the organisation of mutual learning workshops around particular themes, such as foresight evaluation.[15] Both projects are in the initial start-up phase and therefore it is premature to review their progress and impact in promoting mutual learning in foresight here; however the initial results are promising.

A related initiative aimed at promoting networking and mutual learning between national foresight programme managers in the enlarged EU, is the ERANET ForSociety.[16] The Project provides an enabling environment for foresight programme managers "to co-ordinate their activities and on the basis of shared knowledge on relevant issues, methodologies, legal and financial frameworks, regularly develop and implement efficient trans-national foresight programmes that significantly enrich both the national and the European research and innovation systems". The results to date highlight the challenge of coordinating the different perspectives, approaches and interests of 19 contractor organisations from 15 European countries. ForSociety's action-oriented, learning-driven approach evident in the planned launch of joint foresight activity should provide important insights into the utility, constraints and ideal future design of transnational foresight.

CONCLUSION

Learning and transfer of practice are clearly key elements in the development of foresight practice. We have seen that a number of contrasting situations exist in parallel. Among the key distinctions that could be drawn are those between learning taking place within the context of the design of a foresight exercise that has already begun its preliminary stages and learning which is aimed at creating the capability to run a future unspecified activity. In the first case the operators of the programme are highly motivated to learn but in many cases appear to be constrained by decisions already taken about their own activity. Hence, the scope for learning may be reduced. Time is often highly constrained in such circumstances. It is clear that even when the process is quite systematic, unforeseen problems of transfer can emerge, arising from incomplete transfer of key elements, for example the necessary duration for a satisfactory exercise in the transfer of the Delphi approach from Japan/Germany to the UK, or inappropriate transfer in the sense of failure to anticipate different cultural reactions to highly structured surveys rather than interactive meetings. More broadly, the key factors in determining whether a foresight exercise has a high impact may be misattributed to a particular methodological approach when in fact they are more to do with the institutional setting – for example the exercise is "owned" by a part of government that is sufficiently (or not) influential to be able to implement the recommendations. This links to the complexity of attributing the effects of foresight discussed in the context of evaluation of foresight in Chapter 16.

In the process of transfer itself we may see many potential pitfalls. One is linguistic. Apart from the Delphi-related examples already discussed, there are many other cases where it is easy to confuse the same terminology with the same meaning, when the term is used in different social or political contexts. An example of translation difficulties came in the process of transferring experience from the UK to Hungary. One element of this concerned the briefings given to the senior people who were to be panel members. In the UK, after some discussion it was decided that these people would be offended at the concept of being trained and hence the term "appreciation workshops" was used for the events in which they were introduced to their mission and the methods they would be expected to use. The Hungarian language makes fewer distinctions and hence the same event was translated as "training workshops". Beyond translation, terminology can acquire a specific meaning. In the German Futur exercise efforts to broaden participation led to the involvement of the "interested public". Foreign observers misunderstood this as a rare attempt to engage the general public. In fact this group was much more specific, consisting predominantly of people from the education sector.

Another distinction that can be made is between learning from codified and from tacit knowledge. A comparison of the second and third case studies illustrates this quite well. Within eForesee learning by doing was at the core, certainly within the three national contexts. While innovative efforts were made to codify this learning, for example through recording experiences on video clips made available on the website, the case indicated that despite the Learning Circles approach, the main deficiency remained the international transfer of tacit elements. Mutuality appears to be a useful driver in tacit learning.

The third case study focuses on large-scale efforts to codify the experiences of foresight exercises. The issue here is whether sufficiently powerful organising frameworks can be developed to allow what are essentially, in many cases, databases and case histories to be systematically used to produce a cohort of trained practitioners. University courses in foresight are emerging but at the moment foresight training is anchored much more in the world of executive education. Knowledge platform approaches are aimed more at support for existing practitioners and users. Here the issue is what level of tacit knowledge and expertise is needed by these two categories of people to properly appreciate and use the information. Beyond competence there is also the issue of virtuosity – what is the path to becoming a guru in the practice and theory of foresight?

To conclude this chapter we consider what conditions favour policy learning and transfer. As far as approaches or methods are concerned good documentation accompanied by expert support on its operation are preconditions. Instruments that have already been used in multiple circumstances are more likely to be amenable to further transfer as some of their cultural specificities will have been exposed and possibly reduced.

In terms of predicting the success of a planned transfer, much depends upon the environment for the transfer. It certainly helps if there is a similar institutional setting. The shared conditions of EU accession provide a common context that accelerates the sharing of foresight knowledge. At a higher level, foresight needs senior champions to succeed and hence it is important that it has ideological or philosophical appeal to political sponsors. This may be a matter of timing. Foresight is easier to embed in times of change when new policy ideas are actively being sought. At a more mundane level similar available resource levels and similar available expertise levels (both methodological and in the domains addressed by the exercise) are relevant. Big country approaches involving broad scope, large cohorts of experts and corresponding budgets are often not feasible for smaller countries.

Learning is an essential element of foresight and involves both "learning as a policy" and more specifically, foresight "learning as a policy instrument". A well-managed programme recognises the importance both of vectors of learning, in tandem with systematic approaches and of the need for capture and codification of lessons through evaluation and other processes. The relative infrequency and growing convergence of foresight activity means that this process is much enhanced if international knowledge sharing is successfully enabled.

NOTES

1   See EC, Trendchart Innovation Policy in Europe, http://trendchart.cordis.lu/.
2   See Lundvall, B.-A. (1992).
3   See Rose, R. (1993).
4   See First Report by the High Level Group on Benchmarking, European Commission (1999), http://www.benchmarking-in-europe.com/rpt1hlg.pdf.
5   See Krull, W. *et al.* (2002), *Benchmarking S&T Productivity*, Commission of the European Communities.
6   Also referred to as a genealogical tree by Grupp H., and Linstone H., National Technology Foresight Activities Around the Globe – Resurrection and New Paradigms.
7   See Heraud J-A., and Cuhls K. (1999), Current Foresight Activities in France, Spain and Italy, Technological Forecasting and Social Change 60, pp. 55–70.
8   See Nedeva, M., Georghiou, L., Loveridge, D. and Cameron, H. "The use of co-nomination to identify expert participants for Technology Foresight," R&D Management, 26, 2, April, 1996.
9   See Aicholzer, G., 'Delphi Austria: an Example of Tailoring Foresight to the Needs of a Small Country', ITA Manuscript Vienna 12/2001 ITA-01-02, Austrian Academy of Sciences.

10 See Georghiou, L., Acheson, H., Cassingena Harper, J., Clar, G. and Klusacek, K. (2004), 'Evaluation of the Hungarian Technology Foresight Programme (TEP)' – Report of an International Panel, PREST, University of Manchester.

11 See EFMN Brief 8 Foresight Embedding in Malta at the following URL: http://www.efmn.eu.

12 See http://www.jrc.es/projects/foresightacademy/.

13 See Keenan, M., Scapolo, F. and Zappacosta, M. (2004), *Pilot Programme to Scope the European Foresight Academy*, A joint JRC/IPTS-ESTO Study.

14 The project is known as the European Foresight Monitoring Network (EFMN), details of which are available at http://www.efmn.eu.

15 See http://forlearn.jrc.es/.

16 http://www.eranet-forsociety.net/ForSociety/.

REFERENCES

Barré, R., Georghiou, L., Concalves, M.E., Krull, W., Licht, G., Luukkonen, T., Munoz, E., Polt, W. and Sirilli Tsipouri, L. (2002), 'Benchmarking S&T Productivity', in *Benchmarking National Research Policies*, European Commission, Luxembourg.

Dolowitz, D. (2000), 'Introduction', *Governance*, **13**(1), pp. 1–4.

Dolowitz, D. and Marsh, D., (2000), 'Learning from Abroad; The Role of Policy Transfer in Contemporary Policy-Making', *Governance*, **13**(1), pp. 5–24.

Georghiou, L. (1996), 'The UK Technology Foresight Programme', *Futures*, **28**(4), pp. 359–377.

Havas, A. (2003), 'Evolving Foresight in a Small Transition Economy', *Journal of Forecasting*, **22**, pp. 179–201.

Heraud, J.-A. and Cuhls, K. (1999), 'Current Foresight Activities in France, Spain and Italy', *Technological Forecasting and Social Change*, **60**, pp. 55–70.

Irvine, J. and Martin, B.R. (1984), *Foresight in Science – Picking the Winners*, London: Pinter.

James, O. and Lodge, M. (2003), 'The Limitations of Policy Transfer and Lesson Drawing for Public Policy', *Political Studies Review*, **1**, pp. 179–193.

Keenan, M., Scapolo, F. and Zappacosta, M. (2004), *Pilot Programme to Scope the European Foresight Academy*, A joint JRC/IPTS-ESTO Study.

Kuwahara, T., Yuasa, T., Anan, H., Sato, H., Cuhls, K., Breiner, S., Grupp, H. and Nick, D. (1994), *Outlook for Japanese and German Future Technology – Comparing Japanese and German Technology Forecast Surveys*, NISTEP Report No. 33, April.

Lundvall, B.-A. (1992), *National Systems of Innovation: Towards a theory of innovation and interactive learning*, London and New York: Pinter.

Lundvall, B.-A. and Tomlinson, M. (2001), 'Learning-by-Comparing: Reflections on the Use and Abuse of International Benchmarking', in Sweeney, G., *Innovation, Economic Progress And The Quality Of Life*, Cheltenham: Edward Elgar, p. 122.

Metcalfe, J.S. (2006), 'Innovation Systems, Innovation Policy and Restless Capitalism', in Malerba, F. and Brusoni, S. (eds), *Perspectives on Innovation*, Cambridge: Cambridge University Press.

Nedeva, M., Georghiou, L., Loveridge, D. and Cameron, H. (1996), 'The Use of Co-nomination to Identify Expert Participants for Technology Foresight', *R&D Management*, **26**, 2, April.

Nelson, R.R. (1993), *National Innovation Systems – A Comparative Analysis*, New York: Oxford University Press.

PREST (2004), *Evaluation of the Hungarian Foresight Programme*, Report of an International Panel, May 2004, Hungarian National Office for Research and Technology, http://www.nkth.gov.hu/main.php?folderID=159&articleID=3826&c tag=articlelist&iid=1, last accessed 02 August 2007.

Rose, R. (1993), *Lesson Drawing in Public Policy*, Chatham NJ: Chatham House.

Shin, T. (1998), 'Using Delphi for a Long-Range Technology Forecasting, and Assessing Directions of Future R&D Activities – The Korean Exercise', *Technological Forecasting and Social Change*, **58**, pp. 125–154.

Shin, T.., Park, J.H., Jung, K.H. and Kim, H.S. (1994), *The First Survey for Science and Technology Forecasting; Korea's Future Technology*, (in Korean), Science and Technology Policy Institute (STEPI), Seoul.

15. Scoping and Planning Foresight

Michael Keenan and Ian Miles

INTRODUCTION

Much of the foresight literature simply celebrates the achievements of one or other exercise. More analytic literature tends to focus on technical issues of methodology, or on the location of foresight within broader policy contexts. The organisation and management of foresight exercises remains a black box. Yet these are not things that just happen: they require decisions and strategies. These are neither self-evident nor simple choices: they are very dependent on context. Furthermore, they are, of course, critical to the success of foresight – in terms of yielding timely and relevant results, and in terms of the political reception of the process and results. At the outset of an exercise, the scoping process sets the parameters for the subsequent activities, helping to define procedures for identifying areas to cover, methods to apply, and participants to mobilise, and determining how to implement these procedures. This chapter considers the typical scoping processes used in foresight, and elaborates a framework of 12 interdependent elements – from the initiation of an exercise through to the policy actions that may flow from it – that need to be addressed in constructing foresight exercises (see Table 15.1).[1]

Table 15.1   Framework of elements for scoping foresight exercises

Framing Conditions	Scoping Variables
Locus	Coverage
Broader Milieu	Time Horizon
Aims, Objectives and Outcomes	Organisation and Management
Resources	Participation
	Communication
	Methods
	Product-Process Orientation
	Achieving Impacts

THE SCOPING PROCESS

Other chapters in this volume demonstrate that there are many different types of foresight exercise, varying in their purposes, organisational and policy contexts, focus and methods. Scoping is essentially the attempt to specify what a given foresight activity is aiming at, and how it is to accomplish this. This involves the design of all stages of the exercise. Scoping also implies undertaking activities that will inform this design – for example, reviewing previous practice and relevant current activities (if any), identifying the available options for the conduct of the exercise, perhaps piloting or gathering views as to possible approaches. Scoping should result in the design of an exercise so that it is appropriate to local conditions, including the policy needs that are to be met, the available resources, the cultural and political contexts, the capabilities of civil servants, academics and consultants, and so on. Scoping may be a matter of wide participation (e.g. with great publicity and efforts at consultation) or be a narrower affair (undertaken in closed circles and even with some secrecy). The way in which scoping is itself carried out can set much of the tenor of the subsequent foresight exercise, so scoping is not just a technical matter. Against this broad background, this chapter first seeks to develop a general framework for the understanding of mutual learning and policy transfer and then considers how this may be applied to foresight. Three short case studies of learning in foresight are presented to illustrate these issues.

An informal scoping process is inevitable well before the launch of foresight. Some scoping will have been involved even to germinate the idea that a foresight exercise might be an appropriate way of meeting policy needs. Learning about earlier or ongoing exercises (often in other countries or regions, but possibly in other organisations or government departments in one's locality) is an important trigger, and may be seen as a preliminary and often informal scoping activity. Additionally, since several international bodies have been involved in promoting foresight, early learning about foresight is now readily available. For instance, there is much available material on EC and UNIDO websites[2] and studies like Alsan and Oner (2004) have begun to map out the management structures and processes of different foresight programmes.

More formal scoping processes are usually put into place once the possible relevance of foresight has been determined, and before the full activity has been set out and implemented. It is usually done before any foresight activities get underway. Since some commitment of human and financial resources will be required to conduct a scoping process, the political decision to initiate an exercise may already have been taken. But scoping may also constitute intelligence gathering to see whether foresight is appropriate at all, not just how it should be undertaken. The decision may be taken not to proceed with a foresight exercise – and indeed, this option should in any case be considered in the scoping process. It is in fact one of several options: the time might not be right for foresight in terms of political circumstances; the exercise may need to be much wider than technology foresight; foresight may need to be a more focused activity than the typical wide-ranging assessment; and so on. [3]

Our interest in this chapter is mainly with the formal scoping process, which involves gathering data, eliciting the views of stakeholders, and preparing and assessing options for foresight. The process is formal in that someone has been asked to investigate and/or plan a foresight process. A design needs to be established in order to initiate an exercise. At the same time, the design needs to be responsive to the environment, both in terms of general principles and adaptability in the light of unfolding events during the course of an exercise. Some sort of ongoing scoping process tends to be continuously operating during an exercise. This may be formalised into periodic reviews at key stages that set the future course of an exercise. Much of what follows will apply to such reviews, but at the outset of an exercise, scoping has to play the roles of:

- *Making the case for foresight.* This will involve, typically, a well-written report that explains the purposes and principles of foresight to key decision-makers. It is likely to set out to argue the case for instituting foresight as a means of tackling key policy goals, and to explicate the costs and benefits of different designs of foresight programme. This document can be a powerful tool for convincing others of the merits (and informing them of the limitations) of undertaking an exercise.
- *Gathering background information.* As with any policy initiative, it is only responsible for those initiating foresight to inform themselves about similar activities from which lessons may be drawn, and from where there may be dangers of duplication of activities or creating misunderstandings. Obviously recent local activities are especially relevant, though clues as to good and bad practice, and new approaches, may be gleaned from further away. Sponsors and those charged with designing and implementing foresight may have particular methodological approaches in mind (for example, there was a time when foresight seemed to be synonymous with Delphi studies). In such a case, experience with these approaches should also be researched, and it may be well to explore alternatives, too. [4] Background information is typically gathered from literature reviews

of books, journals, reports, and web sites, and by consultation with knowledgeable informants.

- *Eliciting expert views and advice.* Especially when first undertaking foresight, those involved may undertake one of the "crash courses" of foresight training that are available, and will often seek advice from practitioners and observers of similar exercises (often from other countries, sometimes from other government departments, the corporate world, consultants and academics). This may well go beyond the provision of intelligence about the lessons to be learned from foresight elsewhere to involve more "hands-on" knowledge inputs, in the form of ongoing expert advice in the design and running of a programme, for example.

- *Articulating and presenting options.* Once background information has been gathered and views elicited, options for foresight should be set out in some sort of report. This may be openly published, for example, as a consultation document, or may remain a private document to be circulated only amongst sponsors and key stakeholders. It should set the background and rationale for foresight, highlight examples from other countries, regions, organisations, etc. (whichever is most comparable), and describe a set of possible options for foresight. The scoping elements described later in this chapter constitute one framework for constructing these different options. We would recommend that 3–4 different exercise "blueprints" are generated using these scoping elements and used in further discussions with sponsors and key stakeholders.

- *Eliciting views and advice from stakeholders.* Foresight exercises involve some combination of process and product objectives. Process objectives are those that involve creating "behavioural additionality" – helping to establish and activate networks, stimulating foresight processes in other bodies, and so on. Product objectives involve creating "codified knowledge" – such as lists and rankings of key technologies, scenarios and roadmaps of future developments, and action plans. Whatever the precise combination of objectives aimed at, active involvement of the various stakeholders is a core factor differentiating foresight from more narrow futures and planning approaches. It is accordingly an important factor determining how foresight is organised and managed. Key stakeholders – such as the target audience of a foresight exercise, and those who might be expected to participate in the process and/or to act upon the results – will need to be consulted concerning the design of an exercise, as part of the scoping process. This may be done through means ranging from interviews and bilateral consultation, through scoping workshops (these can combine depth and breadth of analysis) to more outreach in the form of conferences. In practice, more limited discussions with key stakeholders are most common. The aim is to gather ideas, obtain commitment of future support and participation, and to begin the process of securing buy-in to the results of the exercise. The extent and quality of consultation and participation in setting the objectives and designing the process from the outset is bound to be conditioned by local circumstances (though it is usually worth pushing the boundaries here, and not just accepting that only "the usual suspects" should be represented). This can affect, for example, precisely who feels affinity with foresight early on in the planning, and how easy it will be to elicit engagement later on in the implementation.

- *Assigning responsibilities and establishing procedures.* New structures or arrangements will probably be required to conduct a foresight exercise. Apart from a chain of management, various roles may have to be assigned, and capacities for these created. For example, some training may be required for panel members and facilitators; particular tasks may be contracted out with a need for tendering or other ways of recruiting contractors, and so on. Often such arrangements will be accommodated to a greater or lesser extent within the institutions that are already in place. But various features of a foresight process may be challenging to existing structures and routines. The cross-sectoral, cross-professional, cross-disciplinary interchange that is required, the networking and participatory environments that are intended, the creative "out of the box" thinking that is hoped for all, of these may require new arrangements.

- *Broad decisions as to methodology and coordination of activities.* It is necessary to identify the sorts of methods and activities which are required to achieve the objectives of the exercise. Though many detailed questions of method will need to be determined later, early on, the range of tools to be employed (surveys, workshops, panels, expert groups, special studies, and so on) should be established (see Chapter 3). Furthermore, plans for how these might be elaborated, and customised to fit the circumstances are needed. A broader plan should set out how they will be integrated, both in terms of implementation and of processing and applying their outputs and results. This means fitting proposed activities into a viable timetable. The plan cannot be too rigid; foresight needs a good degree of flexibility to be able to respond to significant changes in its environment. The scoping process involves setting a direction for foresight, while avoiding premature decisions that restrict options. [5]

Ultimately, scoping is a matter of developing and sharing understanding of the interdependencies between strategic choices, so that decisions can be based on this understanding. We thus differentiate scoping from the detailed planning of the exercise, which scoping is intended to inform. Scoping should result in specification of the options for planning a foresight exercise.

Explicating the tasks involved in scoping foresight, as we have done above, may make this sound like a very daunting challenge. Spelling out the tasks is not intended to deter people from foresight; the hope is that by making these features explicit we are reducing the need for belated discovery of the necessity to gather information and make decisions about one or other element of foresight. We have been able to undertake this explication precisely because those initiating foresight exercises have in the past had to confront the various challenges that have been identified. This experience shows not only that scoping can be done, it also helps us to understand how it can be done, as we discuss below. But first, who should be doing it? Different elements of scoping may be undertaken by different parties – for example, it has been common to employ academic researchers or consultants to prepare a report on the current state of the art in foresight programmes. But there needs to be overall orchestration of the scoping process. This is often performed by sponsoring organisations and/or foresight "champions" in those organisations. It may be accomplished by the institutions that will be expected to lead the foresight exercise, if these are not themselves the sponsor. In quite a few cases, consultants or academics have been drafted in to lead the scoping process, because they are seen as having already acquired relevant expertise and are more likely to be viewed as neutral players.

ARGUING FOR FORESIGHT

The case for foresight may need to be argued, especially where there are powerful institutions that have their doubts about the process and what it means for them. Some misunderstandings about foresight need to be dispelled. For example, foresight is often confused with prediction. The whole point about foresight is to help shape the future, by anticipating a variety of possible futures and working to create desirable futures through actions today.

Foresight is often subject to a more sophisticated and specific criticism. This is centred on scientific serendipity – attempts to control the course of science are seen as misguided since the eventual long-term applications of new scientific discoveries are extremely hard to anticipate, as will be much of the new knowledge itself. Lasers are often cited as a technology discovered decades ago with few initial applications, but that are now widely used in thousands of products and processes, from consumer electronics to military hardware, for instance. Computers are being used for purposes which their inventors never dreamed of. Surprising applications are the rule, rather than the exception, for technological developments that allow us to effect new and significant transformations of matter and energy.

There is something in this argument – curiosity-driven research may be particularly good at yielding important breakthroughs.[6] But foresight is not about controlling all science. Foresight is not used to prioritise all of the scientific enterprise of a nation state, or even all of the public-funded applied research (though this is where it does have most bearing). The main issue at stake is the identification of emerging areas of research that hold promise for socio-economic and scientific developments. Often these critical developments cross established disciplinary frontiers, and may be overlooked by the traditional disciplinary organisations of science. Additionally, methods of foresight can help science and technology better connect to wider socio-economic goals of public and private actors. This can provide vital legitimacy to research funding when conditions of fiscal constraint mean science budgets may come under threat; investment in science can be defended on grounds of promise, if not immediate pay-back. And foresight provides a forum where scientists can discover new ideas from, and articulate new opportunities with, other scientists and socio-economic actors.

Another argument has particular resonance in the context of globalisation. Even the most powerful nations find it hard to attain leadership in all fields of S&T. Many other nation states – let alone disadvantaged regions and communities – are in extremely dependent circumstances in S&T as in trade and political affairs. This may engender a sense of helplessness which makes a collective social activity like foresight appear irrelevant and difficult to initiate. Foresight is about making informing (constrained) choices – understanding what our room for manoeuvre is, and what trade-offs may be required. Even if one cannot occupy the technology frontier in many areas, there are opportunities to be active – for example in specific application areas, in particular niches, in collaborations, in using research to enhance one's "absorption capacity" (i.e. to make excellent use of knowledge generated elsewhere). Dependency and fatalism about effecting change – common in societies where there is an alternation of governments so that no stable agenda can be pursued – are major problems. For foresight to be used to offset these debilitating mind-sets, it is important to enlist foresight "champions" with authority and vision to initiate and implement activities effectively. They can make the case for foresight by benchmarking and comparing the national experience with that of other countries or regions. If TF is organised in tandem with other broadly-based emancipatory policies, it could make a real difference.

Another argument centres on foresight's "value for money". Little evaluation of foresight has been conducted to demonstrate its effectiveness in a systematic way.[7] However the rapid diffusion of the approach, and the way in which several of the countries that pioneered the approach have continued to support successive exercises, indicates that more informal appraisals have been very positive. Nonetheless, the processes of foresight remain poorly understood and documented, though the situation has improved considerably over recent years.[8] Evidence of the worth of foresight is therefore largely anecdotal and focused mostly upon success stories from various countries or regions. The narratives that can be provided by policy-makers and foresight practitioners can prove very convincing as arguments for conducting foresight. Costs as always can prove to be a major stumbling block. Several planned foresight exercises have been scaled back or postponed due to the unavailability of necessary financial resources. This suggests that, rather than ruling out foresight as inherently over-expensive, scoping might be done best in terms of a menu – supplying project plans that require different levels of funding. However, the limitations of cut-price exercises and the benefits of more extensive programmes should be made plain to prospective sponsors.

The arguments for conducting foresight will be dependent upon the organisations (especially the sponsor) and communities involved. Ultimately, the rationales for foresight will tend to emphasise how things can be done better with the help of foresight. Proponents may point to other places or areas where foresight has been successfully deployed as exemplars. The arguments can cover a rage of objectives (a topic discussed further below). These can include:

- Provision of information to help priority-setting and planning for S&T policy, or the research and innovation strategies of individual organisations or networks of organisations. Foresight may also inform related policy areas, from education to intellectual property. Wider foresight exercises can be used to inform practically any policy areas;
- Creating better understanding among key stakeholders of the nature of the S&T system they are working within and the issues and opportunities that are arising in various areas;
- Formation and energising of networks that link together critical actors in the innovation system, allowing them to share knowledge about major developments, capabilities and strategic choices.

These objectives can of course be pursued by other means. Evaluation studies can throw great light upon the innovation system and the effects of innovation within it. Desk research can produce reports on important trends and SWOT analyses. Existing networks can be funded to engage in more collaborative work. Where foresight adds its own value to these approaches is in its reach and engagement with stakeholders. By actively involving a wide range of actors in creating visions of future possibilities and their implications for action in the present, it allows for the wide embedding of perspectives on the longer-term issues and opportunities, and for the building of constituencies that can act on these perspectives.[9]

FRAMING TF WITHIN THE LOCAL CONTEXT

This section examines those elements of a foresight exercise where there is liable to be less room for manoeuvre. Elements such as the locus of an exercise (national, supranational, sub-national, company, etc.), its desired outcomes (usually politically determined), and the available resources for conducting the exercise are more or less givens, though there may be some room for negotiation.[10] These elements will therefore tend to set the parameters for scoping. In the following section we shall move on to consider elements that are typically more open-ended, and where scoping will go a long way toward providing the basis for key design decisions.

Locus – Who Requests and Sponsors TF?

TF has to date been most prominent at the national level, with national governments in many parts of the world organising wide-ranging exercises that cover wide ranges of technologies and/or fields of application of technology. Such exercises are typically located in science ministries, research councils and/or academies of science, and designed to inform decisions about science policy, research policy, innovation policy, and the like. Sometimes the aim is to inform general decisions about government support for, say, research and development; sometimes the focus has been more on the activities of a government department (for example, what research should be undertaken by a ministry responsible for environmental affairs, transport, energy, health, and so on).

International organisations have more commonly been engaged in narrower technology forecasting and assessment studies (for example, the European Commission's FAST programme during the 1980s and early 1990s), which have also often been pursued by national governments and other parties. Since the late 1990s, the EC and UNIDO have supported foresight activities, especially through providing resources and encouragement for the use of such activities in countries where they have not been established (for example, New Member States of the European Union have been targeted by the EC, while UNIDO has helped TF efforts in several Latin American countries). Typically these have involved providing support to national-level TF in countries that are seen to need help in getting work off the ground.

But the sub-national level has also been interested in foresight and this has been reflected in various ways in support from the EC (Miles and Keenan, 2002; Renn and Thomas, 2002). Since regional authorities have not typically had great influence over research and innovation policy – though this is changing – much regional foresight is not focused primarily on technology. Issues such as business cluster development and democratic renewal have been at the fore in most sub-national foresight, though in some regions, foresight exercises have taken place. These include the Basque region (Spain), Bordeaux Aquitaine (France), Lombardy (Italy) and Liege (Belgium). Most analysis of foresight focuses on one or other specific level of policy-making, usually national governments (but also regional governments and private corporations receive some attention). Among interesting exceptions are studies concerning transnational foresight such as Jewell and Sripaipan (1998), and multilevel (national–regional) studies, like Gertler and Wolfe (2004) and Koschatzky (2005).

Non-governmental actors, such as professional associations and industry federations, are most typically active in technology forecasting or futures studies more generally, although these may sometimes be described as being foresight activities. Techniques such as roadmapping and scenario analysis are widely employed in high-tech sectors and in industries where large investment decisions with long time horizons are concerned. When such exercises involve numerous stakeholders from within the constituency in question, they come close to being foresight as we understand it here – though the results of such exercises are rarely tied to specific collective decisions, and more often aimed at building collective vision and establishing better networks.

The locus of foresight is likely to be a given. The drive for an exercise will tend to come from the main sponsor, though sometimes a government agency will be requested by a higher political authority to run an exercise. The sponsorship of a programme has huge implications for its agenda, for what policy levers it is able to pull, and the like. Yet, even within these confines, there is normally considerable room for choice in the details of an exercise – locus need not completely dictate focus. For instance, a national health ministry may decide to use foresight as a policy-making tool. But there is a huge range of ways in which foresight can then be articulated. The exercise could be focused upon any one (set) of hundreds of disease groups, or upon sites of a particular service delivery, or upon the implications of certain technological developments, e.g. nanotechnology or genomics. The ministry might decide to collaborate in the exercise with other health-related institutions in its own country (or internationally). So, whilst the institutional positioning of foresight has a large effect on its scope and shape, even here there is considerable room for choice.

The Broader Milieu

TF is positioned within an institutional setting, a locus, as discussed above. But such institutions are themselves situated in wider policy milieux and socio-economic cultures. Such settings will need to be taken into account when designing a foresight exercise. For example, if there is a longstanding history of conflict between stakeholders in a particular economic sector or policy area, this has implications: for foresight exercises – we cannot expect the parties to sit down together and work productively from the outset. There are also potentially problems where there is long-standing concord amongst key stakeholders – this may signify cosy relations, collusion or complacency. Foresight can play an important role in any event, if it is constructed so as to:

- Stretch perspectives into the future – if possible, beyond the reach of current disputes, but with the scope of identifying potential emerging areas of conflict;
- Develop mutual understanding of, and respect for, different positions, including differences that are not articulated normally – some of the procedures of foresight can be effective in fostering dialogue rather than set-piece speech-making; and
- Construct and cement the foundations for continuous long-term strategic conversations. By contrast, in areas of complacency, emphasis should be placed upon (a) introducing new perspectives and/or data that call into question current assumptions, and (b) instilling a sense of urgency (or even crisis) that demands immediate collective action.

The milieu can constrain the foresight activity in other ways, too. There are other features that will probably need to be considered when scoping foresight, and these are outlined below.

Cultures of collaboration
Are the different communities used to working together? There may be institutional rivalries (for instance, one national exercise was wrecked by conflict between Ministries of Education and of Research); there may be distrust in cross-disciplinary or cross-professional work. Institutional competition has been observed by the authors in several nation states and

regions (and even within the European Commission), where several agencies compete to be the "authority" on, and organiser of, foresight. Such competition can stall or undermine initiatives. The problem seems to be most acute under conditions of financial resource constraint where there may be competition to be the "owner" of foresight; when resources are less constrained it may be practicable for several institutions to organise their own foresight exercises, a situation commonly found in North-west European states, e.g. Denmark, Finland and the Netherlands.[11]

The presence or absence of a forward-looking tradition
Is tradition seen as politically and intellectually respectable? There may be a need to differentiate foresight from central planning and programming, which have dominated state socialist regimes. Where the futures tradition is more literary, and/or the scientific and cultural elites have possessed an almost aristocratic status, it may be important to demystify the idea that visions of the future are always utopian, or only the product of great genius.

The presence of other policies and programmes that profess to take a strategic view of future developments and actions
There may already be a central strategic planning or policy review agency, or a body engaged in scenario-mapping or horizon-scanning for the public sector. If such activities are well-established, a stand-alone foresight exercise may not be an appropriate choice; it might be better to introduce foresight into these existing strategic processes, to enlarge those elements of foresight that already exist in them, and/or to develop projects in coordination with such activities (as long as this does not mean too much compromising of foresight principles).

Aims, Objectives and Outcomes

It is only recently that foresight-type activities have become fairly routine in a number of countries and regions. As the discussion of rationales highlighted earlier, the rise of foresight largely reflects a crisis in the funding and political perception of S&T. Often, demands for foresight (and/or similar strategic futures activities) emerge in the context of a sense of social or political crisis. The crisis may not be a matter of shortage of resources; rather, concern with technological futures can also reflect awareness of major new technological developments that are emerging, and the anticipation that established trends are being undermined.

While a sense of urgency can be stimulating, feelings of panic can be paralysing. It can be helpful to interpret the situation in terms of challenges, to stress that "crises" are "opportunities".[12] The challenges that are associated with such opportunities can be used to set the main thematic orientations of the foresight exercise. For this to be effective, however, there must be a good measure of shared agreement as to the nature of these challenges – or, at least, how they can be productively handled by a range of stakeholders who may have different "takes" on them. Such a working agreement will need to be established at an early stage in the foresight activity.

Once the challenges have been identified in broad terms, then it is important to consider the extent to which the organisations involved in foresight, be they public or private, are able to influence or respond to such challenges:

- The challenges may be particularly pertinent to a particular organisation, region, etc. Often, that organisation or governmental structure will not possess all the political and technical competence to deal with the issues. Other players will have to be brought on board very early on if the chances of connecting to critical actors and users are to be maximised.
- Some issues are best addressed by the private sector. But this does not preclude public administration from leading or facilitating a foresight exercise. For example, a government agency can constitute a forum to help private businesses reach consensus on what actions they might need to take around particular technological developments.
- Some issues have a global reach. The influence that can be brought to bear by any one country or region may be very limited. The crux will be to consider how foresight can inform us as to what sorts of external contingencies (a partly "imposed future") may need to be coped with, what sorts of alliance can be used to shape trends in more positive ways, and so on.

Questions of competence, prerogative and authority, such as are raised here, are absolutely vital issues. They will need to be taken into account in determining the objectives of a foresight exercise – not least because the scoping process may generate disagreements that could prevent or delay an exercise being launched. Conflict at this stage could be minimised by limiting access to the scoping process, involving only those who have an obviously shared agenda. (For example, only including institutions and parties clearly interested in such specific goals as modernising particular sectors, building specific technological capabilities, protecting certain natural resources, and so on). But this may (a) postpone problems till later, (b) undermine the legitimacy of the exercise, (c) exclude stakeholders whose knowledge may be vital, and (d) exclude stakeholders whose engagement may be key to mobilising social resources required for successful implementation of the exercise. Still, it is often impossible to satisfy everyone, without creating a massive and probably unmanageable exercise. The strategy for engaging various actors, and the tactics for dealing with disputes that arise, will have to be specific to the given situation. In any event, some disappointment and complaints are to be expected as a consequence of scoping foresight – as they will be in the light of any priority-setting. The hope is that the eventual achievements of the exercise will outweigh the reservations about its starting point.

The major issue, of course, is definition of a set of objectives for the exercise that can command sufficiently wide consensus, and that promises sufficiently attractive outcomes, to make for a viable and effective exercise. Objectives may be more or less immediate, and more or less operational. Thus an immediate and operational objective of those managing a foresight exercise is its smooth execution. But the sponsors of foresight – while they will want to see value-for-money – will be most concerned with whether the product and process results of the exercise are in line with the rationales offered for conducting the exercise. Objectives may shift over time, and different actors may hold or emphasise different objectives for a foresight exercise. Nevertheless, it is good programme practice to set verifiable objectives, i.e. objectives where it is possible to assess whether they have been met, and where there is thus some form of performance indicators available with which we can measure progress. All too often, this has not been done, especially when pioneering foresight activities are launched in a location. Spelling out such objectives and indicators can assist programme management by relating specific activities to specific objectives, and by providing information as to whether activities are on the right track (and whether intervention is necessary).

The task of generating performance indicators can be valuable in terms of forcing those involved to consider exactly what they mean by specific objectives – often apparent agreement on generalities masks huge divergence in more concrete expectations. Such practice also makes eventual evaluation of the exercise more effective, and this in turn will contribute to making foresight truly a learning process. These issues are considered further in Chapter 16.

Resources

Besides financial resources, the scope of a foresight exercise will be dependent upon factors such as time, political support, human resources, institutional infrastructure, and the culture in which the exercise is embedded. We will now briefly deal with each of these in turn.

Financial resources
The cost of a foresight exercise depends primarily upon the nature and scale of involvement of participants and its duration. In general, and somewhat obviously, the shorter the exercise and the fewer people involved, the cheaper it is likely to be. The financial burden of foresight activities are typically borne by a wide range of players, not least by the participants themselves, who usually provide their thoughts and time for free or nominal costs (travel expenses are more often met). It is common for the "invisible" costs of foresight, contributed in terms of the time of participants, to be by far the largest single economic resource drawn upon. Though little indicative financial data exists on the costs of foresight exercises in general, we can identify the major costs – which are usually borne by a central sponsor – as resulting from such elements as:

- The running of a project management team (this may also involve, for example, workshop facilitators, domain experts to advise on specific topics being covered in depth);
- The organisation of meetings and events, travel and subsistence of at least some of the participants (some participants may even have to be paid to give up their time for the foresight exercise – this is uncommon, but in some places, it might be necessary);
- The production and dissemination of publicity material (especially newsletters, Panel Reports and websites);
- The operation of extensive consultation processes (e.g. questionnaire surveys, consultation via websites, meetings aimed at dissemination and/or consultation, preparation of material for media); and
- Other activities, both routine and one-off, associated with an exercise;

Economic resources can be provided by various "official" sponsors – from the public or private sectors, as well as from the "third" sector (e.g. trade unions, voluntary groups, etc.). It is not unheard of for foresight to be co-sponsored by all three.

Time
Time is nearly always a resource in short supply in foresight. Whether a public or private sector exercise, the results of foresight are usually required by a particular date to feed into policy and/or investment decisions. Typically, national foresight exercises take 1–2 years to complete, depending upon financial resources and political imperatives. Private sector exercises are normally shorter, which is made possible by their being more focused. Foresight can also become a "continuous" activity, as highlighted in the case

of the UK Foresight Programme (see Chapter 4). The available time will have major implications for an exercise's organisational structure and overall methodology.

Political support

Without the support of those in authority, foresight is unlikely to get off the ground, let alone make a difference. It is therefore essential that foresight receives and is seen to receive political commitment at the outset and throughout the lifetime of an exercise. Political commitment can be demonstrated in a number of ways, for example, through institutionally locating an exercise at the heart of power (e.g. in a Prime Minister's office, within Parliament, etc.). More modestly, it can be helpful if someone in a position of authority (e.g. a government minister or company CEO) opens and attends workshops and conferences; if their signature is attached to a letter requesting informants to complete a survey questionnaire; if their photograph can be meaningfully included in a newsletter article; and so on.

Human resources

Foresight requires domain expertise in the areas under consideration, as well as expertise in the use of foresight methods. Expertise in using some foresight methods is very patchy around the world. Skills in forecasting and planning, usually of quite narrow kinds, are often found in state planning departments and universities. But these are typically restricted in the topics addressed and indicators used (typically restricted to quantitative and often financial data); and in being more oriented to technocratic and efficiency-oriented practice than to participative and deliberative processes. Some consultancies and management researchers, too, have developed skills in strategic analysis that can be applied to foresight processes. Again, these are typically restricted (e.g. to market research or environmental scanning).

Only as foresight approaches have become widely diffused are there many experienced experts that can be drawn on. At first – and still in many countries – it was common for less experienced actors to become involved in facilitating foresight through "on-the-job" trial and error learning. The wide diffusion of foresight has meant that advice may be sought from other countries, and training undertaken through courses and placements with foreign agencies. Written guidance has become more widely available, too, though there are often major problems in translating written advice into operational practice. Moving on to domain expertise, foresight should be informed by the best available experts. "Best" may not always equate with "most famous" (some scholars are better at self-promotion than at generating substantial new knowledge) or even "most respected" (because leading scientists are sometimes proponents of very specific approaches, to the extent that they fail to give other possibilities their due). "Best" implies open-mindedness and wide-ranging interests, and some basic social skills, as well as technical excellence. In some countries, regions, or companies, there may be a need to source experts from outside. But if such expertise is difficult to access, then the focus of the TF should be reviewed.

Infrastructural resources

These refer to the existing institutional landscape around a given area, such as research councils, academies of science, universities, science ministries, professional associations, industry federations, consumer groups, banks, etc. In virtually all countries, infrastructural resources such as the organisation and network capacity of stakeholder groups will vary in strength from area to area. The foresight exercise should then be designed in such a way as to be responsive to different institutional landscapes.

A rich institutional landscape can greatly smooth the way for foresight, providing useful data inputs, knowledgeable participants, and fora for dissemination and implementation of an exercise's findings. But institutional "thickness" can also act as a barrier to foresight. There may be institutional rivalries, "not invented here" suspicion of foresight proposals from "outside", and problems with institutional worldviews that are rather static and difficult to openly question. Moreover, an exercise is far more likely to be subject to intensive lobbying by well-organised groups of interests. Strategies for dealing with such opportunities and threats will have to be informed by insight into the social and institutional, as well as the technological features, of those areas to be covered by the foresight exercise.

Cultural resources

These cover a broad range of rather ill-defined conditions, but these are factors that are likely to have an important impact on the conduct of foresight. They include the propensity to take risks, the extent and degree of collaboration between industry and academia (as well as between competitors within each sphere), and the extent to which actors already understand and position themselves *vis-à-vis* the long term. Where such resources are largely absent, foresight will need to begin the process of building them – indeed, this may be one of the main objectives of foresight, and not just an enabling condition.

SCOPING DECISIONS

Having considered various circumstances that more or less set the parameters for a foresight exercise, in this section we consider some elements where scoping can substantially influence key design decisions. These include the coverage of the exercise, methods to be used, the degree of participation, and the organisational structure of the exercise.

Coverage

Any given TF exercise will have to be selective in various ways in terms of the range of S&T topics to consider. Few countries can hope to attain excellence in all fields of S&T, and some S&T areas are of little relevance in some locations. The focus may be on R&D, or more widely on a range of innovation-related topics (including, for example, education and training, IPR issues, etc.).

A review of national foresight exercises conducted in the last decade shows considerable commonality in many of the areas covered. Among technology areas featuring in almost all such exercises we see ICTs, Biotechnology (primarily applied to healthcare and agriculture) and (in more recent exercises) Nanotechnology, together with Transport Technology and Energy Technology.[13] How the major selections are made is rarely made explicit in the documentation of existing foresight activities.[14]

Methods for selecting the broad set of themes and areas to cover in exercises are probably mainly a matter of small groups deciding on these around a table. They may draw upon existing strategic priorities or SWOT analyses of the national economy, but are probably most influenced by steers from their political masters (and direct or indirect lobbying by interest groups).[15] This is one of the most opaque areas of scoping, one where the key decisions seem to be taken very rapidly yet with little transparency – despite its evident centrality to the process.

The definition of areas to cover is a process where stakeholder consultation could play major roles in identifying themes of concern. As well as bringing more interests to bear, this is likely to increase commitment to later stages in the exercise. Nonetheless, difficult decisions will perhaps have to be taken when there is demand for more themes and/or sectors to be addressed than resources or time will allow.

There are other selection processes later in the course of an exercise, and these are somewhat better elaborated, with some use of priority-setting methods (such as ratings by large numbers of experts on the importance and feasibility of pursuing R&D with particular outcomes in mind). But this takes us out of the arena of scoping, and more into foresight implementation and use. There is no a priori reason why such approaches might not be used in scoping the coverage of an exercise, however.

Time Horizon

Foresight is centrally concerned with increasing the time horizon of planning activities. This is not just a matter of "stretching" existing horizons, extending familiar planning and intelligence-gathering into a longer-term future:

- A major point about the longer term is that it brings into relief trends, countertrends and possible events that are of limited concern in the short term. Such developments may well not be crucially important to one's immediate prospects – but if they are not taken into account until the problems start to be highly manifest, then it may be too late to adapt effectively, or the costs of coping with change may be higher than they would be otherwise. Consider, for example, the question of developing a base of skills to cope with economic or technological change: this is often a matter that will require years to put into place.
- Whereas we can often rely on limited extrapolative and "business as usual" approaches to give us a reasonable grasp of the near-term, the longer term will almost inevitably involve qualitative as well as quantitative change. Models and approaches that assume structural continuity are liable to prove inadequate; deep-seated assumptions and even paradigms are liable to be disrupted. Standard tools are rarely able to point to such developments. Furthermore, the sorts of change we are talking of here often take much longer to mature than is expected by excitable social commentators, and trying to tackle them as near-future phenomena is liable to look unsound.

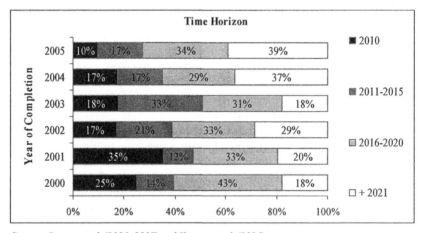

Sources: Popper *et al.* (2005, 2007) and Keenan *et al.* (2006)

Figure 15.1   Time horizons used in recent foresight studies (495 cases)

In practice, the time horizon of foresight activities can vary considerably. What is thought of as the "long term" varies considerably across different issues and different cultures. For example, sectors like the oil and nuclear power industries tend to work with very long time horizons, while retailing has short ones. The average time horizon for national and regional foresight exercises seems to be around 10–15 years, although it may be as long as 30+ or as short as five years. This is supported by time horizon data collected on recently completed foresight exercises by the European Foresight Monitoring Network (see Figure 15.1).

There is some evidence that the time horizons adopted tend, as would be expected, to be related to foresight's objectives and orientation, in other words, to the uses to which foresight is to be put. A long time horizon provides the opportunity to develop a broad vision, and may allow participants to extract themselves from immediate conflicts and hobby horses. But most players are hoping that foresight will result in immediate (and typically short-term) policy and/or investment responses. This is not the paradox it seems – foresight should be instigated in order to think about possible futures, with a view to changing what we do today for the better. The aim is readjustment, in the present, to create more agile organisations, cultures, etc. for the future.

Organisation and Management

Any foresight exercise requires an organisational structure, with roles assigned to various elements of this structure in line with the type of activity planned. Common organisational elements are included below.

A steering committee
A steering committee will have the roles of approving the scoping of the exercise – its objectives, focus, methodology and work programme, to validate the strategy and tools for communication, and help in the promotion of the results. It may undertake much of this scoping itself, or commission it to officials or contractors. It will define/adjust the assessment criteria and review the deliverables, and monitor the quality assurance process for the whole project. The Steering Committee can also be a key actor in terms of raising awareness, nominating and mobilising experts, preparing key reports, etc.

A Project Team
This will manage the project on a daily basis, with tasks such as:

- Maintaining regular contacts with the stakeholders and the Steering Committee, to ensure (and assure them) that the project direction is maintained – in terms both of technical objectives and maintenance of relevance to wider activities, initiatives, and policies;
- Keeping accurate records of costs, resources and time-scales for the project; collecting and collating material that may be required for eventual evaluation purposes;
- Ensuring integration of Management Reports and their presentation to the Steering Committee; and
- Identifying emerging problems and difficulties and developing and implementing strategies for dealing with these.

Foresight champions
Securing high political support early on is an important task, since this will demonstrate that the exercise is taken seriously, and help to protect the exercise in the event of difficult times (sudden budget shortfalls, criticism of foresight in the media, etc.). If key people are first targeted and won over, a momentum can be established, as discussed earlier.

Experts
Sometimes domain experts are used as highly technical project managers, since they can identify (and mobilise) key people, provide a perspective on important themes, supply quality control, etc. In addition, experts are often drawn on for periods of less extensive (but arguably equally intensive) work, organised around expert panels/working groups. Having well-respected experts enlisted in an exercise can be influential in persuading people that the exercise is one to be taken seriously. Expert work is highly significant in terms of:

- Gathering relevant information and knowledge;
- Stimulating new insights and creative views and strategies for the future;
- Helping to build new networks;
- Diffusing the foresight process and results to much wider constituencies; and
- Identifying and acting on ways to maximise the overall impact of foresight in terms of follow-up action.

Organisation needs to be thought through very carefully, since the membership of the various groups set out above will influence the whole exercise. Moreover, the management style and interrelationships of these elements will need to be defined and explicated early on, so as to avoid disagreements flaring up.

Participation

The question of who participates in a foresight exercise, and how, is a central concern of exercise managers.[16] It is vital because the exercise needs to access information and insights that are located in a highly decentralised way; it needs to produce results that can be seen as legitimate, robust and relevant (i.e. not just the received wisdom from the usual suspects); and it must engage many stakeholders to implement its main results. But determining which participants are implied by the call for wide participation will depend on the exercise's objectives, its orientation, the themes/sectors covered and its intended audience. In other words, as we have seen so often, the various elements we need to scope are intimately interrelated. Some exercises are quite limited in their breadth of participation, both in terms of actual numbers and the types of actors engaged. Others, on the other hand, have set out to directly involve widely disparate groups, including citizens.

"Stakeholder analysis" has been developed as a tool for participatory planning.[17] It involves attempting to identify critical stakeholders and the interests they have in the activity. Lists of stakeholders are constructed from a mixture of general guidance about types of actor likely to be engaged, empirical investigation of specific actors, and "inside knowledge" from key participants. A simple starting point is to consider broad classes of stakeholders – for instance, scientist, governmental, non-governmental organisation (NGO), industry, other professional and citizen groups. It is important not to be too restrictive in identifying, for example, the sort of government department or firm that should play a role. Different levels (national, regional) and sizes of organisation might be relevant. What is important is to recruit gifted and open-minded individuals, who are prepared to learn and share, and not just present their organisation's official positions. They are recruited as knowledgeable actors, not strictly as representatives who are unable to think for themselves and voice opinions on topics where their organisation has yet to form an official line.

Methods for locating such individuals involve search through databases and web resources, or seeking advice from other informed people. Representative approaches can involve asking scholarly, professional and industry organisations for names. Again, we stress that the people sought are not to act solely as representatives of their bodies; rather, they are being recruited to give a good range of opinion and knowledge, and to be able to carry the foresight message into diverse quarters. Reputational approaches are also commonly deployed in foresight exercises. For example, questionnaires asking informed sources to nominate particularly knowledgeable people in required areas of expertise have been widely used in foresight – examples being snowball surveys and co-nomination methods. Such methods can be limited by the choice of initial informed sources, so it is important to cast the net widely here. The more formal methods are important for reaching beyond the "usual suspects", but approaches such as co-nomination are time-consuming. If the area under consideration is large, many new names may be generated by such approaches. In smaller areas, there may already be little to learn, since most players are likely to be already well-networked. Another set of considerations arises from equity concerns, as well as legitimacy ones. It may be important to ensure representation of women (gender balance is often highly skewed in foresight activities) and ethnic minorities, people from non-metropolitan regions and smaller businesses, etc.

Identifying participants is, of course, only part of the picture – how they are actually engaged in the foresight exercise is of paramount importance. Such engagement can be thought about along two dimensions: the "frequency" of participation and its "reach". Considering "frequency" first, an exercise might be largely desk-based with views from stakeholders (other than those gleaned from literature review) actively elicited infrequently, at discrete points in the process (for example, during a set of Delphi surveys or of consultation events). Alternatively, an exercise might largely constitute an ongoing dialogue or "strategic conversation" between stakeholders, with panels and working groups set up for an indefinite period of time to deliberate on the future of an area (or some other topic). Foresight activities "naturally" offer a number of opportunities to consult stakeholders – it is up to project managers to decide how to take full advantage of these. Turning to "reach", this leads us to consider who is to be consulted at each round of consultation. While a total pool of participants may be identified, it is likely that different stakeholders will be engaged at different points of the process. Reach can be considered to be either "extensive" or "exclusive", with different methods typically used for different situations. The choices made have implications for the credibility of the outcome of a foresight exercise, for the time needed for its completion, and for its eventual cost.

Stakeholder views on the future (elicited, for example, through Delphi or scenario workshops) are of course of interest. But these are not the only inputs that may be sought. Inputs may be usefully provided during the scoping process, during deliberation on the implications of foresight's results, etc. These can often be the most significant (yet often forgotten) consultation points, since they allow participants to make strategic choices about an exercise, which, in theory, should engender greater ownership of the process and its outputs.

Having identified stakeholders, their orientations to foresight must be considered in order to effectively approach them and make good use of the resources they will hopefully bring to the exercise. Such orientations are usually ascertained from evidence that is available informally (or as the result of pilot exercises or earlier research), but it is possible to commission interviews or even surveys to examine attitudes to foresight. Such stakeholder analysis involves providing answers to such questions as:

- What do various stakeholders specifically expect of the activity? Are these expectations realistic and well-informed, and how have they been formed? Do they have interests or objectives that might conflict with the exercise? What are their attitudes to each other – are there conflicts to resolve or manage?
- What benefits might they experience from the exercise and from active engagement in it? What might be gained by participating in an open-minded way, rather than leaving it up to others to do the work, or undertaking it only to argue for a particular point of view?
- How can effective communication be established with different stakeholders?
- What resources could or should stakeholders contribute?

Communicating with Users and Audiences

Foresight is a participatory process that requires time and commitment from a range of experts and stakeholders. In order to enlist their support, activities must carry a strong stamp of approval: endorsement from leading figures from science, industry and government. The foresight activities must also be clearly explained, transparent and involve the key stakeholders from the outset (including consultation as to the design of the exercise). But care must be taken to be realistic, not to promise too much, and perhaps not to promise too many outcomes to too many players. For a relevant discussion, see von Schomberg *et al.* (2005).

Communication is a key element of foresight. Communication with potential supporters, participants and users is required in order to argue for a foresight activity (and to deal with counter-arguments), and to argue for an activity of a particular sort; to supply guidance as to how to participate effectively and discussion of what sorts of input are required; to disseminate results, implement strategies, provide assistance for others to use the outputs of the exercise, and so on.[18] Various tools can be used to promote widespread appreciation of, and participation in, foresight activities, including:

- Publications and traditional communications tools (databases, newsletters, etc.) aimed at widespread promotion of the activities to be carried out (and thus, identification of players interested in participating), and for disseminating results;
- An online communications "forum" designed to disseminate information and promote the activities carried out and completed by foresight. Websites are being used to increasingly good effect in foresight activities, and can provide an important way of reaching people remotely;
- Initiatives aimed at encouraging participation, such as conferences, workshops, and other meetings. These may be mainly oriented toward dissemination of decisions already taken and preliminary results, or they may constitute more active consultation as to the aims and activities of foresight. They may be tied to the actual work of foresight in terms of generating visions and gathering knowledge. It is often helpful to work together with specific intermediaries and sectors of activity (academies of science, trades unions, research centres, industry associations, government ministries, etc.), whose aim is to encourage participation and promote a more active and knowledgeable involvement among their members or clients;
- Illustration of foresight "success stories" in organisations and/or areas characterised by similar problems and objectives.

Methods

As this Handbook is given over to summarising some of the main methods used in foresight exercises (see Chapter 3), we only discuss these here in terms of the scoping issues that are raised. The particular methods to be employed will be a function of objectives and other scoping decisions (plus likely influence from what technical support capabilities are available, whether particular people approached have acquired a bad name, and so on). What is also important for scoping is to consider how methods can be used together, whether in parallel or in sequence, to constitute a coherent exercise.

Moreover, if we consider some of the key steps in foresight and the methods that might prove useful in these, we can see that foresight methodology is far broader than forecasting and futures methods. It necessarily includes methods for structuring and managing critical tasks such as stakeholder analysis and recruitment, coalition building, project scoping, organisation and management, implementation, etc.

We have already discussed how (many of) these tasks need to be considered in scoping. Most "futures" methods can be used in a variety of ways in foresight exercises.[19] Selection of methods will depend upon several factors, most notably available time and financial resources. Where it comes to methods for systematic thinking about the future, the issues for scoping to address include the following:

- What is the mix between quantitative approaches (with the advantages of ready visualisation, relatively easy ability to check for consistency of elements and to assess extent of change as compared to trends, comparability across different sectors/regions/scenarios) and qualitative approaches (with the advantages of narrative and accessibility)? (See Table 3.3 in Chapter 3.)
- What is the mix between expert-based analysis (with the advantages of specialist knowledge and quality control) and the use of more participatory methods (offering ability to draw on a wider range of stakeholders, and also, hopefully, to mobilise them through the process)? (See Figure 3.3 in Chapter 3.)
- What effort will be put into developing new data and analysis concerning the present (even in countries with elaborate statistical and planning systems, there is at least likely to be a need to "translate" available material to make it relevant to the task at hand)?

Product–Process Orientation

Many commentators have noted a fundamental issue in contemporary foresight exercises as lying in the balance between what are typically referred to as "products" and "processes".[20] National programmes have sometimes placed more stress on one or the other, and most seek to synthesise the two.

Product-orientation focuses on tangible outputs, typically reports – scenario analyses or Delphi results; a "critical list" of key technologies or a hierarchy of priorities for R&D expenditure, etc. The purpose is usually seen as one of improving central decision-making about S&T issues. Tangible outputs are often what some people refer to as "codified" knowledge, in that the knowledge generated through the process has been turned into information that can be circulated widely, without necessarily requiring face-to-face interaction. Highly product-oriented approaches often involve small expert groups, and/or highly formalised methodologies for eliciting and combining expert opinion (most notably, Delphi). French and German national exercises have taken this form, for example.

Process-orientation focuses on enhancing networking and exchange of opinions among actors. Often the purpose has been described as one of "wiring up the national innovation system" (Martin and Johnston, 1999). The idea is that a shared focus on longer-term developments will help those involved to identify emerging issues and the carriers of relevant knowledge about these issues, to share understanding about each others' expectations and the strategies that are liable to be pursued, and to forge enduring networks for collaboration. The Dutch and the second UK exercises are examples. Such "soft" outputs are more difficult to grasp, because these typically take the form of knowledge embodied in people's practices and approaches to issues. Though these may be harder to identify and quantify than documentation, they represent a very important aspect of the benefits of foresight.

A mixed orientation spans and synthesises the two above. Each is valued in its own right, and seen as supporting the other. Thus, the creation of products is also seen as a helpful device to encourage people to work together and network effectively, and as a legitimating tool to convince auditors that money is being spent well. In turn, networking provides a wider range of inputs to improve the quality of products, and this wider participation itself gives social legitimacy to the process.

Achieving Impacts: Policy Intervention

Project managers are often preoccupied with getting the foresight process "right", and assume that the sponsors will ensure that the results of foresight are used. But while getting the process "right" is a necessity for successful follow-up action, it is not a sufficient condition. The possibilities for follow-up action should be considered from the outset of an exercise, in other words, as a part of scoping. There are several elements to this, but let us here focus just on two: (a) planning for the ways in which results and conclusions are presented; and (b) managing expectations as to what can be achieved through foresight.

Taking point (a), the scoping process should include some notion of how results are to be presented. As always, this has to involve some flexibility, partly because it is unrealistic to expect that we will know all of the results of foresight in advance, or even the form these are likely to take; and partly because good creative ideas about how to effectively present and implement results may be generated in the course of the exercise. Nevertheless, we can suggest that planning to present results in a concise and accessible way, that makes their links to action highly explicit, is in general going to be more effective than more diffuse, highly technical, and/or abstract accounts. Just how such high-quality reports are to be prepared should be addressed in the scoping process.

One common strategy in foresight exercises is to produce Action Plans as a major part of the main report of the exercise, or as a stand-alone accompaniment to a series of more detailed reports. Action Plans are, simply, lists of actions that should follow from the identification of problems and possible solutions through foresight. They are not "wish lists", nor should they simply specify end points and objectives. They should indicate actions and responsible agents, ways of monitoring progress, and indicators with which to assess the degree of success attained ("verifiable objectives"). Foresight can produce other sorts of output, of course – there may well be issues where it is hard to identify the responsible agent (often the exercise will have to conclude that further study of such issues is the important action to undertake next). But outlining key actions, key targets, key agents, is vital. The Action Plan should also provide resources – hyperlinks, references to other reports – that justify and provide access to more detailed explanatory and analytic material underpinning the proposed actions.

It is important to link actions to the people responsible for executing them, and considerable presentational skill and political nous may be required to formulate Action Plans in terms that can be accepted by decision-makers. The goals have to be realistic, but also challenging ones ("stretch targets"). Reiterating excessively familiar calls (e.g. for improved training, resources for basic research, university–industry links) is unlikely to do more than disappoint. It will be easier to successfully link decision-makers with actions if they have been involved in the foresight process. Having members of the staff of a particular agency or organisation involved in the exercise means that they can provide the political nous mentioned above – indicating what language to use and what to avoid, suggesting existing policies and programmes that could be modified or adapted to take on board results of foresight, and so on.

Moving on to point (b), as discussed above, the outcomes desired from an exercise are liable to vary across actors – some may hope for a focus on certain types of policy agenda (e.g. training, support for higher education…), some for attention to particular technologies (e.g. biotechnology, nanotechnology…), to particular sectors of the economy (e.g. technology-related business services, small firms…), or particular social groups (e.g. neglected regions, socially excluded reservoirs of talent…), and so on. Some expectations as to outcomes can be unrealistic. Those setting up and "selling" foresight exercises may be inclined to promise too much in order to win the support of a wide range of actors. If this leads to misleading views as to how great an emphasis will be placed on certain issues, how far decision-makers are liable to heed the inputs from foresight in dealing with such issues, and how rapidly to expect change there is liable to be disappointment and loss of legitimacy.

For these reasons, a reasonably clear and realistic notion of the sorts of benefit that can reasonably be expected needs to be developed and conveyed as part of the scoping activity. It needs to be communicated by capturing relevant information, and putting it into a form suitable for stakeholders to examine. Later, as the foresight activity proceeds, better understanding will be gained as to what it can and cannot hope to accomplish. Feedback should be provided that allows for modification of these expectations, too.

Gaps in implementation can be very discouraging. These may occur where recommendations have been prepared, but there has been no mechanism to check on their follow-up; and where networks that were working productively have been allowed to dissolve. Foresight is not a matter of free-floating visions. It is a participatory process of constructing better understanding of what desirable and feasible futures could be, and how different socio-economic partners need to work together to create them. This is a demanding task, inherently linked to action. It cannot be achieved without serious inputs of time and effort from many parties. Perhaps the most crucial message in managing expectations is the following: foresight is not a quick fix.

CONCLUDING REMARKS

This chapter has sought to introduce some key elements for scoping foresight that can be used at national, regional or company/organisation levels. These scoping elements have already been employed widely in foresight exercises across Europe and underpin recent European guidelines on the use of foresight (see Bernabei and Prakke, 2004). We have also sought to raise awareness of foresight's limitations, arguing that expectations should be realistic. Planned appropriately, and with sufficient political support, foresight can be a very positive force, bringing transparency and clarity to decision-making, at the same time as grounding it more thoroughly in longer-term considerations. But foresight is never easy, and those who wish to pursue the use of such policy instruments need to be prepared for the long haul.

Foresight should not be used if there is no possibility to act on the results that it will generate. "Wishful thinking" is not enough to sustain a foresight exercise; those involved are likely to feel that their expectations have been raised unduly, and their time wasted. A minimal degree of political, economic or cultural leverage is required – even if it is recognised that the foresight activity is likely to have to battle with entrenched opposition to achieve any significant impacts.

Nor is "me too" a good basis for foresight. The simple imitation of issues and methods (not to mention the uncritical "borrowing" of results) from elsewhere is liable to be counterproductive. For example, a predominantly rural agricultural region or state cannot "foresight" its way to becoming a high-tech nanotechnology or even biotechnology hub. Neither can a foresight activity that has been designed for a region or state that is accustomed to wide public participatory debates necessarily be (immediately) deployed in one in which public opinion is handled through more traditional routes – surveys, press, political party representation, etc.

If there is no possibility for careful preparation and tailoring of foresight to specific national or regional characteristics, then it probably should not be implemented. We should be explicit in acknowledging that foresight cannot solve all of the social, economic or political problems that beset a state, region or organisation. Foresight can generate visions. Ideally, large elements of these will be shared visions, and ones that are well-founded on knowledge of the relevant developments in social or technological affairs. This ideal is not as utopian as it may at first seem; some national and regional exercises have succeeded in achieving quite widespread consensus behind their results.

But foresight is not a "magic wand" with which to impose consensus in situations where there are profound disagreements. Political discretion also needs to be exercised in cases where conflict is inevitable between certain sectors on highly contentious issues. Skills at mediating conflictual discussions are liable to be required! In some situations, unfortunately, there is a strong probability that the conflict-resolution powers of foresight methods will be insufficient, and that conflict may even be exacerbated by embarking upon an exercise at this moment. In such cases, foresight should not be undertaken, or at least taken up in a very cautious way. Foresight may help find areas of agreement shared between opposing factions, but it can become mired in disputes between entrenched antagonists, especially when the focus of foresight is on topics that divide these groups – which will often involve issues of social welfare, governance, and the like.

Finally, we should reiterate that foresight should not be seen as a "quick fix". A foresight exercise may provide the information (e.g. a priority list) needed for a particular policy to be implemented. But the sorts of longer-term analyses that foresight involves, and the new networks and capabilities that it can forge, cannot be expected to achieve results overnight. Often the processes of interacting around ideas of what opportunities might be seized, how particular challenges might be confronted, etc. will take a long time to result in widely-accepted notions of the way forward. The problems we wish to address have often matured over many years. Effecting significant change, then, is often going to require long preparation, and considerable groundwork to prepare people for the change.

NOTES

1   The chapter builds on earlier efforts to discuss managerial as well as methodological aspects of foresight in Miles and Keenan (2002) and Keenan *et*

al. (2003), and draws on other reviews of the topic such as RAND Europe (2006) and Woodling (2004). A list of relevant documents and links to them is available online at: http://cordis.europa.eu/foresight/tools.htm, and a number of useful accounts of foresight organisation (mainly focused on regional foresight) is at http://cordis.europa.eu/foresight/conference2002_background.htm.

2   EU material is available at http://cordis.europa.eu/foresight/ (especially relevant are the FOR-LEARN pages) and UNIDO material at http://www.unido.org/doc/5216, see also the governmental foresight network material at http://www.eranet-forsociety.net/ForSociety/index.html.

3   Much of the discussion in this chapter is relevant to foresight in general, not just TF exercises. For an introduction to organisational use of strategic planning, see Bryson (1995).

4   Staying with the Delphi example, it is remarkable what a limited range of the vast spectrum of possible Delphi approaches has been brought into play in existing TF programmes. Some other methods seem to have been completely overlooked, as other chapters of this Handbook make clear.

5   More recently, Eriksson and Weber (2006) have usefully introduced the concept of "adaptive foresight" to capture the need for flexibility in conducting TF exercises.

6   Though it remains for this to be demonstrated – and in any case, curiosity is often likely to be informed by a sense of the importance – and applicability – of solutions to specific problems. And it could well be that areas of work identified as particularly interesting in foresight studies could equally well yield serendipitous consequences.

7   See for instance Cuhls (2004), Cassingena Harper and Georghiou (2005), and references in Chapters 16; also Bernabei and Prakke (2004) on EU activities, and regional foresight studies reviewed at http://cordis.europa.eu/foresight/conference2002_background.htm.

8   In addition to the studies documented here, especially the guides cited at the outset of this chapter, one should consider the new journals *Foresight* and *International Journal of Foresight and Innovation Policy*, plus the treatment of foresight in established journals such as *Futures*, *International Journal of Forecasting*, and *Technological Forecasting and Social Change*.

9   See Von Schomberg *et al.* (2005) for excellent discussions of the relations between Foresight quality and arguments for organising and using Foresight.

10   For instance, more parties might be brought on board as co-sponsors of the exercise; some desired outcomes might be determined to be unrealistic; while others might usefully be added, and so on.

11   For a study examining how foresight is effective within more and less loosely knit organisational cultures, see the work of the FORMAKIN project, e.g. Brown *et al.* (2001).

12   This derives from the Chinese translation of the two terms as being identical. The point has been frequently noted in the foresight field, perhaps because futures work is often turned to at times of crisis. The veteran futurist Johan Galtung did much to disseminate awareness of this significant convergence of meanings, and the proponent of strategic prospectives, Michel Godet, has even used it as the title of a book (Godet, 2005).

13 The first three represent areas of continuing rapid advance in scientific knowledge and technology development – each has been hailed as involving a technological revolution where new knowledge can be applied on a very pervasive scale. The latter two technology areas are ones of considerable relevance to national economic development and environmental affairs, and are problem areas in many countries.

14 For instance, (much of) the public sector is rarely taken up in TF exercises, and the case for not doing so remains obscure. Clearly the concern of most exercises with competitiveness leads people to focus on private industry; but public sector performance is important in underpinning and facilitating that of the private sector, and public services are often pioneers in using new technologies, setting standards, forming major users, and so on. The rationale for excluding some areas, including others, could fruitfully be made more explicit.

15 For example, concerted effort from those concerned with marine science and engineering led to this sector being introduced into the first UK TFP, though it had not been among the original set of panels.

16 See for example Gustafsson (2000) and Pirttimäki (2006).

17 See Slocum (2003) for a wide-ranging review of participatory methods. In addition, Elias *et al.* (2002) examine stakeholder analysis in the context of R&D programmes.

18 For an interesting elaboration of the use of scenarios in Foresight work see Berkhout and Hertin (2002).

19 One type of method that has received considerable attention, in terms of alternative approaches to participation and use, is scenario approaches: see, for example, Ringland (1998), Roubelat (2000), and Van der Heijden (1996).

20 Presumably inspired by the product–process innovation distinction in innovation studies; Martin (1995) has previously noted the close intersection of foresight and innovation research.

REFERENCES

Alsan, A. and Oner, M.A. (2004), 'Comparison of National Foresight Studies by Integrated Foresight Management Model', *Futures*, **36**.

Berkhout, F. and Hertin, J. (2002), 'Foresight Futures Scenarios: Developing and Applying a Participative Strategic Planning Tool', GMI newsletter.

Bernabei, G.C., Prakke, F., Aichholzer, G., De Lattre-Gasquet, M. and Salminen, K. (2004), *Mid-Term Assessment of Foresight Activities*, Independent Expert Panel Report to Directorate General Research of the European Commission, available online at ftp://ftp.cordis.europa.eu/pub/foresight/docs/mta_finalreport.pdf.

Brown, N., Rappert, B. and Webster, A. (2001), *Foresight as a Tool for the Management of Knowledge Flows and Innovation (FORMAKIN)*, available from Science and Technology Studies Unit, The University of York, UK, see http://www.york.ac.uk/org/satsu/Projects/Formakin.htm.

Bryson, J.M. (1995), 'Strategic Planning for Public and Nonprofit Organizations', *A Guide to Strengthening and Sustaining Organisational Achievement*, Revised Edition, New York: Jossey-Bass, Wiley.

Cassingena Harper, J. and Georghiou, L. (2005), 'The Targeted and Unforeseen Impacts of Foresight on Innovation Policy: The eFORESEE Malta Case', *International Journal of Foresight and Innovation Policy*, 2(1), pp. 84–103.

Cuhls, K., (2004), 'Futur – Foresight for Priority-setting in Germany', *International Journal of Foresight and Innovation Policy*, 1(3–4), pp. 183–194.

Elias, A.A., Cavana, R.Y. and Jackson, L.S. (2002), 'Stakeholder Analysis for R&D Project Management', *R&D Management*, 32(4), pp. 301–310.

Eriksson, E.A. and Weber, K.M. (2006), 'Adaptive Foresight: Navigating the Complex Landscape of Policy Strategies', paper presented at the Second International Seville Seminar on Future-Oriented Technology Analysis, Seville, September.

Gertler, M.S. and Wolfe, D.A. (2004), 'Local Social Knowledge Management: Community Actors, Institutions and Multilevel Governance in Regional Foresight Exercises', *Futures*, 36, pp. 45–65.

Godet, M. (1985), *Crises are Opportunities*, Montreal: Gamma Institute Press.

Gustafsson, T. (2000), *Participatory Foresight Processes For Finnish RTD Programmes*, Master's Thesis submitted in partial fulfilment of the requirements for the degree of Master of Science in Technology Department of Electrical and Communications Engineering, Espoo: Helsinki University of Technology.

Jewell, T. and Sripaipan, C. (1998), 'Multi-country foresight – issues and challenges: a paper based on a foresight study on the future for water supply and management in the APEC Region to 2010', presented to the Conference of the International Association of Technology Assessment and Forecasting, New Delhi, India.

Keenan, M., Butter, M., Sainz, G. and Popper, R. (2006), *Mapping Foresight in Europe and Other Regions of the World: The 2006 Annual Mapping Report of the EFMN*, report to the European Commission, Delft: TNO.

Koschatzky, K. (2005), 'Foresight as a Governance Concept at the Interface between Global Challenges and Regional Innovation Potentials', *European Planning Studies*, 13(4), June, pp. 619–639.

Martin, B.R. (1995), 'Foresight in Science and Technology', *Technology Analysis and Strategic Management*, 7(2), pp. 139–168.

Martin, B.R. and Johnston, R. (1999), 'Technology Foresight for Wiring up the National Innovation System: Experiences in Britain, Australia and New Zealand', *Technological Forecasting and Social Change*, 60, pp. 37–54.

Miles, I. and Keenan, M. (2002), *Practical Guide to Regional Foresight in the United Kingdom*, Luxembourg: European Commission, EUR 20478, ISBN 92 894 4682 X.

Miles. I. and Cunningham, P. (2006), *Smart Innovation: A Practical Guide to Evaluating Innovation Programmes*, Luxembourg: European Commission, at: ftp://ftp.cordis.lu/pub/innovation/docs/sar1_smartinnovation_master2.pdf.

Miles, I., Keenan, M. and Kaivo-oja, J. (2003), *Handbook of Knowledge Society Foresight*, European Foundation, Dublin, available at http://www.eurofound.eu.int/transversal/foresight.htm.

Pirttimäki, A. (2006), *Foresight in a Research and Technology Organisation*, Master's Thesis submitted in partial fulfilment of the requirements for the degree of Master of Science in Technology, Department of Electrical and Communications Engineering, Espoo: Helsinki University of Technology.

Popper, R., Keenan, M. and Butter, M. (2005), *Mapping Foresight in Europe and other Regions of the World: The EFMN Annual Mapping Report 2004–2005*, report prepared by PREST and TNO to the European Commissions' DG Research.

Popper, R., Keenan, M., Miles, I., Butter, M. and Sainz, G. (2007), *Global Foresight Outlook 2007: Mapping Foresight in Europe and the rest of the World,* The EFMN Annual Mapping Report 2007, report to the European Commission, Manchester: The University of Manchester/TNO.

RAND Europe (2006), 'Improving the science/policy relationship with the help of Foresight: a European perspective', *European S&T Foresight Knowledge Sharing Platform,* available online at: ftp://ftp.cordis.europa.eu/pub/foresight/docs/ ntw_using_report_en.pdf.

Renn, O. and Thomas, M. (2002), *The Potential of Regional Foresight - Final Report of the STRATA-ETAN Expert Group: "Mobilising the regional foresight potential for an enlarged European Union – an essential contribution to strengthen the strategic basis of the European Research Area (ERA)",* Luxembourg: Office for Official Publications of the European Communities, EUR 20589, available at: ftp://ftp.cordis.europa.eu/pub/foresight/docs/regional_foresight_en.pdf.

Ringland, G. (1998), *Scenario Planning: Managing for the Future,* Chichester: John Wiley.

Roubelat, F. (2000), 'Scenario Planning As A Networking Process', *Technological Forecasting and Social Change,* **65**(1).

Slocum, N. (2003), *Participatory Methods Toolkit: a practitioner's manual,* King Baudouin Foundation and Flemish Institute for Science and Technology Assessment (viWTA) in collaboration with the United Nations University – Comparative Regional Integration Studies (UNU/CRIS); available at http://www.kbs-frb.be, http://www.viWTA.be, and http://www.unu.cris.edu.

Van der Heijden, A. (1996), *Scenarios: The Art of Strategic Conversation,* Chichester, UK: John Wiley.

Von Schomberg, R., La Guimarães Pereira, A. and Funtowicz, S. (2005), 'Deliberating Foresight-Knowledge for Policy and Foresight Knowledge Assessment', Luxembourg: European Commission, working paper, EUR 21957 at: ftp://ftp.cordis.europa.eu/pub/foresight/docs/deliberating_foresight2.pdf.

Woodling, G. (2004), 'Foresight, Futures & Vision', in *EU-US Seminar: New Technology Foresight, Forecasting & Assessment Methods,* Seville, May.

16. Evaluation and Impact of Foresight

Luke Georghiou and Michael Keenan

INTRODUCTION

From the perspective of a new user of foresight, the key evaluation questions concern additionality, essentially what difference has it made and how sustainable is this difference? This is true in both of the ways in which foresight can be encountered in policy-making, foresight "for policy" and foresight "as policy". In the former case the aim would be to use it as a tool to inform and develop policy in any area or to "join up" policy across domains. Technology foresight is normally applied in this way in areas of policy with a strong science/research input. In the latter case the aim is to use foresight as an instrument to implement budgetary, structural or cultural changes in the domain of research and/or innovation policy. For each of these contexts the policy-maker needs to know whether foresight is achieving the desired effects and how it may be made more effective. Exploring the answer to these questions is the territory of evaluation.

In several of the country chapters reference has been made to full or partial evaluations of foresight activity. It is already a complex issue to consider the relationship between foresight and evaluation. In the array of strategic policy approaches available to the policy-maker, the two are often grouped together, with the simple distinction being made that evaluation is looking backwards at what has occurred and foresight looking forward at possible futures. A European working group sought to combine these activities, along with technology assessment, in the domain of technology policy into a common concept of distributed strategic intelligence but largely they remain institutionally distinct (Kuhlmann et al., 1999). Yet, in other ways, they are not so easily separated, as most evaluations include some form of formative perspective on the future and often have to consider the future implications of the measures they seek to assess. In turn, foresight activity generally needs to be informed by a thorough understanding of the past and present.

The foresight community is quite introspective and self-evaluation forms one tradition. For example, Durand drew 12 key lessons from the French Key Technologies 2005 exercise (Durand, 2003) and Havas also sought to draw generalisable lessons from the experiences of the Hungarian TEP foresight programme with particular emphasis upon the linkages to policy-makers (Havas, 2003). Recent experiences in the evaluation of foresight have been reviewed by two of the authors (Georghiou and Keenan, 2006) from whose article this chapter in part derives.[1]

MATCHING EVALUATION

Against the background of variation in foresight practice described in this volume, it is important to refer back to the rationales and definitions discussed in Chapter 1. A key message emerging there was that foresight could be summarised as an interactive approach producing shared visions of the future and joint actions in consequence. This perspective avoids what it will be argued below is a common trap, the treatment of foresight and its implementation as separate processes without serious attempts to build bridges between or to link the two. Establishing this connection provides an important backdrop to effective evaluation. This is more than the obvious statement that evaluation of foresight should be concerned with its impacts. Rather, the point here is that foresight is part of a broader set of influences in most of the effects it seeks to achieve in terms of public policy or the strategy of firms. Furthermore, once a foresight output has been produced and enters the environment for implementation the question may be asked of how is it different from other types of policy information emanating from, for example, lobbying, evaluation, strategic studies, or from the influence of historic commitments and budgetary analyses? Probably the answer lies in a longer time horizon, and elements of creativity or commitment through participation, but all of these elements can also come from other sources. The implication is that evaluation of foresight must include understanding of the interaction of foresight outputs with the strategic behaviour of policy and economic actors.

There are also some normative issues involved. Foresight is not always tuned to the needs of recipients and hence, to extend the analogy, the signal may be obscured by noise and not picked up. Information needs to be presented in such a way that policy/strategy mechanisms can receive and absorb it. One moderating factor is that of timing. This needs to synchronise with policy and strategic cycles. Keenan (2000) has shown that the results of the first UK Technology Foresight Programme were delivered at the wrong time of year in respect of the priority-setting mechanisms of the Research Councils (the funding bodies in the strongest position to implement them), and hence that their impact was both delayed and diminished. Furthermore, the level of recommendations needs to match available funding or capacity for reform. However, foresight cannot always work within the status quo and occasionally it is the policy/strategy structure that needs to change in the light of disruptive foresight information.

The common space and joint ownership elements associated with foresight (Cassingena Harper, 2003) imply that it should not be viewed as being in a linear or sequential relationship with implementation but rather that it should move into the "implementation space". In other words, the conduct of foresight itself moderates the implementation of emerging findings (or at least the conditions for their implementation), and the environment for implementation affects the way in which foresight ought to be conducted – foresight and implementation are interactive activities. The implication is that, on the one hand impact-oriented evaluation must closely consider the foresight process if impacts are to be fully accounted for and explained, while on the other hand also understanding the drivers of the strategic behaviour of the implementing bodies.

Implicit in the generational models of foresight discussed in Chapter 1 are very different approaches to evaluation (Georghiou, 2001). For first generation foresight the key issues are accuracy of prediction and diffusion of results (to non-experts). In the second generation the take-up of priorities and establishment of networks among the industrial and academic participants become key evaluation issues, while the third generation implies the involvement of stakeholders in evaluation and looks for evidence of the emergence of a foresight culture.

The need for conditioning the evaluation approach to foresight may be illustrated by considering the difference in the unit of aggregation at which foresight is evaluated. For example foresight may be presented as a policy, a programme or as practice. Each of these demands a different evaluation approach. In a policy evaluation, issues of rationale for public action predominate and the interaction of foresight with other policies becomes a topic of focus. In the more conventional format of programme evaluation the programme objectives become a primary focus, mostly in terms of objectives achievement but also in terms of the appropriateness of the objectives – the link to policy evaluation. Foresight as practice has as its focus the methods and structures used. These may be evaluated both in their own terms and in terms of whether they were fit for purpose. For example, it is a very different question to ask whether a Delphi survey allows an expert consensus to emerge than it is to judge whether it was the right way to consult a particular group of participants. Again, in a real situation we may see combinations of these elements, albeit with different emphases.

EVALUATION OF FORESIGHT: SCOPE AND PURPOSE

Since Foresight is a policy instrument consuming time and resources, it is reasonable to expect that it should be subject to evaluation of a comparable rigour to other tools. In a generalised evaluation framework, three basic tests could be applied:

1. *Accountability* – with questions such as whether the activity was efficiently conducted and proper use made of public funds;
2. *Justification* – with questions such as whether the effects of foresight justify its continuation and extension; and
3. *Learning* – asking how foresight can be done better in particular circumstances.

Each of the above can singly or in combination form the basis of a motivation to evaluate a foresight action and each carries with it implications for the conduct of that evaluation. Although it is the most traditional basis for evaluation, accountability is probably the most problematic in this context, if only because foresight is unlikely easily to yield an answer to the type of question that is being asked here. Accountability generally follows the lines of resources and governance but foresight presents difficulties for evaluators in both cases. Some resources for foresight come centrally from government but in a "soft" policy instrument of this type, the most important resource input is the voluntary contribution of the participants. This is also the case for governance since "ownership" of a foresight exercise is not always clear. Indeed, foresight exercises tend to be distributed activities across (and beyond) communities of actors with various needs and objectives. The practical implication is that evaluators cannot expect to impose demanding questions on a volunteer taskforce that is working without a contract and hence with potentially flexible objectives.

A further difficulty for an evaluation originating in accountability considerations is that there will be pressure to apply standard tests that compare it with other publicly funded activities. Again the specific nature of foresight, as already described, makes this difficult. What happens may ultimately depend upon the degree of institutionalisation of foresight. First efforts in a country or region often represent a "heroic age" in which creativity dominates at the expense of good organisation but where mistakes are forgiven on the basis that this is a prototype. By the time a second cycle is entered institutionalisation may have reversed the proportions of creativity and organisation and brought in its wake requirements for evaluation against more conventional criteria.

The motivation of justification is a common one in evaluation and should not be seen as pejorative. The typical situation is one in which the operators of a programme are approaching a key decision point regarding extension or repetition of an activity that is likely to be novel and possibly not strongly institutionalised. In these circumstances they desire evidence to support their belief that their efforts have been worthwhile. The independence of view afforded by an evaluation meets this need.

In reality, the justification motive is frequently combined with the third major group, that of learning. Evaluations of foresight are almost inevitably formative (Scriven, 1991), and given the novelty and sensitivity to context of the instrument, lessons are a likely outcome. These tend to relate to process rather than outcome because of the timing issue which is discussed below.

In a standard evaluation approach, it is important to define the scope and purpose of what is being evaluated at an early stage (Gibbons and Georghiou, 1986). The variety of forms of foresight has been discussed in the previous section. Another dimension in which foresight has to be delineated is that of location in time. The key question is where does a foresight activity begin and where does it end. In a first national effort, the beginning is usually clear as the process is initiated with a decision to commit resources and often to establish some sort of secretariat. The end is frequently much less clearly delineated. Where the aim is a report or list of priorities, publication and launch marks some kind of termination though dissemination and other implementation activities may well follow. The launch of networking activities is far less likely to offer a clean break, as these are likely to persist for some time after the foresight activity has ended. An arbitrary decision may need to be made on when to demarcate the cut-off point by when foresight outputs cease to provide a distinct or influential voice in policy discourse. Furthermore, this reinforces the point made in the previous section, that to understand the effect of foresight it is necessary to locate it in a broader strategic and policy context. The evaluation will have to explore the period in which foresight emerged and its interaction with other elements of the system.

The timing issue is also linked to the type of question being asked. If a linear or sequential view of foresight is taken, process issues are best pursued while the activity is still under way. However, many outputs and outcomes will not be clearly visible at this time and will need to be investigated ex post. Here the problem becomes one of attributing effects. Over time, phenomena of "diminishing signal" and "knowledge creep" frustrate the evaluator looking to disentangle the effects of foresight exercises from other influences on outcomes. The evaluator needs to move away from such reductionist approaches to a more holistic and systemic framework that views foresight as complementing the use of other policy tools.

Finally, if accuracy of the future visions is an issue the ex post delay corresponds to the foresight period. This may be less of a problem with short horizon five-year critical technology exercises, but it requires a remarkably stable system if the issue is to be usefully pursued for foresight on, say, a 15-year timescale. Only the Japanese STA/NISTEP forecasts have been properly assessed on this basis (see Chapter 8). In this case the method adopted was to use expert panels assembled for the current iteration of Delphi to review the extent to which topics had been "realised" on the expected timescale. The first assessment of this kind was reported in the Fifth Delphi Survey published in 1992 (NISTEP, 1992). The panels judged that 28 per cent of the topics had been realised and a further 36 per cent "partly realised". Kuwahara has noted that accuracy in the time dimension is not necessarily the major benchmark as use of the results may have changed the direction of research and technology and affected realisation (Kuwahara, 1999). Other issues may be raised – accuracy may not necessarily coincide with relevance or utility. Partial realisation is also a difficult concept to enumerate. Data of this kind are best seen as an input to evaluation rather than an evaluation in itself.

CRITERIA AND ADDITIONALITY

The classic criteria of evaluation are:

- *Efficiency of implementation,* otherwise known as process evaluation and focused upon managerial and logistical issues. These are not necessarily trivial or only of bureaucratic concern. Process evaluation covers topics such as organisation and management, and would for example ask: Were the "right" people involved in an exercise? Did expert panels (if used) receive adequate support? Was the exercise adequately linked to decision-making centres? It may also address the question of the appropriateness and efficiency of methods used, for example: Should a Delphi have been used? Were scenario workshops properly facilitated? A well-conducted process evaluation can cast light upon the dynamics of foresight. As noted above it should be conducted in real-time or immediately after an activity is complete to ensure that the findings are not distorted by hindsight or obscured by loss of data;
- *Impact and Effectiveness,* often at the core of policy-makers' concerns, these criteria deal with what has been produced by foresight in terms of outputs and outcomes. Probably the most important observation here is that outputs measure only activity and not its significance. Hence, while it may be useful to know numbers

participating in meetings or surveys, reports disseminated, meetings held, website hits and so on, none of these measure the effects of these contacts or their contribution to outcomes. Numbers may even be misleading; the number of "new networks" formed disguises variation in their novelty, size, significance and durability. Outcome evaluation is normally made far more difficult by the problem of attribution, discussed more extensively below;

- *Appropriateness* as a criterion links back to the earlier discussion of policy level evaluation by engaging with questions of rationale. For national foresight activities this includes the issue of state intervention but also raises questions of what the alternatives would have been (including the counterfactual).

Pursuing this issue of counterfactual, a key question in the evaluation of any public policy intervention is that of additionality – the extent to which the activity would have taken place without a public intervention. This in effect examines the rationale discussed earlier. Under this framework, the questions which should be asked about a foresight or TF activity are:

- Would foresight have happened without the policy intervention?
- Is TF done differently/better because of the policy intervention?
- Are the resulting actions better because of foresight?
- Have persistent changes been achieved (e.g. foresight culture)?

Within the field of evaluation, recent thinking has moved away from treating additionality as a binary stop–go item. In a "systems of innovation" framework, temporary financial interventions are seen as less important than efforts to change the innovation system for the better in a lasting way. If it is accepted, as discussed above, that foresight is correcting an inherent tendency to have excessively short-term horizons and a difficulty in forming new networks around technologically and socially innovative activities, then foresight may be best evaluated ultimately in terms of its ability to change values and behaviour in these directions.

This perspective on additionality – known as behavioural additionality – has implications for how evaluation is done (Buisseret *et al.*, 1995; Georghiou 2002). For example, there is an interest in the persistence of effects – have new routines or practices (including networks developed and the capacity to continue doing foresight) been adopted by the participants? The behavioural additionality perspective is now being applied in many OECD countries to examine the effects of R&D grants for industry. However, it also provides an appropriate framework for evaluating foresight and indeed was used in practice as the underpinning concept for the evaluation of the third cycle of the UK Foresight Programme. To answer questions of this type we once more return to the need to explore the interactions of foresight with the strategies of the participating organisations. This will require both the practical knowledge of how foresight interacts with and penetrates the organisation, and an understanding of the other influences upon strategic decision-making that contribute to a particular outcome.

The evaluation of foresight should also beware of potential traps. Traditionally foresight is seen as a process of building commitment among stakeholders –an important element for example in Martin's "5Cs" (Martin and Johnston, 1999). However, from an evaluation perspective this also creates risks when trying to assess the additionality of foresight. One risk is that of the self-fulfilling prophecy when the "owners" of a foresight activity (for example a sponsor ministry) also control the distribution of resources at the implementation phase. There may be a tendency in this situation to cause foresight priorities to have a stronger influence in the implementation environment than may be justified in terms of the rigour and merit of the exercise. At a more methodological level, stakeholding and consensus may be seen to some extent as a trade-off with creativity and insight. It may be somewhat easier to get "buy-in" to a set of views that are already commonly held than for a really novel or disruptive idea.

SOME EVALUATION EXPERIENCES

The table below (Table 16.1) shows some recent experiences of foresight evaluation, and serves mainly to emphasise that a consistent and comparable approach has not emerged. However, it also demonstrates that almost every country that has undertaken a national foresight exercise has also seen some need for evaluation. Three of the countries mentioned can be discussed further as case studies. The aim is not to report the result of these evaluations but rather to indicate the types of evaluation approach adopted (and sometimes dropped) and to make some observations about the influence exerted by the evaluations. In effect these are brief meta-evaluations.

Case 1: UK Foresight Evaluation Experiences

Rationale and objectives have shifted substantially through the three cycles of the UK Foresight Programme and thinking about evaluation has to some extent tracked this shift. The first cycle was launched as a key policy instrument to reorient the country's science base in the direction of wealth creation and quality of life. The initial expectation was that this would be achieved through establishing priorities for science funding but as time passed and the difficulties of implementing this wish became clear, the secondary objective of building industry–science networks around new opportunities came to predominate. Hence, the Programme could be seen to be addressing a failure in the innovation system. In the current third cycle these "second generation" aims survive to some extent, but the main focus is upon engagement with stakeholders in government to address policy issues which cut across government structures and have a high science content. The foresight culture objective is also stressed.

Table 16.1   Evaluation of national foresight activities

Country	Evaluation Effort
Austria	Internal assessment of impacts by Science Ministry
Colombia	Panel evaluation 2007/8 addressing process and impact with national and international Validation/Evaluation Committees
Germany	Delphi 98 evaluation questionnaire; FUTUR evaluated during 2002 and again in 2004
Hungary	Panel evaluation 2003/4 addressing process and impact
Japan	Assessment of realisation of results some 15–20 years after identification in STA forecasts. Also foresight evaluated as a part of broader evaluations of its host institute NISTEP
Malta, Cyprus and Estonia	"Light" expert evaluation of the eForesee project, examining the achievements of an EU-funded project that linked the foresight activities of these three small countries
Netherlands (OCV)	Self-evaluation, PhD study, Masters thesis, evaluation by Advisory Council for Science & Technology (AWT)
Sweden	Process (and not the impacts) evaluated continuously by an Evaluation Committee. New evaluation in 2005
United Kingdom	For the first cycle, sub-critical *ad hoc* studies; some limited external (and independent) scrutiny, e.g. by Parliament, a PhD study, etc. OSI conducted a self-evaluation of the second cycle. External evaluation conducted of the third cycle.

Table 16.2 UK evaluations relating to foresight

Year	Evaluation Activity	Implemented	Impact
1995	OST/PREST survey of panellists	Yes	●
1995	OST draft comprehensive evaluation proposals that remain unfulfilled	No	●
1995–1999	OST sponsors PhD studentship at PREST on evaluation of programme (Keenan, 2000)	Yes	●
1996	Panels asked to draft performance indicators	No	●●
1996–1998	Research Councils and other Government Departments asked to account for foresight implementation	Yes	●●●
1997	Royal Academy of Engineering case study and survey work (RAE, 1998)	Yes	●
1997–2000	Academic evaluations at York (Brown *et al.*, 2001) and Brunel (Henkel, 2000) universities	Yes	●
1997	Parliamentary Office of Science and Technology produces review of foresight and its impacts (POST, 1997)	Yes	●●●●
1997	OST consultation about Lessons from First Round published as Second Round Blueprint (OST, 1999)	Yes	●●●●
1998	Consultants Segal Quince Wickstead contracted to develop impact indicators	No	●
2000	PREST/WiseGuys/SUPRA contracted to develop evaluation framework for 2nd Foresight cycle	No	●
2001	Chief Scientist's Review of 2nd cycle (OST, 2001)	Yes	●●●●
2005	Full evaluation of the 3rd cycle carried out (PREST, 2005)	Yes	●●●

Notes: Little/no impact [●], some impact [●●], significant impact [●●●], major impact [●●●●]

Table 16.2 illustrates the main efforts in foresight evaluation in the UK. It illustrates that without a consistent, credible central approach to evaluation, the likely result is a proliferation of activity. Much of the work listed in Table 16.2 was at a sub-critical level, or else relied very heavily on anecdotal and potentially prejudiced evidence. The Table indicates whether the evaluation got past the stage of planning the methodology and also gives the authors' rating of the impact on the Programme. It may be seen that the operating ministry for foresight, the Office of Science and Technology (OST) was the main driver of activity but, despite commissioning a number of methodological studies and some fieldwork, it has taken more than a decade for a full independent evaluation to be commissioned (see PREST, 2005). Impact does not necessarily indicate the quality of the evaluation, only whether its conclusions were implemented. Other interested bodies such as the Parliamentary Office of Science and Technology (POST) were able to put forward more critical and insightful views but lacked the resources to follow up in terms of extensive collection of evidence (POST, 1997). The POST report was, nevertheless, unusual in having an impact, despite not being "owned" by OST, partly because of the high status of its Parliamentary home and partly because it was a well-researched and balanced report. A parallel theme was growing government enthusiasm for performance indicators in all aspects of public sector activity. Notwithstanding the comments made above about the limitations of output indicators, the constant pressure was to capture the effects of foresight in terms of key indicators.

During the second cycle the authors were asked to develop an indicator-driven evaluation framework. A particular difficulty, as noted earlier, with this accountability-style of approach is that foresight depends heavily upon the unpaid involvement of panellists and other contributors who do not take kindly to being monitored. A "softer" evaluation approach was therefore suggested, which relied upon participants to collect and analyse a significant part of the data, while other items would be compiled centrally. The organising principle was to separate process from impact and in the latter case to identify the five main stakeholder groups: the science base, industry and commerce, the voluntary sector, government, and education, training and public understanding of science. From the framework a set of key indicators was derived that attempted as far as possible to cover the full range of outcomes and outputs. Some of these indicators are outlined in Table 16.3. In this case impact was curtailed by the early termination of that phase of the programme. The stress on accountability may also be seen in the review of the UK's second cycle, which superseded the performance indicators framework. The context was a growing realisation that the Programme was not on a trajectory likely to offer insights or impacts comparable with its first cycle predecessor. The Government Chief Scientific Adviser was the senior official ultimately responsible for the programme. Recognising the situation, he instituted a high-level review. This was an internally conducted evaluation based upon soliciting views from stakeholders but without any attempt to codify a systematic approach or to present detailed evidence. However, the conclusions were powerful and resulted in the major change of direction described in Chapter 4.

Table 16.3   Suggested indicators for the second UK Foresight Programme

Item measured	Indicator
Level of awareness of foresight and foresight culture in the industry	SME survey questions
Commitment of participants	Consistency of attendance at Panel and Task Force meetings
Consultation exercise	Number and quality of responses received to consultation documents
Cross-Panel communication	Documented contacts and joint activities. Cross-references in reports to issues from other Panels
Influence on Government Departments spend	Additional resources committed to foresight activities
Influence on Government Departments coordination	Frequency of Foresight on Agenda of Ministerial Science Group
Influence on Science Base spend	Proportion of new programmes and initiatives which are clearly aligned with foresight priorities
Influence on the formation of new industry–science networks	Persistence of groups founded by Panels or Task Forces
Contribution to quality of life goals	Engagement of voluntary sector in Foresight activities. Take-up of recommendations by regulatory or standards-setting bodies in areas such as environment protection, health and safety, etc.
Regional engagement	References to Foresight in Regional Innovation Number and extent of regional foresight groups

It was not until the Third Cycle of UK foresight that a formal evaluation was commissioned (PREST, 2005). The terms of reference of this study were to provide answers to questions on the impact of the programme and the projects; the cost effectiveness of the programme; and the way in which foresight is run. On this basis the evaluators were to deliver a report with findings and recommendations on the above which provided a sound basis for the programme management and/or senior stakeholders to understand the costs and benefits of the programme and to identify any relevant lessons or improvements.

The principal instrument of the evaluation was a series of semi-structured interviews with eight members of the Foresight team and 28 key stakeholders up to the level of Minister. Interview guides were derived from a logic chart (Figure 16.1) and the evaluation criteria. A logic chart approach was used to structure the evaluation. This links higher and Programme objectives to the actions taken and the effects. Effects are spread over time, immediate being those manifested while the programme/project is under way, intermediate those which are evident at the end of the programme/project (sometimes called outputs) and ultimate those which are manifested some time after completion (sometimes called outcomes). Logic charts have three roles in evaluations (Kellog Foundation, 2004; Yampolskaya *et al.*, 2004):

1. To set out objectives, actions and impacts in a hierarchical structure as a guide to the effects that can reasonably be expected;
2. To check the logical consistency of a programme – that is to say whether the actions can in any circumstances achieve the desired effects; and
3. To form an agreed basis for evaluation between evaluators and those commissioning the evaluation or those being evaluated.

In this case, several observations followed:

1. The Programme objectives were consistent with the overall objectives of the parent ministry;
2. Programme activities fall into three major groups – project selection, analysis and foresight, and consulting with stakeholders;
3. Project selection itself creates effects;
4. Key outputs are state of the art reviews, action networks, visions, consequential actions, and innovation in forms of engagement;
5. Expected areas of influence are in the science base, business and government policy;
6. Action networks could be expected to continue further project-related activities.

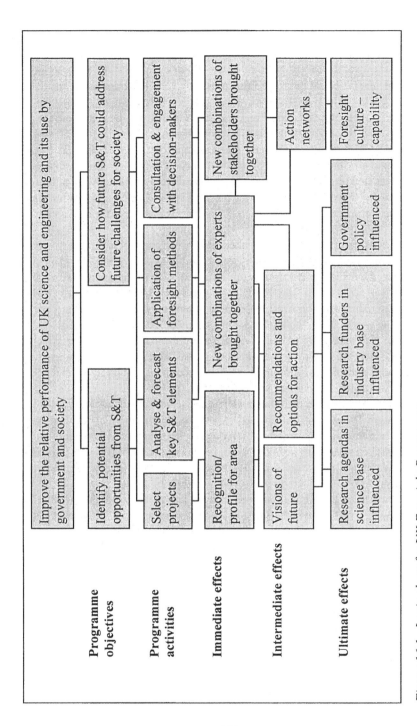

Programme
objectives

Programme
activities

Immediate effects

Intermediate effects

Ultimate effects

Figure 16.1   Logic chart for UK Foresight Programme

389

The logic chart is based upon the intended actions as far as they could be derived from documents. For foresight, as for many programmes, the outcome shows additional degrees of richness. For example at the programme objectives level it could be argued that "providing challenging visions of the future to ensure effective strategies now" has *ex post* become an objective. Activities include the full range of what is needed to implement a project, from scoping, to appointing experts, commissioning studies, organising workshops, and communicating findings to stakeholders.

Intermediate impacts *ex post* include stakeholder/sponsor ownership of action plans and ultimate impacts could include developing a knowledge base which can be drawn upon when future unanticipated needs arise (what we shall later term the "reservoir model"). The evaluation therefore sought evidence both for targeted and for unforeseen effects.

Key areas explored during the interviews with stakeholders were firstly the pre-foresight conditions to establish a baseline against which the project took place. Immediate impacts were investigated – identifying perceptions of key outputs, effects on the knowledge base and on communities and networking. This was followed by an assessment of ultimate impacts – the programme's effects on policy or strategy, whether there were alternative sources of information other than foresight to inform these, the broader effect on use of scientific evidence and what follow-ons were taking place. Finally, management and process effectiveness were examined, including the value-for-money obtained for public expenditure.

Additional evaluation activities undertaken included a Forum with most of the Foresight Directorate and invited experts in which emerging evaluation findings and hypotheses were explored in a structured environment. An international comparison was undertaken, though the value of this was limited by the distinctive nature of the UK programme, which meant that direct comparison with similar activities elsewhere was problematic.

The evaluation concluded that the Programme had achieved its objectives of identifying ways in which future science and technology could address future challenges for society and identifying potential opportunities. It was regarded as a neutral interdisciplinary space in which forward thinking on science-based issues can take place. The prior engagement of senior stakeholders (with a Minister or similarly senior person as sponsor) was very important. A critical phase was identified when responsibility for follow-up is handed over to stakeholders. The evaluation therefore recommended more resource for "aftercare".

It was concluded that some impacts take time to materialise and only become apparent when the policy agenda catches up with the content. This we termed the "reservoir model" – also seen in the evaluation of the Hungarian TEP programme described below. There was a risk of unintended policy effects, for example implicit prioritisation for the areas selected for foresight projects. Such unintended effects are harder to detect but in some cases are more profound (Cassingena Harper and Georghiou, 2005).

Case 2: Evaluation of German FUTUR Initiative

The German FUTUR initiative had a much more specific objective than its UK counterpart in that it sought to introduce fresh ideas into the research funding priorities of one ministry, BMBF, by adding to the traditional mechanisms of agenda setting and prioritisation a third generation approach involving broader stakeholder groups.

The recent evaluation of FUTUR (Cuhls and Georghiou, 2004) was commissioned by the responsible ministry, BMBF, and was largely a process evaluation, focusing upon:

- The objectives of FUTUR, which are assumed to summarise the central assumptions upon which the exercise is based;
- The different instruments and methods with regard to their effectiveness, efficiency and interplay: and
- The process in general.

In terms of the issues described above, this was an evaluation motivated by justification and learning with the results used both as an input to the decision to continue the initiative into a new phase and in terms of making some adjustments to the process. The limitations of this exercise were too little time and resources available and, in terms of timing, the fact that the exercise was conducted too early to pick up outcomes. However, several process-related recommendations were made and an impetus was gained for the continuation and improvement of the activity. This was an example of foresight being used to bring new actors into the strategic debate and hence one measure of success could be that of participation. On the other hand the quality of that participation and the eventual influence it had on policy outcomes was also an evaluation issue. In this regard a key finding was that the participants felt disconnected from the implementation process and to a lesser extent the programme managers responsible for implementation lacked a sense of ownership of FUTUR, responding more readily to traditional policy inputs and influences. Hence the emphasis upon understanding the implementation environment and its interaction with foresight processes was highly relevant in this case.[2]

The terms of reference of the FUTUR evaluation were presented as six key questions:

1. Were the pursued objectives reasonable?
2. Were the pursued objectives reached?
3. Were the individual steps taken in the process appropriate to accomplish the goals?
4. Is the process in general appropriate to achieve the goals?
5. What can be improved to optimise the achievement of the objectives?
6. Were the applied instruments effective and efficient with regard to the objectives?

To answer these questions the evaluation was tasked to cover four segments:

1. The objectives of FUTUR (meta targets) which summarise the central assumptions the process is based on;
2. The different instruments and methods with regard to their effectiveness, efficiency and their interplay with each other, including the basic process principles, the methods of selecting participants and topics as well as the implemented activities;
3. The process in general; and
4. The results in form of the comparison of the topics developed in the lead visions with the big research programmes already being executed by the BMBF.

In this evaluation, the information was collected *ex post*. It was difficult to identify definitive documentation as the there were many versions of concepts kept on file. A more timely collection of information would have been helpful but was impossible due to the process procedure and its timing.

Before going into further detail, some limitations of the evaluation have to be mentioned. Because of the early point in time the first evaluation took place (which was necessary in order to learn for the next phase of FUTUR), no questions concerning the outcomes of the implementation of the process results could be answered. The limited resources for evaluation, temporal and financial, also have to be taken into account.

Organisation of the evaluation of FUTUR
An international expert panel was nominated by the BMBF. This panel consisted mainly of international foresight experts who in most cases also had considerable experience in evaluation. In addition, two members were drawn from the German industrial and scientific constituencies. Thus the framework approach for the panel was "modified peer review" (Gibbons and Georghiou, 1986) and provided a further example of the extension of the peer review into

a new domain (Guston, 2003). It may be remarked that the peer element of the panel, that is the group of other foresight practitioners, was exclusively foreign, while the stakeholder element was exclusively German and did not extend to representation from the broader target groups engaged in FUTUR. However, the panel was entirely independent from the sponsors. This issue will be further discussed in the conclusions.

The panel was assisted by a scientific secretariat from the Fraunhofer Institute for Systems and Innovation Research (ISI), a team which provided a bridge between the panel and the consortium. Potentially, there could have been a conflict of interest but the panel was satisfied that sufficient "Chinese walls" were in place to ensure independence, including the additional presence of the senior member of staff who had no involvement in FUTUR.

The secretariat designed a methodology for the evaluation (see Figure 16.2) to answer the questions posed above. The concept was based upon examination of the central assumptions (meta-targets) of FUTUR, the instruments and methods used, the process in general and the results achieved. These areas were expressed as 25 hypotheses which formed the basis for four questionnaire surveys to actors in the process: participants, BMBF staff, facilitators and coordinators. The core of the information gathered resulted from an electronic (e-mail) survey asking participants their views after the last events had finished, and a survey of other actors involved in the process (conducted by ISI). The questionnaire for this survey was prepared by some interviews and pre-tested. A separate small survey with similar but additional questions was directed at the consortium including especially the mediators and so-called "subject advisors" (Themenpaten) in the different working groups. In both cases, short in-depth interviews followed to verify the results, and to better understand and check ambiguous answers. Other material from the process (minutes, especially "observers" during the different events, protocols, the database with addresses, letters and others) was analysed and summarised to give additional information.

Following an initial informal briefing of the Panel in 2003, members were provided with a package of documents which contained a description of FUTUR, the survey results, the samples of FUTUR, the hypotheses and questions to be asked including some answers from the surveys, and other information relevant for an assessment of FUTUR in 2002.

The panel met once formally to hear evidence and cross-examine individually the member organisations of the consortium that managed FUTUR and the sponsoring department from BMBF. Following the meeting, the Panel felt that it was lacking a fully independent user-perspective from the Ministry itself. To address this gap, the Panel Chair was invited to interview the State Secretary of BMBF.

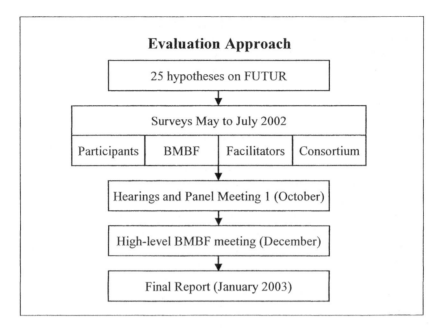

Figure 16.2   The general evaluation approach

The time frame of the evaluation was short but had the advantage that a first assessment (based on the surveys) was available when the lead visions were ready for handing over and the official report of the panel, which was scheduled later, could be used for improvements and for the follow-up and optimisation of the FUTUR process. A presentation of the intermediate results of the FUTUR evaluation was made at an international workshop in Berlin (Georghiou, 2003; Banthien *et al.*, 2003).

The official evaluation report was handed over to BMBF Secretary of State at the end of May 2003 at a closed seminar for senior Ministry officials. The text of the report has not been officially published but was taken into account when the decision about the continuation of FUTUR was made and when designing phase 2 of FUTUR which started in March 2003. One might question how an unpublished evaluation could be reconciled with the aim of the programme to get maximum engagement from its stakeholders. In reality this decision reflected a politically precarious position for the programme in which misinterpretation by its critics of what was actually a positive evaluation could have had potentially terminal effects. The second evaluation was carried out in a climate less favourable to FUTUR owing to political and personnel changes and it may be seen as having contributed to the demise of the programme in its then incarnation.

Case 3: Evaluation of TEP (Hungarian TF Programme)

The Hungarian Technology Foresight Programme (TEP) was a holistic exercise with a structure based on sectoral and thematic panels and the use of a Delphi survey, macro-scenarios (an innovative feature in this context) and workshops. The original rationale for the programme was firmly set in the systems of innovation perspective. According to the programme manager the problem faced was a highly fragmented innovation system during a time of fundamental economic and social change (Havas, 2003). The Programme aimed to bring together business, the science base and government to identify and respond to emerging opportunities in markets and technologies. The evaluation (PREST, 2004) was commissioned by the successor body to the ministry that had sponsored the programme (the Research and Development Division of the Ministry of Education) three years after completion of the main phase of the TEP. Its objectives were:

- To answer the question as to what extent TEP has achieved its objectives; and
- To help orient decision-making on future foresight activities in Hungary.

In so doing it was to address:

- Appropriateness of the original objectives;
- Whether the original objectives were adequately formulated;
- Organisational structure and performance of the management of the exercise;
- The extent to which direct and indirect outputs have been achieved;
- Justification of the exercise in terms of value for money;
- Barriers to implementation of the recommendations;
- Recommended directions for the organisation of future foresight activities in Hungary.

The approach adopted was again a light one, with an international panel designing a questionnaire for stakeholders (two thirds of whom were panellists or members of the overall Programme Steering Group, with the rest being experts or government officials) and hearings during which 22 participants and users were interviewed by the panel.

Again this could be seen as a mixture of learning and justification, though another motivation was to revive interest in foresight after a political hiatus and in the formative element of the evaluation to assist in the design and adoption of a new type of foresight activity. The later timing was very beneficial as many of the effects had taken much longer than expected to materialise and were recent in terms of the evaluation. TEP provides a good example of the importance of policy context, as the initial failure to have a significant impact resulted mainly from a lack of a clear client base – the sponsor organisation OMFB was not at that time well-connected to propagate results across government and the situation was worsened by political change.

In the event, the evaluation found that that the most important effects were mainly in the area of cultural change – in establishing longer-term perspectives and in introducing greater interdisciplinarity. This was seen as a welcome introduction of longer-term thinking during a period when the country was dominated by a short-term agenda (partly because of economic difficulties but also as a reaction against long-term planning of a previous era). Questionnaire respondents were quite negative about the effects achieved in terms of the original objectives, particularly in influencing the research directions of industry or the public sector. As noted above, the panel found that effects on public policy had been substantial but had materialised through a slow and non-linear process.

CONCLUSIONS

Examining foresight against the type of framework presented by evaluation raises several issues. If we consider the range of situations in which foresight is applied, ranging from priority-setting in technology strategy, through network-building and participative approaches to broader restructuring of the science system it is evident that there is no "one-size-fits-all" evaluation approach. Furthermore, the evaluation approach is affected by a series of other factors, including the motivation for evaluation, its timing in relation to foresight activity and the unit of aggregation at which foresight is approached.

The conceptual basis of evaluation helps to place some order. A systematic understanding of the rationale behind a particular foresight intervention can in turn lead to an evaluation framework. However, the systems rationale also leads us to the realisation that foresight cannot be evaluated independently of its context. This falls out at two levels: the need to understand the relative signal strength of foresight compared with other influences in determining the attribution of impacts, and the interaction of foresight with the strategies of the organisations it seeks to affect. Evaluation has to steer a difficult course between under- and over-attribution.

The three national case studies discussed illustrate the variety of evaluation approaches but also indicate the immaturity of evaluation in this context. Process issues and early impacts are generally well taken care of by what is effectively peer review but much remains to be done in understanding and measuring impacts, particularly in the longer term. As many of the countries discussed here move towards the fourth generation "distributed" model of national foresight, some new evaluation issues appear. On the one hand the political pressures surrounding a national initiative of high prominence are less likely to be present, which in principle should make evaluation easier to carry out. On the other hand, there is an increased complexity as it is likely that policy-makers will want to understand the linkages between different future-oriented exercises and their sources of information and expertise.

One cross-cutting theme emerging from the findings is that evaluation is apparently much simpler when there is a clear "client" for the outputs of foresight but, paradoxically, the more the relationship is one of client rather than partner, the less space there is for broader process-related effects to be manifested. Rather, participants will see themselves as instruments serving the end goal, not as active beneficiaries.

Another trend has been the emergence of several exercises in Europe which aim to systematise the exchange of knowledge about foresight and its effects, either through information platforms or through explicit approaches to benchmarking (see Chapter 14). However, attaching credibility and validity to such activities places rigorous evaluation as a prerequisite. Foresight can only be strengthened by the emergence of a more systematic and rigorous body of knowledge to assist learning and improvement. Furthermore, as with foresight itself, evaluation can offer a forum for the exchange of knowledge that allows some of the more tacit but essential lessons to diffuse through the community.

NOTES

1. This chapter is partly based on Georghiou and Keenan (2006), Copyright 2006.
2. Issues of impact were pursued in more detail in a second evaluation of Futur, following broadly the same methodology with a reconstituted panel that met in 2004 and reported in 2005.

REFERENCES

Banthien, H., Cuhls, K. and Ludewig N. (2003), 'Introduction of Futur – The German Research Dialogue, About the Futur Process', in Cuhls, K. and Jaspers, M. (eds), *Participatory Priority-Setting for Research and Innovation Policy – Concepts, Tools and Implementation in Foresight Processes*, Proceedings of the international expert workshop in Berlin, 13 and 14 December 2002, Karlsruhe/Berlin/Bensheim.

Brown, N., Rappert, B., Webster, A., Cabello, C., Sanz-Menendez, L., Merkx, F. and van der Meulen, B. (2001), *Foresight as a Tool for the Management of Knowledge Flows and Innovation*, Final report of the FORMAKIN project, University of York

Buisseret, T.J., Cameron, H.M. and Georghiou, L. (1995) 'What Difference Does it Make? – Additionality in the Public Support of R&D in Large Firms', *International Journal of Technology Management*, **10**(4–6), pp. 587–600.

Cassingena Harper, J. (ed.) (2003), *eFORESEE Malta ICT and Knowledge Futures Pilot*, report to the European Commission (January).

Cassingena Harper, J. and Georghiou, L. (2005), 'The Targeted and Unforeseen Impacts of Foresight on Innovation Policy: The eFORESEE Malta Case Study', *International Journal of Foresight and Innovation Policy*, **2**(1).

Cuhls, K. and Georghiou, L. (2004), 'Evaluating a Participative Foresight Process: Futur – The German Research Dialogue', *Research Evaluation*, **13**(3), December, pp. 143–153.

Durand, T. (2003), 'Twelve Lessons from Key Technologies 2005: the French Technology Foresight Exercise', *Journal of Forecasting*. **22**, pp. 161–177.

Georghiou, L. (2001), 'Third Generation Foresight – Integrating the Socio-economic Dimension', in *Proceedings of the International Conference on Technology Foresight – the approach to and potential for New Technology Foresight*, NISTEP Research Material 77, March.

Georghiou, L. (2002), 'Impact and Additionality of Innovation Policy', in Boekholt, P. (ed.), *Innovation Policy and Sustainable Development: Can Innovation Incentives make a Difference*, Brussels: IWT-Observatory 2002, pp. 7–22.

Georghiou, L. (2003), 'Evaluation of Futur – Intermediate Results', in Cuhls, K. and Jaspers, M. (eds), *Participatory Priority-Setting for Research and Innovation Policy – Concepts, Tools and Implementation in Foresight Processes*, Proceedings of the international expert workshop in Berlin, 13–14 December, Karlsruhe/Berlin/Bensheim.

Georghiou, L. and Keenan, M. (2006), 'Evaluation of National Foresight Activities: Assessing Rationale, Process and Impact', *Technological Forecasting and Social Change*, **73**(7), 1 September, pp. 761–777.

Gibbons M. and Georghiou L. (1986), *Evaluation of Research – A Selection of Current Practices*, Paris: OECD.

Guston, D.H. (2003), 'The Expanding Role of Peer Review Processes in the United States', in Shapira, P. and Kuhlmann, S. (eds), *Learning from Science and Technology Policy Evaluation: Experiences from the United States and Europe*, Cheltenham and Northampton MA: Edward Elgar.

Havas, A. (2003), 'Evolving Foresight in a Small Transition Economy', *Journal of Forcasting*, **22**, pp. 179–201

Henkel, M. (2000), 'Academic Responses to the UK Foresight Programme', *Higher Education Management*, **12**(1), pp. 67–84.

Keenan, M. (2000), *An Evaluation of the Implementation of the UK Technology Foresight Programme*, Doctoral Thesis, Manchester: The University of Manchester.

Kellog Foundation (2004), *Logic Model Development Guide*, W.K.Kellog Foundation, January, http://www.wkkf.org/Pubs/Tools/Evaluation/Pub3669.pdf.

Kuhlmann, S., Boekholt, P., Georghiou, L., Guy, K., Héraud. J.-A., Laredo, P., Lemola, T., Loveridge, D., Luukkonen, T., Polt, W., Rip, A., Sanz-Menendez, L. and Smits, R. (1999), *Improving Distributed Intelligence in Complex Innovation*

Systems, Final Report of the Advanced Science and Technology Policy Planning Network, Frauenhofer Institute, Karlsruhe: Systems and Innovation Research.

Kuwahara T. (1999), 'Technology Forecasting Activities in Japan', *Technological Forecasting and Social Change*, **60**, p. 12.

Martin B.R. and Johnston R. (1999), 'Technology Foresight for Wiring Up the National Innovation System – Experiences in Britain, Australia, and New Zealand', *Technological Forecasting and Social Change*, **60**(1), 2 January, pp. 37–54.

NISTEP (1992), *2nd Policy-oriented Research group, 5th Technology Forecast Survey – Future Technology in Japan*, NISTEP Report No. 25, Tokyo: NISTEP.

OST (1999), *Blueprint for the Next Round of Foresight*, available at http://www.foresight.gov.uk/previous_rounds/foresight_1994__1999/, last accessed 19 August 2005.

OST (2001), *Foresight Review – A summary report*, Department of Trade and Industry.

POST (United Kingdom Parliamentary Office of Science and Technology) (1997), *Science Shaping the Future: Technology Foresight and its Impacts*, London: POST.

PREST (2004), *Evaluation of the Hungarian Foresight Programme*, Report of an International Panel, May 2004, Hungarian National Office for Research and Technology, http://www.nkth.gov.hu/main.php?folderID=159, last accessed 02 August 2007.

PREST (2005), *Evaluation of the United Kingdom Foresight Programme*, Office of Science and Technology, http://www.foresight.gov.uk.

Royal Academy of Engineering (1998), *A Preliminary Assessment of the Impact of Foresight*, London: RAE.

Scriven, M. (1991), *Evaluation Thesaurus*, (4th ed.) Beverly Hills, CA: Sage.

Webster, A. (1999), 'Technologies in Transition, Policies in Transition: Foresight in the Risk Society', *Technovation*, 19, pp. 413–421.

Yampolskaya, S., Nesman, T., Hernandez, M. and Koch, D. (2004), 'Using Concept Mapping to Develop a Logic Model and Articulate a Program Theory: a Case Example', *American Journal of Evaluation*, **25**(2), pp. 191–207.

17. New Frontiers: Emerging Foresight

Ian Miles, Jennifer Cassingena Harper, Luke Georghiou, Michael Keenan and Rafael Popper

INTRODUCTION

The many faces of foresight that we have seen in these chapters have reflected an evolving and adaptive phenomenon that is recognisable to most who are concerned with strategy and policy for technology and associated social change. In the five countries and five regional groupings we have focused upon a range of experience emerges which is characterised both by its variety and by the common thread that we were searching for at the start of this volume. This is not a thread constrained by definitions but rather by a shared set of problems, approaches and values and a social network that shades into other communities – policy-making, futures, business, research, and so on, but which has its own identity nonetheless. In some chapters we have sought to codify the explicit knowledge involved in designing and implementing foresight but only through the totality can we access the tacit element which forms the basis of this community of practice. Three core elements of the meaning of foresight are: its position in the policy- or strategy-making process, the future of foresight methods, and the community of actors who perform and use foresight. In concluding we discuss and consider the future of these elements, and draw out some tentative principles of good practice.

POSITIONING FORESIGHT

Despite our limited focus upon technology foresight, it is apparent that such activities are in fact conducted across a wide range of locations and at different levels, including various sites at the national level (e.g. ministries, research councils, and the like), in sub-national and supra-national regions, and in organisations (e.g. national laboratories, large companies, etc.). At the national level (the main focus of this book), foresight has moved well beyond the boundaries of traditional S&T actors in many countries, and is now regularly carried out by a variety of ministries and agencies across several domains of government. This is most apparent in the chapters on France and the Nordic countries, though it can also be seen in most other Western countries. The degree of connectedness between sites and levels of activity is minimal, however, with foresight landscapes typically described as "fragmented" with little collaboration between different foresight exercises. This is hardly unexpected while foresight remains a largely punctual, as opposed to a continuous, activity. Under these circumstances, cooperation is likely to be rare and opportunistic, with linkages mostly confined to some recycling of foresight products and to a few instances of methodological learning. By contrast, continuous activities would offer the time and stability for more profound cooperation to develop.

There remain differences as to where in an organisation or "system" foresight is "located", i.e. coordinated and managed, with little discernible pattern according to country/region or foresight rationale. There are many options available here, including in-house, semi-detached, and outsourced configurations. While the pros and cons of these different options are sometimes discussed in the literature, mostly around notions of autonomy and connectivity (with an apparent trade-off between the two), the country/region chapters tend to pay little consideration to this tension. For sure, foresight is seen as providing a "space" for the sorts of discourse, analysis and creative visioning that are normally absent in day-to-day operations, or even in more long-term strategic planning. This needs to be a "safe" space, however, if foresight is to be open and adventurous, where the unthinkable can be thought about and discussions are not wholly dominated by current controversies. While this creates a natural need for some disconnection from the rough and tumble of day-to-day policy and decision-making, the challenge has always lain in reconnecting with policy arenas. This has been achieved usually through participation of major stakeholders in the foresight process itself, reflecting an increasingly common belief that foresight is more likely to impact on policy through the agenda-setting and mobilisation of actors – rather than through the provision of some new and enlightened codified facts.

There are a growing number of instances where foresight is embedded in existing strategic processes, linking it ever closer to policy and decision-making, and making it (perhaps) more difficult to discern as a distinct and stand-alone activity. Some would argue that such foresight runs a greater risk of being compromised through its embeddedness. This is probably true, but it would be unrealistic to expect all foresight activities to conform to a particular "gold standard" or to specific organisational forms.

Experimentation will no doubt continue, and we are likely to see foresight being used in a wider variety of settings and in combinations with other decision-support tools and policy instruments.

In fact, in some academic STI policy writing, foresight is increasingly viewed as one instrument in a distributed, strategic policy intelligence "toolbox" (sometimes abbreviated to SPI) that also includes evaluation, technology assessment and various other strategic approaches. Conceptual work on how such tools might be combined in such a way as to provide policy makers with readily available "strategic intelligence" has been funded by the European Commission, e.g. the Advanced Science and Technology Policy Planning (ASTPP) network (see Kuhlmann *et al.*, 1999) and the more recent RegStrat project.[1] It is worth noting that some of this thinking sees foresight as a practice that can be combined with other SPI tools and, in fact, there is probably considerable untapped potential in embedding foresight into practices such as evaluation, although there remains little evidence of many multi-tool approaches being developed for use in policy-making at the current time. Another view around SPI – and perhaps one that is more widely held – sees foresight as a pool of knowledge in a wider, distributed landscape of knowledge production. In other words, there is greater emphasis on the concrete outputs (products) of foresight than on the practice (processes). From this perspective, the knowledge generated by foresight exercises is combined with knowledge from other SPI tools to inform policy-making processes. Undoubtedly, this could be beneficial, but it also assumes a rather narrow instrumental view of the uses of foresight, limited to foresight as information input into quasi-rational decision-making processes.

This brings us to the topic of use of foresight and its impacts. Many of the case study chapters are a little coy on the subject of impacts, either not mentioning them at all, or else claiming that it is still too early to say what the effects have been. Some of the challenges facing evaluation of foresight were highlighted already in Chapter 16, and it is clear that accounting for the uses and impacts of foresight will never be an easy task. But an expanded and more sophisticated view of foresight's uses could be emerging – and there is evidence from the chapters that points to a wider repertoire of acknowledged uses that reflect the rationales set out in Chapter 1. Thus, we see foresight being presented as a tool for positioning actors, for agenda-setting and problem-framing, for managing relations at the science–society interface, and so on. This is a far cry from the original, and perhaps rather simplistic, rationales of priority-setting and networking seen in earlier foresight exercises.

These shifting rationales reflect the generations of foresight presented earlier in Chapter 1. Simply put, the early priority-setting rationales were the product of fiscal crises within states, as well as the need to manage the ever-growing scientific estate. It quickly became apparent, however, that many of the issues around science and technology were connected to an innovation deficit – particularly in Europe – and that firms needed to conduct more R&D or at least be better connected to centres of techno-science knowledge production. Foresight therefore assumed a more networking and community-building function, particularly during the 1990s, and sought to serve a variety of innovation system actors beyond a sole public funding agency. A greater emphasis upon the relations of S&T with society also began to emerge later in the 1990s, with many governments (and the European Commission) establishing or strengthening their policies and capabilities in this area. Again, in many places, foresight adapted to this new emphasis, particularly in Germany, UK and Japan (the Nordic countries already had a strong tradition in this area, which shaped their foresight activities somewhat earlier). More recently, an increasingly fragmented landscape is emerging, with foresight activities carried out across an extensive, multi-scalar and multi-level space – and with a variety of rationales.

But the overall picture is a mixed one. The evolution of rationales described here (and their associated stylised generations) is largely confined to those countries where foresight has been practiced for some time, particularly in Western Europe. The picture is somewhat different elsewhere, for example, in industrialising countries and the transition countries of Central and Eastern Europe. In the latter case, for instance, foresight continues to be (almost) solely viewed as an approach for "picking winners" through processes of S&T priority-setting. Programme managers might sometimes extol the process benefits associated with foresight, but these are rarely, if ever, appreciated by S&T policy-makers, who are normally the funders of national technology foresight. It could be that there is a learning curve – as suggested in Chapters 11 and 12 – characterised by imitation, assimilation and adaptive innovation, which will see an evolution and maturation in rationales. Even so, it would be presumptuous to assume that other parts of the world will follow the same Eurocentric trajectory.

What is apparent from the chapters is that national exercises are tending to become more complex in their scope and design.[2] This might be attributed to learning and a better understanding of foresight dynamics, as well as a growing confidence in the use of some of the methods and frameworks – such as the Foresight Diamond presented in Chapter 3. At the same time, the multiple "layers" of rationales for exercises – since societal dialogue rarely substitutes priority-setting, for example, but is instead an additional objective – often necessitates more complex exercises. There is a danger here, however, of overloading foresight with too many objectives, as highlighted in the German and UK chapters, where rounds of national foresight activity have collapsed under the weight of multiple expectations.

Yet, the needs for foresight, as well as the likely range of applications, are expected to continue to grow. In the field of techno-science alone, there are many newly emergent frontiers opening up that will require an active shaping if future problems are to be managed. These include issues around environmental degradation, energy supply, various forms of human-enhancement, and the convergence of nanoscience, biotechnology, information technology and cognitive science (NBIC), to name but a few. How foresight will be used to address these, and other grand challenges, remains to be seen. But they will need to be addressed and foresight practitioners will need to rise to the challenge.

THE FUTURE OF FORESIGHT METHODS

The future prospects for TF methods will be intimately bound up with the future of foresight itself. One possibility is that the "foresight" brand may become submerged into the more general futures field, with TF programmes and projects being hard to distinguish from many other futures studies – perhaps losing the emphasis on participatory approaches and feeding vision into planning processes. It is also possible that particular lines of work – like technology forecasting or roadmapping – will come to be seen as the gold standard in TF work, which might happen if such approaches receive substantial endorsement from some key players (corporate giants, leading countries, etc.). Chapter 2 stressed that foresight differs from much futures work by combining emphasis on prospective analysis with attention to participative approaches and policy-informing applications. On the assumption that foresight continues to have these three major elements, we identify trajectories of methodology development relevant to each. Chapter 3 explored the range of methods commonly employed in TF: now we seek to consider how methods may evolve and diffuse in the future. Most of the foresight exercises reviewed in earlier chapters are using a fairly limited set of tools, though we can see progress within some of these tools – for example, the online surveys used in Latin American studies (see Chapter 12) and the not altogether successful Knowledge Pool of the second UK programme (Chapter 4).

NEW METHODS

For each of the three elements of TF, the most important developments over the past years involve new IT, and this is likely to remain true into the medium-term future.[3] The key underpinning trends here are:

- Greater access to a growing range of media for interfacing with information systems; and
- Increasing power of information-processing and hyper-realism of data presentation (visualisation and more user-friendly tools for content development and programming – together with growing familiarity with such systems, their potentials and limitations).

The continuation of these trends into the very long term has been debated, mainly in terms of how long Moore's Law and its numerous analogues in related areas can be sustained. But even if the pace of IT-related change were to slow down, there is still considerable scope for further adopting and exploiting available technologies. Of greater long-term significance are the issues raised by the debate among futurists as to whether we can expect a "singularity" in the foreseeable future, when machine intelligence reaches levels such that innovation and even scientific and technological understanding begin to move out of human grasp.[4] Already, of course, high-tech methods may be hard for many non-experts to comprehend. Apart from the scope for utilising new IT in improving methods used in foresight, there is also scope for importing (often with modifications) methods being developed in parallel fields, such as social and environmental impact analysis, ecosystem modelling,[5] participatory planning and deliberative democracy systems, management information and knowledge management systems, risk management, and so on.

Given the many possible avenues of methodological development, it is inevitable that the discussion below will be rather partial, but hopefully it will provide a first glimpse into the main trajectories. Further analysis could involve reviewing material on new methods and approaches in such journals as *Foresight, Futures, International Journal of Foresight and Innovation Policy, International Journal of Forecasting, Technological Forecasting and Social Change* – and also in the many journals on strategic planning, knowledge management, and related topics.

Prospective Methods

Some key methodological developments involving prospective (forecasting and anticipatory) analysis are likely to include:

- Use of new IT in scanning for technological developments and emerging areas of science, and exploring emerging markets and social issues associated with these. We would expect to see considerable development of web-based scanning methods: for example, the use of intelligent software agents to "crawl" the Internet and identify emerging and hot topics; the application of data and text mining tools to analyse individual documents and bodies of literature and commentary; statistical aids for network analysis and visualisation of network structures (for instance, to identify key ideas and protagonists); co-word and co-citation methods (for instance, to provide guidance as to new clusters of ideas in scientific areas).[6] Such methods could be applied to the determination and analysis of drivers of change, of links between primary and secondary drivers, and of how drivers impact on each other;

- Modelling, where increasing access to powerful computer capabilities has made it much more practicable to apply established methods like econometric analysis and system dynamics. Equally and perhaps more significantly, many new approaches are developing – for instance, agent-based and fuzzy approaches. The development of new quantitative approaches has been heralded by some RAND staff[7] as meaning that "robust decision-making can harness the heretofore unavailable capabilities of modern computers to grapple directly with the inherent difficulty of accurate long-term prediction that has bedeviled previous approaches ..." Computer tools are employed to create and visualise a large set of scenarios, drawing on different assumptions about how the world evolves (how different these assumptions really are is an interesting question) and what alternative strategies might be brought to bear. The claim is that robust policies can be developed and used to deal with the actually emerging future;[8]

- Technology roadmapping is attracting wide interest, and sophisticated computer aids are becoming available for organizing and visualizing the process and its results;[9]

- Improvements in ways of eliciting expert opinion – such as Delphi technique – will also draw upon computers and networking, e.g. web-based questionnaire surveys and more discursive approaches (see the discussion of networking below). With less emphasis on IT, we might also anticipate some return to relatively neglected ways of using Delphi that were demonstrated in early studies using the technique but which have been overshadowed by the emphasis in most studies on prediction of when things will happen. Several studies have demonstrated the scope for applying Delphi to explore the implications of various lines of technology development for social goals, for instance;[10]

- Expert systems to guide practitioners and managers through the process of carrying out foresight – from decision support for managing the foresight process, to aids for application of specific techniques.

Planning and Decision-making Methods

Significant methodological developments may be expected concerning planning and decision-making tools. These include:

- Application of more formal tools – such as MCA (Multicriteria Analysis)[11] and AHP (Analytical Hierarchy process) – for the comparative assessment of options and the definition and selection of priorities.[12] These tools confront several problems, not least being the simplistic rationalism they often involve (experts frequently bridle against the results of, say, multiplying a judgement of probability of an outcome by the value of that outcome; and policy-makers may well feel that they have knowledge about political feasibility and other features of key decisions that has not been encompassed within the analyses). The tools are often cumbersome to implement, with elaborate methodologies that sometimes require users to make large numbers of structured choices. One danger here is that the process of making such decisions may take so much time and energy that debate and interchange is marginalised. The effort to establish consistency and rigour may limit opportunities for learning, clarification of thinking, and the creation of shared insights. IT applications here may allow for some form of distribution of the judgemental effort across larger numbers of experts – freeing up more time for each individual, or for innovations that increase the ease of the tasks and reduce the need for repetitive choices;

- Introduction and elaboration of Management Information and Knowledge Management Systems, to support better sharing of information and knowledge. Such systems can allow for location of expertise and of "libraries" of solutions to specific problems that might be generalised to new problems. The "Knowledge Pool" of UK Foresight's second cycle – see Chapter 4 – was not altogether successful as an early instance of this, but we can expect better experiences to accumulate as Foresight imports and learns from methods and tools developed in other contexts.

Participative and Networking Methods

Significant methodological developments involving the enhancement of participation and networking are reviewed in Slocum (2003). Among these we would highlight:

- "Groupware" for supporting face-to-face interaction in scenario workshops and similar settings, as well as in "virtual meetings" where participants are remote from each other. Again, such methods can be overused, so that the keyboard tends to displace direct interaction with those around you. But accumulation of experience with such methodologies is improving understanding as to how to make them supportive rather than intrusive. The tools are fairly self-explanatory, so that participants are unlikely to be alienated or mystified by them (unlike in the case of some highly complex IT applications in forecasting and planning). Considerable development of such tools can be anticipated for the future;
- Qualitative information is hard to capture and process effectively, and there are often suspicions that verbal inputs are being used in a selective fashion. Thus tools for speech capture, transcription of dialogue, content analysis of debates, and "mindmapping"-type techniques to explicate main lines of argument, are liable to be sought after;
- Finally, we may well see elements of foresight exercises built into "deliberative democracy" activities that are emerging within e-government and e-democracy initiatives.

METHODOLOGY AND SOCIAL DIALOGUE

One problem in the foresight process lies in the tension between human stakeholders (the participative dimension) and technical expertise (employed in prospective studies and some planning tools). The issue is that increasingly sophisticated methodologies of futures analysis and planning may be hard to integrate with more participative activities.

Already in the days of *Limits to Growth* we saw computer modelling used to persuade people that a particular vision of the future was correct "because the computer says so" (see Chapter 2). Many people remain prepared to accept the results of (at least some sorts of) expert analysis – without exploring whether these results might reflect partisan or selective approaches It does not help that it can be difficult to find clear explanations of how many techniques work, or what the implications of technical and more loaded assumptions may be. On the other hand, across the Western world (at least) there is decreasing trust in technical as well as political elites. This is reflected in suspicion that experts can be paid to come up with whatever results their paymasters require. The "new age" and postmodern critiques of technical rationality, together with the claims of fundamentalists of all stripes, surface in many debates. Many influential commentators and leaders use such arguments to deny the validity of data and analysis that does not conform to their local experience – or prejudices. This is apparent in the vociferous arguments against climate change, where indeed there is good reason to be suspicious about the motives of some corporate groups and the experts they fund.

The present discussion is focusing more on methodology (such as the models and use of data in climate change analysis) rather than the palatability of the results they arrive at. So far it is only a minority of those involved in methods such as simulation, systems analysis and decision theory who have been prepared to engage with the issues of public understanding of methodologies. Foresight needs to be steered between the rocks of uncritical acceptance and uninformed rejection of particular methodological approaches. It is thus important to have those responsible for designing and implementing such approaches to undertake much more demystification of the tools and the underpinning assumptions associated with them.[13] This may mean developing more simplified versions of some approaches, allowing for "counterexperts" or devil's advocates to explore the implications of different assumptions (technical or otherwise), and investment in public/stakeholder engagement. These sorts of issues have been taken up over the years in "socially responsible" and/or "critical" science, technology management and operations research movements. Insights and approaches developed there might be valuable. "Constructive technology assessment" is one point at which foresight work overlaps with such approaches. One important topic for methodological development then, will be bridging these approaches.

THE FORESIGHT COMMUNITY

Expertise and Participatory Approaches

We have seen argued several times that participation is a cornerstone of the foresight approach. This puts the spotlight on the actors in foresight. These can be seen broadly as falling into three groups: experts in foresight methods and organisation; experts in the domain(s) addressed by a particular exercise; and the eventual users of the outcomes of the exercise. Within these groups, sub-categories of actors are emerging which include specialists in corporate foresight as opposed to public sector activity, and one can distinguish between local, national, regional and global players. Users nowadays extend from those actively engaged to those who sponsor foresight activity and those who watch and learn from afar. This lends a level of complexity to the term community since there are multiple networks, formal and informal, linking foresight practitioners and would-be foresighters worldwide. Tracing the knowledge flows is no easy task and silent participants (who watch exercises from afar) also play a role unwittingly in disseminating results and widening the impact of an exercise.

The exercises in Part Two present a mixed picture. Domain experts are present in all cases but expertise in foresight methods is often acquired in the context of a first encounter with foresight – the responsible officials taking it upon themselves to engage in more or less systematic learning through training courses and direct contact with other exercises, often in other countries as we saw in Chapter 14. A minority of these self-taught experts move on to become internationally recognised authorities in foresight while others are absorbed into the management functions of their domain. Those who join the foresight community on a permanent basis more often come from the public sector than the private. On this basis we could examine the contributors to this volume. The editors are four academics, all of whom engage regularly in policy advice and even management at different levels, and one policy-maker with a further identity as one of those mentioned above who have gained authority through experience and participation in the community. The contributors present a similar picture: several working in applied research organisations with a national role, a group of academics who have moved in or out of government, and a couple of knowledge-based consultants who have also moved across public–private barriers. The most common thread here is that all are mobile individuals used to working at the kind of interfaces involved in foresight.

Actors from many domains and sectors become engaged with foresight – we saw energy as a recurrent theme in Nordic foresight, professional trainers in Thailand and those concerned with flood management in the UK. A key issue is whether these are existing communities or whether foresight itself is part of the process of building new interaction around a problem area. There is a role in the first case – normally renewal – but the second is much more central to recent rationales for foresight. In both large and small countries we may see these linkages being put in place – the UK's project on flooding brought together expertise on climate change, environmental planning and regulation, construction and even the insurance sector. The eForesee exercise in Cyprus (Chapter 10) explicitly sought to resolve policy conflict between four trade unions in the agricultural sector. The additionality of foresight often lies in the ability to bring new actors into an area of strategy formulation and what is innovative depends very much on context. For one country bringing private industry into the process of deciding national research priorities may be revolutionary, while for another such consultation is the norm. Sometimes actors who feel their voice is not sufficiently heard use foresight to gain an entrance to the debate. The Futuris exercise in France

(Chapter 5) was initiated by an industrial association even though it succeeded in getting government buy-in.

We have already hinted in our discussion of methods that there is an inherent tension between sophistication of approach and the possibility of wider engagement (interaction in the Foresight Diamond – Chapter 3). It is salutary to note that despite the rhetoric of participation, the message from the national chapters is that foresight has succeeded in widening the circle beyond that of futurologists but that the widening has clear limitations. The active stakeholders in science and technology are present, mainly those who would be direct beneficiaries of any research and innovation funding that might follow from foresight recommendations. The science community itself, industry and certain parts of government sit comfortably in what we called the second generation model. In the third generation we do see some presence of social stakeholders (for example, voluntary organisations and consumer associations) but these too are normally members of the decision-making circle through other activities such as lobbying and committee membership. Another group is made up of social research domain experts contributing knowledge on demographic trends, market developments, social change, entrepreneurship and attitudes to innovation. Even where an explicit attempt was made to bring in the public as in the German Futur experience, in reality active stakeholders remained dominant and even the public presence turned out to be what some might see as a proxy group under the heading of "interested public".

In general one might say that the public have been seen, more or less globally, as a passive audience to be addressed by communication strategies and through the media. This in itself is commendable – efforts in Japan to produce cartoon images capturing future visions are an example of the kinds of device being used. The UK's current foresight programme involves a "launch" which seeks maximum press coverage (despite the use of a term implying a beginning, the launch is actually of the finished product). The PREST evaluation of the third cycle saw some irony in a Ministry (the OSI) committed to promoting public engagement with science falling back solely on expert views, particularly when some of the issues addressed interact strongly with public opinion.

It would be quite feasible to argue the opposite view, to say that the public has little to offer beyond a reflection of the immediate issues of the day, with the media playing a strong role in influencing their views. To the extent that the public were involved in Futur, the evaluation of that exercise perceived them as a conservative influence on visions. The counter to this view comes not from foresight but from the related family approaches generally known as technology assessment (TA). Here there is a long history of "confronting" expert and lay opinion with the aim of building in public concerns and priorities at the formative stages of a technology's development and diffusion. In Chapters 2 and 7 the role of the now defunct US Office of Technology Assessment is presented as a foresight-like activity and we have chosen to include some core methods of TA such as citizens' panels or juries as elements available to foresight practitioners. Barré's distinction between analytic and societal models of foresight (Chapter 5) goes to the heart of the matter and in the latter case presents foresight in the tradition of extended TA associated with "dialogic" democracy, but later in the same chapter he raises concerns about falling below the standards of professional practice in social science.

This would take us to the position of treating the general public view as a data item or driver rather than as a means of building commitment (this being difficult given the low numbers in a population likely to be engaged). Such a perspective would definitely be the case in corporate foresight where the aim is to identify trends that could impact upon consumer tastes and attitudes in the market which might also manifest themselves through regulation. This would leave us with public engagement and participation in a double loop mode – feeding in on selected issues where public attitudes are deemed to be relevant and then engaging with the results at the "end-of-pipeline".

Ethics and Quality Control in Foresight

"Responsibility without foresight is blind, but foresight with responsibility is dangerous" (Fuller and Tilley, 2005).

From a superficial reading of the chapters, the reader may easily conclude that foresight has the properties of a malleable fabric, which can conveniently stretch to take on multiple forms and faces to suit different contexts. But is foresight being stretched too thinly at the risk of losing its inherent qualities and features? This is a commonly expressed concern when one enters into a discussion on foresight definitions and its essential features. There are however deeper concerns and issues at stake when confronted with the prospect of the many faces of foresight and its more long-term evolution. Futures tools have often been used in the past to alert the social and human conscience and as the means for raising the alarm on key global concerns associated with fast change in demographics, technological development, climate, and sustainability, thereby emphasising our long-term responsibilities towards future generations. Contrasting images of utopia and dystopia evident in the positive and negative future scenarios developed through foresight exercises help to shape, manipulate and even mobilise different stakeholders to action or inertia. As highlighted by each of these chapters, foresight exercises are motivated by some form of political agenda, whether explicit or implicit. But it is the extent to which they also embrace some form of societal concerns which gives them a human face and somehow renders them more ethical.

Capturing and respecting the multitude of perspectives on future developments in science and technology and their impacts on society places a heavy burden on foresight and its practitioners, sponsors and participants, current and potential. "The research community has a responsibility to analyze different perspectives on the opportunities and dilemmas that flow from the advance of science and technology, and to consider how these points of view might be reconciled within a global context."[14] Foresight, whether it takes this role seriously or not, has important implications in terms of setting long-term trends and orientations in policy, promoting particular styles of decision-making, restricting/opening up the range of alternative futures considered, and respecting the perspectives of the full range of stakeholders. From the design phase of a foresight exercise to implementation, ethics and quality control have a highly significant role to play in shaping the activity and giving it direction. The choice of a particular foresight approach or mix of approaches in a particular context often embrace ethical dilemmas and concerns, while the professional implementation of the approaches and the skilled use of the tools involve equally important issues of quality control. To date the foresight community, despite acknowledging the importance of these issues, has shied away from addressing them directly. Yet it is clear that foresight, as a skill or discipline, would benefit enormously from more discussion on ethical and quality control guidelines.

In an age when broad stakeholder engagement and active citizen involvement have become key aspects of an open, democratic process of governance, foresight exercises often provide a convenient backdrop for exploring and validating government policy. It is critical that laypeople are given appropriate room for manoeuvre in order to express their opinion and that their participation in foresight activity does not work against their own interests. The foresight practitioner has a critical role to play in extending the foresight space and political room for manoeuvre, beyond the sponsors' expectations and/or demands. This task requires skills, tact and experience in handling situations which can make or break a foresight exercise. Stretching the flexibility and patience of the foresight sponsors to accommodate the real interests and needs of the stakeholders and real purpose of the exercise is one of the key hidden challenges of a foresight exercise. Such flexibility and patience is one of the key hidden success factors of a foresight exercise.

Building and Sustaining the Foresight Community

Setting some form of guidelines for the optimum design and implementation of foresight exercises has emerged as an underlying common if (underplayed) unspoken thread in the different chapters. The newly initiated should not feel unduly discouraged at the outset since codifying appropriate foresight practice through the setting of common standards comparable across a range of contexts, whether transnational, transregional, national, regional, local, sectoral, organisational, is the indicated way forward, despite the complication of context specificities and contingency. To ensure that foresight practitioners can meet these standards, the current fragmented approach to foresight training needs to be addressed, through the development of tertiary (secondary and primary) level foresight education and improved professional training programmes. Understanding core foresight skills, together with further interdisciplinary international research and dialogue on the theoretical approaches to foresight, will help to provide the framework for commonly agreed definitions and approaches to foresight education and training and subsequent activity. International and interregional benchmarking and evaluation of foresight activity, together with the publication, transfer and learning from the results, will provide the additional fuel for driving the quality control system forward.

An activity like foresight, which requires careful configuration to its application, is not easily reduced to a formulaic statement of good practice. But we could propose as a starting point the following meta-principles, such that foresight should be:

- *Contextualised*: Foresight needs to be rooted in the context within which it is to be implemented be it national, regional, local, corporate, organisational;
- *Credible*: The robustness of the evidence and the reputation of those presenting and validating it should be such that results are treated as credible;
- *Diversed*: Foresight must keep an ear open to unpopular views and not rush to a consensus; relevant (and seemingly less relevant) stakeholders should be engaged wherever possible, either in the exercise itself or in pre- and post-foresight activities;
- *Systematic*: A foresight exercise should develop and follow a systematic approach which can easily be replicated. Methods should as far as possible allow comparisons/benchmarking and yield reproducible results;

- *Far-sighted*: There is little point to foresight which does not include a creative element that is explicitly future-oriented and moves beyond mere *zeitgeist*;
- *Transparent*: The objectives of an exercise should be clear to all; the design of the process, the sources of information and the means used to analyse data should all be available to those expected to participate and make use of the results;
- *Embedded*: Foresight's impact endures where a culture for foresight is able to spread;
- *Engaged*: The commitment of those capable of acting upon the results should be secured in advance;
- *Efficient*: In its use of public (or private funds) foresight should be carried out with due economy and efficiency but be adequately resourced to be effective;
- *Adaptive*: Foresight should be adaptive, drawing upon lessons from previous and current activity to meet evolving demands.

No doubt readers will be able to distil their own mix of principles and those undertaking foresight may find innovative ways forward; this is after all the consequence of being adaptive and evolutionary in approach. Nonetheless, it is our sincere hope that those who have navigated this Handbook will feel better equipped in their chosen task, be it commissioning, designing, implementing, comprehending or studying foresight. In so doing they will join the contributors and those they write about in a community which is striving to make the creation of policy or strategy for the future less precarious and more systematic, engaged and inclusive.

NOTES

1   See http://www.regstrat.net for further information and a guide to developing and using SPI.
2   Though at the same time, the growing popularity of foresight has also seen pressure for more rapid processes to be developed. This includes instances of "mini-foresights", such as 24-hour success scenario workshops, which are relatively simple but effective for high-profile stakeholders (see Miles, 2005).
3   See also the discussion in the conclusions of Chapter 7.
4   See the discussion in the special issue of *Technological Forecasting and Social Change*, **73**(2), 2006.
5   For example, Clark *et al.* (2001), Potting and Bakkes (2004), among many others.
6   The Science Foresight project at http://www.sussex.ac.uk/spru/foresight/ is one major development here, and co-citation is playing a major role in the current round of the Japanese Foresight programme. Chapter 8 describes Japanese foresight's use of co-citation methods.

7   Lempert *et al.*, (2003).
8   We have also seen a revival of claims about being on the edge of developing a viable "social physics", and repeated calls for displacement of expert opinion with supposedly robust methods of extrapolation and fitting of logistic curves and the like to data.
9   E.g. Phaal *et al.*, (2003).
10 E.g. Loveridge *et al.*, (2004), Popper and Miles (2005).
11 Applied, for example, by a team studying climate change and other long-term futures issues, including new technologies, to carbon sequestration possibilities, in Gough *et al.*, (2002).
12 This approach is currently attracting interest from Japanese foresight teams; it has also been considered and in some cases critiqued by European foresighters such as Ahti Salo. For an application to research priority-setting, see Braunschweig (2000).
13 Understanding the technical issues involved in network analysis or multidimensional scaling, for instance, and the impacts of choosing particular thresholds or assumptions as to data structures, is an arcane art to all but a few data analysts. To take a more prevalent example, much economic forecasting is based on models that assume equilibrium to be the norm – while there are substantial lines of argument arguing that this is never a valid assumption, let alone one that makes much sense when we are examining circumstances characterised by rapid technological change.
14 See "Key Technologies: The Social Sciences and the Humanities (SS&H)" report prepared by George Gaskell (July 2005), available at ftp://ftp.cordis.europa.eu/pub/foresight/docs/kte_social_humanities.pdf.

REFERENCES

Braunschweig, T. (2000), *Priority Setting in Agricultural Biotechnology Research: Supporting Public Decisions in Developing Countries with the Analytic Hierarchy Process*, ISNAR Research Report No. 16, at: http://www.isnar.cgiar.org/publications/pdf/rr-16.pdf.
Clark, J.S., Carpenter, S.R., Barber, M., Collins, S., Dobson, A., Foley, J.A., Lodge, D.M., Pascual, M., Pielke, R., Pizer, W., Pringle, C., Reid, W.V., Rose, K.A., Sala, O., Schlesinger, W.H., Wall, D.H. and Wear, D. (2001), 'Ecological Forecasts: An Emerging Imperative', *Science*, **293**, 27 July, pp. 657–660.
Fuller, T. and Tilley, F. (2005), 'Corporate Ethical Futures; Responsibility for the Shadow on the Future of Today's Ethical Corporations', *Futures*, **37**, pp. 183–197.
Gough, C., Shackley, S., Melvin, G. and Cannell, R. (2002), *Evaluating the Options for Carbon Sequestration*, Tyndall Centre (North), UMIST, Manchester, at: http://www.tyndall.ac.uk/research/.
Lempert, R.J., Popper, S.W. and Bankes, S.C. (2003), *Shaping the next one hundred years: new methods for quantitative, long-term policy analysis*, RAND Pardee Center, MR-1626 at: http://www.rand.org/.

Loveridge, D., Miles, I., Keenan, M., Popper, R. and Thomas, D. (2004), *European Knowledge Society Foresight: The EUFORIA Project Synthesis Report*, European Foundation, Dublin available at: http://www.eurofound.europa.eu/publications/.

Miles, I. (2005), 'Scenario Planning', in *UNIDO Technology Foresight Manual*, Volume 1 – Organization and Methods, pp. 168–193, Vienna: UNIDO.

Phaal, R., Farrukh, C. and Probert, D. (2003), *Technology Roadmapping: linking technology resources to business objectives*, Centre for Technology Management, University of Cambridge Institute for Manufacturing, http://www-mmd.eng.cam. ac.uk/ctm/.

Popper, R. and Miles, I. (2005), *The FISTERA Delphi Future Challenges, Applications And Priorities For Socially Beneficial Information Society Technologies*, Report prepared for FISTERA project, available at http://prest.mbs.ac.uk/prest/FISTERA/delphi_results.htm and http://fistera.jrc.es/ docs/RP_The_FISTERA_Delphi.pdf.

Potting, J. and Bakkes, J. (eds) (2004), *The GEO-3 Scenarios 2002-2032: Quantification and analysis of environmental impacts*, Nairobi: UNEP/DEWA/ RS.03-4 and Bilthaven: RIVM 402001022.

Slocum, N. (2003), *Participatory Methods Toolkit. A Practitioner's Manual*, Bruges, UNU-CRIS available at: http://www.cris.unu.edu/.

Index

Printed and bound by CPI Group (UK) Ltd, Croydon, CR0 4YY

27/10/2024

14580409-0005